Environmental Microbiology

METHODS IN BIOTECHNOLOGY™

John M. Walker, Series Editor

18. **Microbial Processes and Products**, edited by *Jose Luis Barredo, 2005*
17. **Microbial Enzymes and Biotransformations**, edited by *Jose Luis Barredo, 2005*
16. **Environmental Microbiology: *Methods and Protocols***, edited by *John F. T. Spencer and Alicia L. Ragout de Spencer, 2004*
15. **Enzymes in Nonaqueous Solvents: *Methods and Protocols***, edited by *Evgeny N. Vulfson, Peter J. Halling, and Herbert L. Holland, 2001*
14. **Food Microbiology Protocols**, edited by *J. F. T. Spencer and Alicia Leonor Ragout de Spencer, 2000*
13. **Supercritical Fluid Methods and Protocols**, edited by *John R. Williams and Anthony A. Clifford, 2000*
12. **Environmental Monitoring of Bacteria**, edited by *Clive Edwards, 1999*
11. **Aqueous Two-Phase Systems**, edited by *Rajni Hatti-Kaul, 2000*
10. **Carbohydrate Biotechnology Protocols**, edited by *Christopher Bucke, 1999*
9. **Downstream Processing Methods**, edited by *Mohamed A. Desai, 2000*
8. **Animal Cell Biotechnology**, edited by *Nigel Jenkins, 1999*
7. **Affinity Biosensors: *Techniques and Protocols***, edited by *Kim R. Rogers and Ashok Mulchandani, 1998*
6. **Enzyme and Microbial Biosensors: *Techniques and Protocols***, edited by *Ashok Mulchandani and Kim R. Rogers, 1998*
5. **Biopesticides: *Use and Delivery***, edited by *Franklin R. Hall and Julius J. Menn, 1999*
4. **Natural Products Isolation**, edited by *Richard J. P. Cannell, 1998*
3. **Recombinant Proteins from Plants: *Production and Isolation of Clinically Useful Compounds***, edited by *Charles Cunningham and Andrew J. R. Porter, 1998*
2. **Bioremediation Protocols**, edited by *David Sheehan, 1997*
1. **Immobilization of Enzymes and Cells**, edited by *Gordon F. Bickerstaff, 1997*

METHODS IN BIOTECHNOLOGY™

Environmental Microbiology

Methods and Protocols

Edited by

John F. T. Spencer

and

Alicia L. Ragout de Spencer

Planta Piloto de Procesos Industriales Microbiológicos (PROIMI)-CONICET, San Miguel de Tucumán, Argentina

HUMANA PRESS TOTOWA, NEW JERSEY

999 Riverview Drive, Suite 208
Totowa, New Jersey 07512

www.humanapress.com

This publication is printed on acid-free paper. ∞
ANSI Z39.48-1984 (American Standards Institute)

Permanence of Paper for Printed Library Materials.

Cover illustration: Figure 1A from Chapter 35, "Techniques for Manipulating the Bacterial Endophyte *Bacillus mojavensis*," by Charles W. Bacon and Dorothy M. Hinton.

Production Editor: Jessica Jannicelli
Cover design by Patricia F. Cleary.

For additional copies, pricing for bulk purchases, and/or information about other Humana titles, contact Humana at the above address or at any of the following numbers: Tel.: 973-256-1699; Fax: 973-256-8341; E-mail: humana@humanapr.com; or visit our Website: www.humanapress.com

Printed in the United States of America. 10 9 8 7 6 5 4 3 2 1

eISBN: 1-59259-765-3

Library of Congress Cataloging-in-Publication Data

Environmental microbiology : methods and protocols / edited by John
F.T. Spencer and Alicia L. Ragout de Spencer.
p. ; cm. -- (Methods in biotechnology ; 16)
Includes bibliographical references and index.
ISBN 1-58829-116-2 (alk. paper)
1. Microbial ecology.
[DNLM: 1. Environmental Microbiology. 2. Bacteria--isolation & purification. 3. Microbiological Techniques. QW 55 E607 2004] I. Spencer, J. F. T. II. Ragout de Spencer, Alicia L. III. Series.
QR100.E584 2004
579--dc22

2004008392

Preface

The methods included in *Environmental Microbiology: Methods and Protocols* can be placed in the categories "Communities and Biofilms," "Fermented Milks," "Recovery and Determination of Nucleic Acids," and the review section, containing chapters on the endophytic bacterium, *Bacillus mojavensis,* the engineering of bacteria to enhance their ability to carry out bioremediation of aromatic compounds, using the hemoglobin gene from a strain of *Vitreoscilla* spp., and the use of chemical shift reagents and ^{23}Na NMR to study sodium gradients in microorganisms, all of which should be of interest to investigators in these fields.

The subjects treated within the different categories also cover a wide range, with methods ranging from those for the study of marine organisms, through those for the investigation of microorganisms occurring in ground waters, including subsurface ground waters, to other types of environmental waters, to as varied subjects as the biodiversity of yeasts found in northwest Argentina.

The range of topics described in the Fermented Milks section is smaller, but significant for investigators in areas concerned with milk as an item of foods for infants, small children, and even adults.

The section on recovery and determination of nucleic acids and other compounds affecting, and affected by, microorganisms also covers a considerable range, as well as including methods for some of the enzymes produced by plant pathogens and methods for obtaining microbial species tolerant of some inhibitors, such as heavy metals. Thus, there is something for most investigators concerned with microorganisms in their native environments.

There is one aspect of *Environmental Microbiology: Methods and Protocols* concerned with the special problem of microorganisms in the environment, that of the so-called "non-culturable" forms. This problem has been solved, at least partially, by growing the organisms in a two-chambered system in which the organisms of interest are grown in a central chamber, separated by a semipermeable membrane from an outer compartment, in which a mixed culture of an unidentified group of organisms is grown, and produces unidentified (as yet) growth factors for the organisms in the central chamber. This problem has existed for at least 50 years, for soil microbiologists and probably others. The method is described in a recent issue of *Science*. Unfortunately, the editors were unable to find an author willing, or able to take the time, to write about this subject.

The editors wish to acknowledge the considerable help in producing this volume, by several people. We would especially like to thank Dr. Faustino Siñeriz, Director of PROIMI and Head of Secretaría de Ciencia y Técnica (CIUNT) for CONICET in Tucumán, for the use of the facilities of PROIMI and authorship of some of the chapters in this volume, and encouragement of us in the work. We also wish to thank Pharm. María Laura Tereschuk and Dr. María Alejandra Martínez for their important efforts in checking and correcting the writing, and Dr. Javier Ochoa for much technical assistance. Finally, we would like to thank all of the authors who gave their time and expertise in the writing of the chapters.

John F. T. Spencer
Alicia L. Ragout de Spencer

Contents

Contributors

CARLOS M. ABATE • *Planta Piloto de Procesos Industriales Microbiologicos (PROIMI)-CONICET, Facultad de Bioquimica, Quimica y Farmacia, Universidad Nacional de Tucumán, Tucumán, Argentina*

JULIANA ADELMANN • *Biomass Chemo/Biotechnology Laboratory, Department of Pharmacy, Federal University of Paraná, Curitiba-PR, Brazil*

SPIROS N. AGATHOS • *Unit of Bioengineering, Université Catholique de Louvain, Louvain-la-Neuve, Belgium*

VIRGINIA H. ALBARRACÍN • *Planta Piloto de Procesos Industriales Microbiologicos (PROIMI)-CONICET, Facultad de Ciencias Naturales e Instituto Miguel Lillo, Universidad Nacional de Tucumán, Tucumán, Argentina*

ROSALÍA ALESSO • *Laboratorio Central, SanCor Cooperativas Unidas, Argentina*

MARÍA J. AMOROSO • *Planta Piloto de Procesos Industriales Microbiologicos (PROIMI)-CONICET, Facultad de Bioquimica, Quimica y Farmacia, Unviersidad Nacional de Tucumán, Tucumán, Argentina*

MARÍA C. APELLA • *Centro de Referencia para Lactobacilos (CERELA)-CONICET, Tucumán, Argentina*

M. CARINA AUDISIO • *Centro de Referencia para Lactobacilos (CERELA)-CONICET, Tucumán, Argentina*

CHARLES W. BACON • *Russell Research Center, Toxicology & Mycotoxin Research Unit, ARS, USDA, Athens, GA*

CHRISTOPHER E. BAGWELL • *Environmental Biotechnology Section, Westinghouse Savannah River Company, Aiken, SC*

PAUL W. BAKER • *Biological Sciences, Kent State University, Kent, OH*

ALICIA BARDÓN • *Universidad Nacional de Tucumán, Tucumán, Argentina*

NICOLÁS G. BARENGO • *Department of Biology and Biochemistry, University of Houston, Houston, TX*

CLAUDIA S. BENIMELI • *Centro de Referencia para Lactobacilos (CERELA)-CONICET, Tucumán, Argentina; Department of Biology and Biochemistry, University of Houston, Houston, TX*

JULIANA M. BENITO • *Planta Piloto de Procesos Industriales Microbiologicos (PROIMI)-CONICET, Tucumán, Argentina*

D. K. BUTTON • *Institute of Marine Sciences, University of Alaska, Fairbanks, Alaska*

ADRIANA P. CHAILE • *Obras Sanitarias Tucumán, Tucumán, Argentina*

LUCÍA I. C. DE FIGUEROA • *Planta Piloto de Procesos Industriales Microbiologicos (PROIMI)-CONICET, Facultad de Bioquimica, Quimica y Farmacia, Universidad Nacional de Tucumán, Tucumán, Argentina*
CRISTINA A. DE LACERDA • *Chromatography Laboratory, Institute of Chemistry of São Carlos, University of São Paulo, São Carlos-SP, Brazil*
ANNE-MARIE DELORT • *Laboratoire de Synthese et Etude de Systèmes d' Intéret Biologique, Université Blaise Pascal-CNRS, Aubière Cedex, France*
JOSÉ C. DEL RÍO • *Instituto de Recursos Naturales y Agrobiología, CSIC, Sevilla, Spain*
PETER F. DUNFIELD • *Max-Planck-Institute for Terrestrial Microbiology, Marburg, Germany*
SAID EL FANTROUSSI • *Unit of Bioengineering, Université Catholique de Louvain, Louvain-la-Neuve, Belgium*
LAURENT EYERS • *Unit of Bioengineering, Université Catholique de Louvain, Louvain-la-Neuve, Belgium*
MARTA E. FARÍAS • *Facultad de Bioquimica, Quimica y Farmacia, Universidad Nacional de Tucumán, Tucumán, Argentina*
JOSÉ DOMINGOS FONTANA • *Biomass Chemo/Biotechnology Laboratory, Department of Pharmacy, Federal University of Paraná, Curitiba-PR, Brazil*
GRACIELA FONT DE VALDEZ • *Centro de Referencia para Lactobacilos (CERELA)-CONICET, Tucumán, Argentina*
EVELYNE FORANO • *Laboratoire de Microbiologie, INRA, Centre de Recherches de Clermont-Ferrand-Theix, Saint-Genes-Champanelle, France*
MARK E. FULLER • *Shaw Environmental Inc., Princeton Research Center, Lawrenceville, NJ*
MARISA S. GARRO • *Centro de Referencia para Lactobacilos (CERELA)-CONICET, Tucumán, Argentina*
GENEVIEVE GAUDET • *Laboratoire de Microbiologie, INRA, Centre de Recherches de Clermont-Ferrand-Theix, Saint-Genes-Champanelle, France*
ERIC S. GILBERT • *Department of Biology, Georgia State University, Atlanta, GA*
SILVIA N. GONZÁLEZ • *Centro de Referencia para Lactobacilos (CERELA)-CONICET, Tucumán, Argentina*
E. ANNE GREENE • *Department of Biology, University of Calgary, N. W. Calgary, Alberta*
ANA GUTIÉRREZ • *Instituto de Recursos Naturales y Agrobiología, CSIC, Sevilla, Spain*
NATSUKO HAMAMURA • *Marine Biotechnology Institute, Kamaishi Laboratories, Heita, Kamaishi City, Iwate, Japan*
SHIGEAKI HARAYAMA • *Department of Biotechnology, National Institute of Technology and Evaluation, Tokyo, Japan*

ELVIRA M. HÉBERT • *Centro de Referencia para Lactobacilos (CERELA)-CONICET, Tucumán, Argentina*
ANAVELLA GAITAN HERRERA • *Planta Piloto de Procesos Industriales Microbiologicos (PROIMI)-CONICET-Biotecnologia, Tucumán, Argentina*
MARTHA DINA VALLEJO HERRERA • *Universidad Nacional de San Juan, San Juan, Argentina*
DOROTHY M. HINTON • *Russell Research Center, Toxicology and Mycotoxin Research Unit, ARS, USDA, Athens, GA*
NOBUO KAKU • *Marine Biotechnology Institute, Kamaishi Laboratories, Heita, Kamaishi City, Iwate, Japan*
JAY D. KEASLING • *Department of Chemical Engineering, University of California Berkeley, Berkeley, CA*
FERNANDO MAURO LANÇAS • *Chromatography Laboratory, Institute of Chemistry of São Carlos, University of São Paulo, São Carlos-SP, Brazil*
WERNER LIESACK • *Max-Planck-Institute for Terrestrial Microbiology, Marburg, Germany*
MÓNICA M. LOCASCIO • *Centro de Referencia para Lactobacilos (CERELA)-CONICET, Universidad Nacional de Tucumán, Tucumán, Argentina*
MANUEL SIÑERIZ LOUIS • *Planta Piloto de Procesos Industriales Microbiologicos (PROIMI)-CONICET, Tucumán, Argentina*
CHARLES R. LOVELL • *Department of Biological Sciences, University of South Carolina, Columbia, SC*
GUSTAVO A. LOVRICH • *Centro Austral de Investigaciones Cientificas, Tierra del Fuego, Argentina*
MARY LOWE • *Department of Physics, Loyola College, Baltimore, MD*
MARÍA C. MANCA DE NADRA • *Facultad de Bioquímica, Química, y Farmacia, Universidad Nacional de Tucumán y Centro de Referencia para Lactobacilos (CERELA)-CONICET, Tucumán, Argentina*
MARCELO MARASCHIN • *Laboratory of Plant Tissues Culture, Federal University of Santa Catarina, Florianopolis-SC, Brasil*
ÁNGEL T. MARTÍNEZ • *Centro de Investigaciones Biológicas, CSIC, Madrid, Spain*
M. ALEJANDRA MARTÍNEZ • *Planta Piloto de Procesos Industriales Microbiologicos (PROIMI)-CONICET, Tucumán, Argentina*
SCOTT R. MILLER • *Division of Biological Sciences, University of Montana, Missoula, MT*
VILMA I. MORATA • *Universidad Nacional de Cuyo, Mendoza, Argentina*
KRISHNA R. PAGILLA • *Department of Chemical and Environmental Engineering, Illinois Institute of Technology, Chicago, IL*
MAURICIO PASSOS • *Biomass Chemo/Biotechnology Laboratory, Department of Pharmacy, Federal University of Paraná, Curitiba-PR, Brazil*

Raúl O. Pedraza • *Facultad de Agronomia y Zootecnica, Universidad Nacional de Tucumán, Tucumán, Argentina*
Garret W. Perney • *Institute of Marine Sciences, University of Alaska, Fairbanks, Alaska*
Luiz Pereira Ramos • *Departamento de Química, Universidade Federal de Paraná, Curitiba, Paraná, Brasil*
Raúl R. Raya • *Centro de Referencia para Lactobacilos (CERELA)-CONICET, Tucumán, Argentina*
Juan C. Díaz Ricci • *INSIBIO, Facultad de Bioquimica, Quimica y Farmacia, Tucumán, Argentina*
Betsy R. Robertson • *Institute of Marine Sciences, University of Alaska, Farbanks, Alaska*
Fabiana M. Saguir • *Facultad de Bioquimica, Quimica y Farmacia, Universidad Nacional de Tucumán, Tucumán, Argentina*
Graciela Savoy de Giori • *Centro de Referencia para Lactobacilos (CERELA)-CONICET, Tucumán, Argentina*
Faustino Siñeriz • *Planta Piloto de Procesos Industriales Microbiologicos (PROIMI)-CONICET, Facultad de Bioquimica, Quimica y Farmacia, Universidad Nacional de Tucumán, Tucumán, Argentina*
John F. T. Spencer • *Planta Piloto de Procesos Industriales Microbiologicos (PROIMI)-CONICET, Tucumán, Argentina*
Alex Spiro • *Department of Physics, Loyola College, Baltimore, MD*
Benjamin C. Stark • *Department of Biological, Chemical, and Physical Sciences, Illinois Institute of Technology, Chicago, IL*
Anne O. Summers • *Department of Microbiology, University of Georgia, Athens, GA*
María Inés Torino • *Centro de Referencia para Lactobacilos (CERELA)-CONICET, Tucumán, Argentina*
María Eugenia Toro • *Universidad Nacional de San Juan, San Juan, Argentina*
Carina Van Nieuwenhove • *Centro de Referencia para Lactobacilos (CERELA)-CONICET, Tucumán, Argentina*
Fabio Vazquez • *Universidad Nacional de San Juan, San Juan, Argentina*
Liliana Villegas • *Planta Piloto de Procesos Industriales Microbiologicos (PROIMI)-CONICET, Tucumán, Argentina*
Gerrit Voordouw • *Department of Biology, University of Calgary, N. W. Calgary, Alberta*
Kazuya Watanabe • *Marine Biotechnology Institute, Kamaishi Laboratories, Heita, Kamaishi City, Iwate, Japan*
Dale A. Webster • *Department of Biological, Chemical, and Physical Sciences, Illinois Institute of Technology, Chicago, IL*
Joy Wireman • *Department of Microbiology, University of Georgia, Athens, GA*

I

Communities and Biofilms

1

Isolation and Molecular Characterization of Seawater Bacteria

Juliana M. Benito, Gustavo A. Lovrich, Faustino Siñeriz, and Carlos M. Abate

1. Introduction

A wide diversity of microorganisms, including bacteria, archaea, fungi, and microalgae, are found in marine habitats. Although there has been much progress in describing the diversity of marine bacteria and archaea, we still know very little about the "role" of these microbes *(1)*.

Jannasch and Jones *(2)* noted a discrepancy of several orders of magnitude between the number of cells that can be seen in seawater samples by direct observation and the number of colonies that grow on agar plates. Fundamentally there are two explanations for this discrepancy: bacterial communities are composed of known species that are capable of forming colonies on agar plates, but do so with low efficiency *(1)*; or bacterial communities are composed of unknown species that do not grow on common microbiological media *(3)*. The recent application of molecular biological techniques to investigate the diversity of marine bacterial communities has revealed 16S rDNA sequences of previously unsequenced and possibly uncultured new bacteria *(4)*.

Connon and Giovannoni *(5)* obtained isolates of many novel microbial strains, including members of previously uncultured groups that are believed to be abundant in coastal seawater, using high-throughput culturing (HTC) methods. These methods would enable a large number of extinction cultures to be identified so that the efficacy of this approach could be assessed with a larger sampling of isolates.

From: *Methods in Biotechnology: Environmental Microbiology: Methods and Protocols*
Edited by: J. F. T. Spencer and A. L. Ragout de Spencer

Eilers et al. ***(6)*** suggested that future cultivation attempts could consider (1) filtration (pore size, <1.2 μm) of the inoculum to remove large, highly active, particle-associated bacteria, (2) dilution to favor dominant bacteria ***(7)***, and (3) colony isolation in semiliquid (soft-agar) medium and subsequent subculturing in liquid medium for bacteria unable to grow at the air–water interface.

Characterization of the 16S rRNA gene is now well-established as a standard method for the identification of species, genera, and families of bacteria, taking into account that rRNA genes are essential for the survival of all organisms and are highly conserved in the bacterial and other kingdoms ***(8)***. The relationship between 16S rRNA gene similarity and percent DNA–DNA reassociation is a logarithmic function in which the sequence similarity within a species (>70% relatedness) is expected to be more than 97% ***(9)***.

Sequence information has also become available on the 16S–23S intergenic spacer region (ISR) and suggests that considerable variation can occur between species in both the length and the sequence of this region. The facts that many bacteria have multiple copies (alleles) per genome of the rDNA operon and that the 16S–23S intergenic region may encode tRNAs depending on the bacterial species, raise the possibility that spacer heterogeneity in both length and nucleotide sequence between strains, species, and genera may be used for identification and typing purposes ***(10)***.

The aim of this chapter is to provide new protocols for isolation of marine bacteria and the subsequent molecular differentiation and characterization of the isolated strains.

2. Materials

2.1. Sampling

Seawater collected in a sterile disposable plastic vessel.

2.2. Culture Media

1. Marine R2A agar medium (in 75% seawater): 0.5 g/L yeast extract; 0.5 g/L Proteose Peptone (Difco); 0.5 g/L casamino acids; 0.5 g/L dextrose; 0.5 g/L soluble starch; 0.3 g/L sodium pyruvate; 15 g/L agar ***(11)*** (*see* **Note 1**).

2.3. DNA Isolation, Purification, and Visualization

1. Lysis buffer: 20 m*M* EDTA, 400 m*M* NaCl, 750 m*M* sucrose, 50 m*M* Tris-HCl, pH 9.0, and 2 mg/mL lysozyme (*see* **Note 2**).
2. 10 mg/mL Proteinase K in distilled water.
3. 20% SDS (w/v).
4. 3 *M* Sodium acetate pH 4.8.
5. Phenol-chloroform-isoamyl alcohol (25:24:1) ***(12)***.

6. Chloroform-isoamyl alcohol 24:1 (v/v).
7. Isopropanol and 70% ethanol.
8. RNAase A solution (stock 10 mg/mL) in distilled water *(12)*.
9. 1X TAE buffer: 0.04 *M* Tris-acetate, 0.001 *M* EDTA, pH 8.0 *(12)*.
10. 0.8% Agarose in 1X TAE buffer.
11. Ethidium bromide staining solution: 1 µg/mL in 1X TAE buffer *(12)*.
12. Molecular weight markers: K562 High Molecular Weight (Promega).
13. Loading buffer (Gibco BRL).

2.4. PCR Amplifications and Purification of PCR Products

1. Primers (final concentration 0.5 µ*M*). For the eubacterial domain, 16S rDNA 27F, 5'-AGAGTTTGATCMTGGCTCAG-3'; 1492R, 5'-GGTTACCTTGTTAC GACTT-3' *(13)* (*see* **Note 3**). For ISR amplification, ISR-1494 (5'GTCGTA ACAAGGTAGCCGTA 3') and ISR-35 (5'CAAGGCATCCACCGT 3') *(14)* (*see* **Note 4**).
2. *Taq* polymerase and 10X STR buffer (Promega).
3. Thermal cycler (e.g., Gene Amp PCR System 9700, Applied Biosystems, CA).
4. TAE buffer (1X): 0.04 *M* Tris-acetate, 0.001 *M* EDTA, pH 8.0 *(12)*.
5. 1% and 2% agarose gel in 1X TAE buffer.
6. Ethidium bromide staining solution in 1X TAE buffer *(12)*.
7. Molecular weight markers: 1 kb and 100 bp DNA Ladders (Promega).
8. Loading buffer (Promega).
9. Sterile-distilled water.

2.5. Sequence Analysis

1. Wizard PCR Preps DNA Purification System (Promega).
2. BLAST tools at http://www.ncbi.nlm.nih.gov.

3. Methods

3.1. Strain Isolation

1. Collect the seawater samples from different locations in sterile containers and transport to the laboratory as soon as possible. Store at 4°C until processing (*see* **Note 5**).
2. Spread subsamples (100 µL each) onto 10 marine R2A agar plates. Incubate at 4–5°C in the dark (*see* **Note 6**).
3. Examine the plates periodically and streak the colonies onto R2A plates after approx 7 d, over the course of 1 mo (*see* **Note 7**).

3.2. DNA Isolation, Purification and Visualization (see Note 8)

1. Transfer 2 mL of an overnight culture to a microcentrifuge tube and centrifuge at 10,000*g* for 2 min. Discard the supernatant (*see* **Note 1**).
2. Resuspend the pellet in 2 mL of lysis buffer.

3. Vortex 1 min and incubate at 37°C for 30 min.
4. Add SDS (final concentration 1% wt/vol) and proteinase K (final concentration 100 µg/mL)
5. Vortex 10 s and incubate at 55°C for 2 h.
6. Add an equal volume of phenol-chloroform-isoamyl alcohol and mix by inverting the tube several times.
7. Centrifuge (10,000*g* for15 min). Transfer the aqueous phase (upper) to a new tube and repeat phenol extraction once.
8. Transfer the aqueous phase to a clean tube and add an equal volume of chloroform-isoamyl alcohol. Again mix well and centrifuge (10,000*g* for 5 min). Repeat this extraction two times.
9. Transfer the aqueous phase to a new tube and precipitate the DNA by adding 1/10 vol of 3 *M* sodium acetate and 0.6 to 1 vol of 2-propanol. Mix gently and incubate at –20°C from 1 h to overnight.
10. Centrifuge (10,000*g* for 15 min). Discard the supernatant and wash DNA with 500 µL of 70% ethanol. Centrifuge (10,000*g* for 5 min), carefully discard the ethanol and dry until ethanol has been removed (*see* **Note 9**).
11. Resuspend DNA in 20–30 µL of double-distilled sterile water and 0.1–0.2 µL RNAase A. Allow to dissolve at 37°C at least 3 h.
12. To visualize the extracted DNA run 5 µL of the sample in a 0.8% agarose gel electrophoresis using 1X TAE electrophoresis buffer and high-molecular-weight marker. Electrophorize at 10 V/cm for 1 h at room temperature; gels should be stained with ethidium bromide solution and observed under UV light (*see* **Note 10**).

3.3. PCR Amplifications

3.3.2. PCR Conditions for 16S rDNA Amplification

Prepare the reaction mixture to make a final volume of 25 µL (*see* **Note 11**): 100 ng isolated DNA; 2.5 µL of 10X STR buffer; 0.2 µL of *Taq* polymerase (1 U); 0.2 µL of each primer: 27F and 1492R (0.5 µ*M* final concentration); and double-distilled sterile water to 25 µL.

Amplification is performed with an initial denaturation at 94°C for 5 min; 25–30 cycles of 94°C for 1 min, 55°C for 1 min, and 72°C for 2 min; and a final extension at 72°C for 10 min (*see* **Note 12**).

3.3.2. PCR Conditions for ISR Amplification

Prepare the reaction mixture, final volume 25 µL (*see* **Note 11**): 100 ng isolated DNA; 2.5 µL of 10X STR buffer; 0.2 µL of *Taq* polymerase (1 U); 0.2 µL of each primer: ISR-1494 and ISR-35 (0.5 µ*M* final concentration); and double-distilled sterile water to 25 µL

The following temperature profile is used: initial denaturation at 94°C for 5 min; followed by 30 cycles of 94°C for 1 min, 55°C for 3 min, and 72°C for 2 min; and a final extension at 72°C for 10 min.

For the evaluation of PCR fingerprint: PCR amplified products are separated by 2% agarose gel electrophoresis using 1X TAE electrophoresis buffer and appropriate markers in the range of 100 bp to 2000 bp, e.g., 1 kb and 100 bp DNA Ladders (Promega). Electrophoresis is carried out at 20 V/cm for 2–4 h at room temperature; gels should be stained with 1 µg/mL ethidium bromide solution and observe under UV light (*see* **Note 13**).

3.4. Sequence Analysis

16S rDNA amplified products are recovered from agarose gel using a Wizard PCR Preps DNA Purification System (Promega). Sequencing is carried out in a DNA sequencer (e.g., ABI 373 Stretch). Retrieved sequences can be compared to bacterial rDNA sequences present in the GeneBank, and similarities searched using BLAST tools. Phylogenetic trees can be constructed with the programs available at http://rdp.cme.msu.edu.

4. Notes

1. Marine R2A is a complex medium with higher carbon content than the natural content of seawater; therefore, it is more likely to isolate heterotrophic bacteria capable of thriving at higher organic carbon concentrations than seawater, known as copiotrophs.

   The same medium but without agar is used to obtain liquid cultures of the isolated strains. Incubation times will depend on the particular strains; in general they range from 24 to 72 h.
2. The lysis buffer without lysozyme can be aseptically prepared and stored at room temperature for long periods of time. The lysozyme is added just before using.
3. In addition to the primer 1492R, others may be used to amplify a shorter fragment of the 16S rDNA gene (e.g., 518R, 5'-CGTATTACCGCGGCTGCTGG-3'). Sequences obtained from this type of fragment are as informative as the complete sequences of the 16S rRNA genes.
4. One issue that is critical for the successful detection of spacer variation is the choice of the PCR primers. Conserved regions in the 16S rRNA gene can be identified accurately, but it is not possible to do the same with the less well characterized 23S rRNA gene. Gurtler and Stanisich ***(10)*** suggested that region 2 of the 16S rRNA gene (nt 1390–1407) and region 10 of the 23S rRNA gene (nt 456–474) are the regions of choice for the construction of primers, the former because of its proximity to the spacer region and the latter because of the high level of sequence conservation among the species analyzed.
5. Samples should be processed as soon as possible or kept at 4°C until they are used. Ferguson et al. ***(14)*** showed that the bacterioplankton community of confined seawater at 25°C changes significantly within 16 h of collection.
6. The number of plates to be spread depends on the objective of the work. We recommend an intermediate number of 10 in order to isolate as many different colonies as possible. The incubation temperature is also variable. If the aim of the

work is to isolate psychrophilic bacteria, then 4–5°C is a good temperature. If this is not the case, then we suggest 15°C.

7. Pure cultures can be stored in glycerol at –20°C. Nevertheless, it is always recommended to maintain the cultures in agar plates at 4°C.
8. When manipulating DNA, wear gloves to minimize the risks of DNase contamination. In the case of PCR preparation, gloves should be powder free because powder inhibits DNA polymerases. Be extremely careful when handling dangerous solutions such as phenol, chloroform, and ethidium bromide.
9. This step removes any residual salt or isopropanol, and should be repeated once.
10. Concentration and quality of the DNA samples obtained should be determined by gel electrophoresis and/or absorbance measurements ***(12)***. Also, concentration of all samples should be adjusted to be similar, especially when differentiating strains with ISR patterns.
11. It is convenient to prepare a master mix with all the reaction components except the DNA, allowing for the total number of samples, including a positive control plus an additional control tube that will not include template DNA (negative control).
12. Visualization of the PCR results is done by running 5-μL aliquots of the products on a 1% agarose gel using a molecular-weight marker—e.g., 1 kb ladder (Promega). Electrophoresis is run at 10 V/cm for 1 h. Bands corresponding to the 16S rDNA fragments (1.5 kb) are visualized after staining the gel with ethidium bromide.
13. Total PCR reaction volume should be loaded to allow detection of all bands, even the less intense ones. Capture image systems are useful if several bands are obtained, to determine precisely the number and size of bands. This procedure is particularly useful when dealing with lots of isolates, as it permits grouping them based on the ISR profiles.

References

1. Kirchman, D. L. and Williams, P. J. leB. (2000) Introduction, in *Microbial Ecology of the Oceans* (Kirchman, D. L., ed.), John Wiley & Sons, Inc., New York, pp. 1–12.
2. Jannasch, H. W. and Jones, G. E. (1959) Bacterial populations in seawater as determined by different methods of enumeration. *Limnol. Oceanogr.* **4,** 128–139.
3. Sherr, E. and Sherr, B. (2000) Marine microbes. An overview, in *Microbial Ecology of the Oceans* (Kirchman, D. L., ed.), John Wiley & Sons, Inc., New York, pp. 13–46.
4. Weidner, S., Arnold, W., and Puhler, A. (1996) Diversity of uncultured microorganisms associated with the seagrass *Halophila stipulacea* estimated by restriction fragment length polymorphism analysis of PCR-amplified 16S rRNA genes. *Appl. Environ. Microbiol.* **62,** 766–771.
5. Connon, S. A. and Giovannoni, S. J. (2002) High-throughput methods for culturing microorganisms in very-low-nutrient media yield diverse new marine isolates. *Appl. Environ. Microbiol.* **68,** 3878–3885.

6. Eilers, H., Pernthaler, J., Glöckner, F. O., and Amann, R. (2000) Culturability and in situ abundance of pelagic bacteria from the North Sea. *Appl. Environ. Microbiol.* **66,** 3044–3051.
7. Button, D. K., Schut, F., Quang, P., Martin, R., and Robertson, B. R. (1993) Viability and isolation of marine bacteria by dilution culture: theory, procedures, and initial results. *Appl. Environ. Microbiol.* **59,** 1707–1713.
8. Woese, C. R. (1987) Bacterial evolution. *Microbiol. Rev.* **51,** 221–271.
9. Stackebrandt, D. and Goebel, B. M. (1994) Taxonomic note: a place for DNA–DNA reassociation and 16S rRNA sequence analysis in the present species definition in bacteriology. *Int. J. Sys. Bacteriol.* **44,** 846–849.
10. Gurtler, V. and Stanisich V. A. (1996) New approaches to typing and identification of bacteria using the 16S–23S rDNA spacer region. *Microbiology* **142,** 3–16.
11. Suzuki, M. T., Rappé, M. S., Zara, W. H., Winfield, H., Adair, N., Strobel, J., et al. (1997) Bacterial diversity among small-subunit rRNA gene clones and cellular isolates from the same seawater sample. *App. Environ. Microbiol.* **63,** 983–989.
12. Sambrook, J., Fristch, E. F., and Maniatis, T. (eds.) (1989) *Molecular Cloning: A Laboratory Manual, 2nd ed.* Cold Spring Harbor Laboratory Press, Cold Spring Harbor, NY.
13. Lane, D. J. (1991) 16S/23S rRNA sequencing. In *Nucleic Acid Techniques in Bacterial Systematics* (Stackebrandt, E. and Goodfellow, M., eds.). John Wiley & Sons, Inc., New York.
14. Daffonchio, D., Borin, S., Frova, G., and Sorlini, C. (1998) PCR fingerprinting of whole genomes: the spacer between the 16S and 23S rRNA genes and of intergenic tRNA gene regions reveals a different intraspecific genomic variability of *Bacillus cereus* and *Bacillus licheniformis. Int. J. Syst. Bacteriol.* **48,** 107–116.
15. Ferguson, R. L., Buckley, E. N., and Palumbo, A. V. (1984) Response of marine bacterioplankton to differential filtration and confinement. *Appl. Environ. Microbiol.* **47,** 49–55.

2

Reverse Sample Genome Probing to Monitor Microbial Communities

E. Anne Greene and Gerrit Voordouw

1. Introduction

Environmental microbial communities are often highly complex. To evaluate community activities, it is desirable to be able to accurately monitor community composition. A variety of methods to monitor environmental microbial diversity exist, e.g., the use of 16S rRNA probes, combined with fluorescence *in situ* hybridization (FISH), or fatty acid methylester analysis. However, no currently available method completely overcomes the difficulties presented by the complexity of environmental microbial communities.

Reverse sample genome probing (RSGP) monitors culturable members of microbial communities. Extracted and labeled total community DNA is used to probe a filter containing chromosomal DNAs from many pure strains. These are selected to have limited cross-hybridization and are referred to as standards. Information on the occurrence of multiple standards is thus obtained in a single hybridization assay. The information obtained is limited to the culturable component of the microbial community. RSGP thus measures microbial diversity in the selected target environment by following the fate of selected culturable community members. It can screen for the presence of specific microbial strains in the target environment under a variety of conditions. Its advantage: once a set of standard microbial strains has been selected, quantifying their relative abundance is straightforward. Its drawback is that environmental microbial diversity is huge. Thus a master filter containing a set of standards may be of limited use outside its intended target environment. Once a filter for a target environment has been obtained, RSGP is a straightforward

From: *Methods in Biotechnology: Environmental Microbiology: Methods and Protocols*
Edited by: J. F. T. Spencer and A. L. Ragout de Spencer

method to evaluate changes in a microbial community over time, and during various treatments (e.g., bioremediation).

RSGP requires isolating and purifying total DNA from microbial community samples. This DNA is radiolabeled and used to probe a master filter that consists of chromosomal DNA from individual standard microorganisms of interest bonded to a nylon membrane **(Fig. 1)**.

2. Materials (see Note 1)

2.1. Designing and Preparing the Master Filter

2.1.1. Isolation of DNA From Pure Bacterial Strains

1. Lysis buffer (0.15 *M* NaCl, 0.1 *M* EDTA, pH 8.0).
2. Freshly made lysozyme (1 mg/mL in lysis buffer).
3. 25% SDS.
4. 5 *M* $NaClO_4$.
5. $CHCl_3$:isoamyl alcohol (24:1, v:v).
6. Tris-EDTA (TE) *(1)*: 10 m*M* Tris-Cl, pH 7.4; 0.1 m*M* EDTA, pH 8.0.
7. 95% and 70% ethanol, ice cold.
8. 10 mg/mL DNase-free RNase *(1)*.
9. 14 mg/mL Proteinase K.
10. TE-buffered phenol *(1)*. Phenol is toxic and should be handled with gloves in a fume hood.

2.1.2. Quantification of DNA

Ethidium bromide is extremely toxic and should be handled with gloves at all times.

1. λ DNA (usually obtained as a 500 ng/μL stock solution).
2. Ethidium bromide plates.
   a. Combine 25 mL of melted 1% (w/v) agarose with 3 mL of 10 mg/mL ethidium bromide and pour the mixture into a 100 × 15 mm Petri plate.
   b. Allow plate to solidify. Plates can be stored in the dark at 4°C for several weeks.

2.1.3. Preparation of Filters

1. λ DNA (**Subheading 2.1.2.**).
2. Purified DNA from microorganisms of interest; at least 100–200 ng/filter.
3. Nylon membrane (e.g., Hybond™-N, Amersham).
4. Sterile-distilled water.
5. 0.5 *N* NaOH.
6. 6X SSC: 20X SSC *(1)* is 175.3 g NaCl, 88.2 g Na_3 citrate per liter, pH 7.

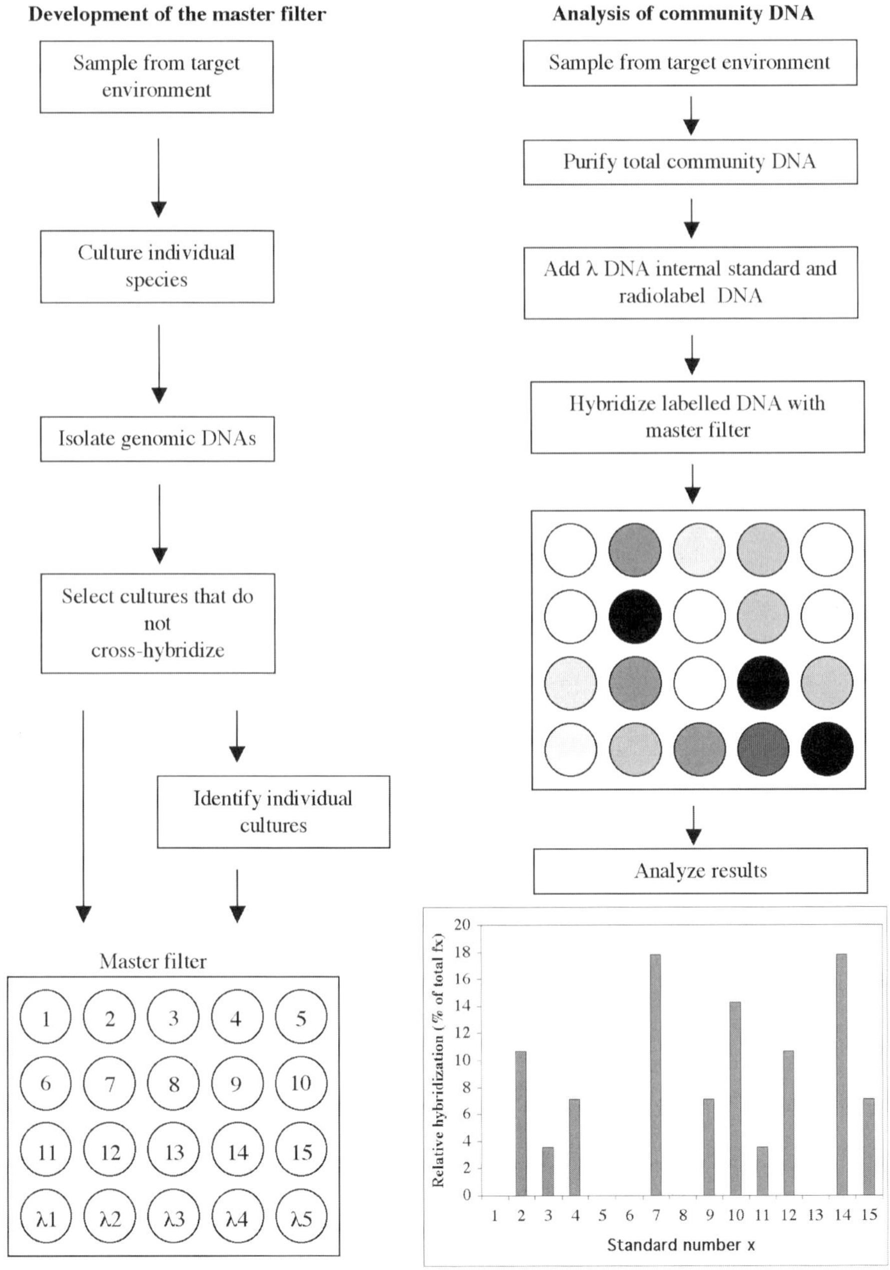

Fig. 1. Overview of the RSGP method. The λ spots represent five different concentrations of λ DNA, which are used to calculate the relative amount of hybridization of labeled community DNAs to various master filter standards.

2.1.4. Generation of Probes and Hybridization With Filters

The work described in this section involves the use of radioisotope (^{32}P); the appropriate precautions should be observed.

2.1.4.1. Generation of Probe

1. DNA (**Subheading 2.1.2.**).
2. DNA samples for generating probe (at least 100 ng of each).
3. Primer extension (PE) mix *(2)*: 44 µL of 0.9 *M* HEPES, 0.1 *M* $MgCl_2$, pH 6.6; 25 µL of 1 *M* Tris-Cl pH 7.4; 10 µL of 0.1 *M* dithiothreitol; 4 µL each of 50 m*M* dATP, dGTP, and dTTP; 10 µL of 10 mg/mL randomly generated hexanucleotides. PE mix can be stored indefinitely at –20°C.
4. DNA polymerase I Klenow fragment, 1,000 U/µL.
5. [$\alpha^{32}P$]dCTP (10 mCi/mL, 3000 Ci/mmol).

2.1.4.2. Preparation of Filters for Hybridization

1. Polypropylene bags and bag sealer.
2. Prehybridization solution *(3)*: 300 mL 20X SSC (**Subheading 2.1.3.**); 50 mL 10% SDS; 100 mL 50X Denhardt's reagent (*[1]*: 5 g Ficoll Type 400), 5 g of polyvinylpyrrolidone, 5 g bovine serum albumin (Fraction V, Sigma, St. Louis, MO) made up to 500 mL in distilled water); 10 mL salmon sperm DNA *(1)*. Prehybridization solution can be stored indefinitely at –20°C.

2.1.4.3. Probing Filters With Labeled DNA

1. 1X SSC (**Subheading 2.1.3.**).
2. 1X SSC, 0.2% (w/v) SDS, preheated to 68°C.

2.2. Isolation and Purification of DNA From Environmental Samples

2.2.1. Isolation of Bacterial Cells

1. Freshly made 0.1% (w/v) $Na_4P_2O_7 \cdot 10H_2O$.
2. Acid-washed polyvinylpolypyrrolidone (PVPP; *[4]*).
3. Mix 300 g PVPP with 4 L of 3 *N* HCl, let stand overnight.
4. Filter with a Buchner funnel apparatus and Whatman no. 1 filter paper.
5. Resuspend PVPP in 20 m*M* KH_2PO_4, stir and filter. Repeat until pH is 7.0–8.0.
6. Air-dry PVPP. Acid-washed PVPP can be stored indefinitely at room temperature.

2.2.2. Cell Lysis and DNA Isolation

1. Lysis buffer (**Subheading 2.1.1.**).
2. Freshly made 300 mg/mL lysozyme in lysis buffer.
3. 25% SDS.
4. 5 *M* $NaClO_4$.
5 $CHCl_3$:isoamyl alcohol (24:1, v:v).

6. 95% Ethanol, ice cold.
7. 70% Ethanol, ice cold.
8. TE-buffered phenol (**Subheading 2.1.1.**).

2.2.3. Removal of Humic Acids From Soil DNA

1. TE (**Subheading 2.1.1.**).
2. Spin columns (based on the protocol of Jackson et al. *[5]*).
3. Soak glass wool in 1 *N* HCl for at least 1 h, rinse with 0.1 *M* KH_2PO_4 until the pH is 7.0–8.0, then air dry overnight (*see* **Note 2**).
4. Pack the bottom 0.5 cm of a 1-mL syringe with acid-washed glass wool, then fill with Sepharose 4B. Centrifuge for 4.5 min at 1000*g*. Refill and pack the material by centrifugation until the syringe is filled to the 0.9-mL mark.
5. Wash with 200 µL of TE, centrifuge for 4.5 min at 1000*g*. Repeat for a total of four washes; after the final wash centrifuge the column again to remove excess TE. Packed columns can be stored up to 2 wk at 4°C with both ends sealed.
6. 0.7% Agarose gel and running buffer for analysis of spin column eluent.
7. Ice-cold 95% and 70% ethanol.

2.2.4. Final DNA Purification

1. 1 mg/mL DNase-free RNase (**Subheading 2.1.1.**).
2. 14 mg/mL Proteinase K.
3. TE-buffered phenol (**Subheading 2.1.1.**).
4. Ice-cold 95% and 70% ethanol.
5. 0.7% Agarose gel and running buffer for analysis of purified DNA.
6. Ethidium bromide plates for DNA quantification.

3. Methods

3.1. Designing and Preparing the Master Filter (see *Note 3)*

3.1.1. Isolation of DNA From Pure Bacterial Strains

Cells should be grown to a high density in appropriate medium. The procedure described below is designed for approx 0.5 g (wet weight) of cells.

1. Collect cells by centrifugation for 20 min at 4°C, 15,000*g*. Some samples may require centrifugation for longer periods of time at higher speeds to collect cells.
2. Resuspend cells in 250 µL of lysis buffer and transfer to a 1.5-mL Eppendorf tube.
3. Add 10 µL of 1 mg/mL lysozyme and incubate at 37°C for 30 min.
4. Add 20 µL of 25% SDS and lyse cells using three freeze/thaw cycles (30 min at –80°C followed by 30 min at 65°C).
5. Add 60 µL of 5 *M* $NaClO_4$ and 300 µL of $CHCl_3$:isoamyl alcohol, then gently mix on a tube roller for 30 min.
6. Centrifuge in a microfuge at maximum speed, room temperature, for 5 min.

7. Transfer aqueous phase to a fresh 1.5-mL Eppendorf tube and add 1 mL of 95% ethanol; incubate at –20°C for at least 1 h. Centrifuge in a microfuge at maximum speed, 4°C, for 15 min. Discard supernatant, rinse pellet once with 500 µL of 70% ethanol, then air-dry pellet.
8. Resuspend in 300 µL TE, then add 15 µL of 1 mg/mL DNase-free RNase. Incubate 30–60 min at 37°C, add 5 µL of 14 mg/l proteinase K, and incubate at 37°C for 30–60 min.
9. Add 100 µL of TE-buffered phenol, mix, and centrifuge (maximum speed in a microfuge) for 5 min at room temperature.
10. Transfer aqueous phase to a fresh 1.5-mL Eppendorf tube and precipitate with 750 µL ethanol as in **step 7**.
11. Resuspend DNA pellet in TE at desired concentration.

3.1.2. Quantification of DNA (see **Notes 4–6**)

Two microliters of known concentrations of λ DNA (5, 10, 15, 20, 25, 30, 40, 50, 60, 80, and 100 ng/µL) are spotted on an ethidium bromide plate, followed by 2 µL of sample DNA. The plate is allowed to "develop" for 1 h, then DNA spots are visualized on a UV light box. Standards must be spotted at the same time as samples.

3.1.3. Preparation of Filters (see **Note 7**)

Filters can be prepared using various amounts of DNA; 20–1450 ng per spot have been used successfully ***(6,7)***. A useful concentration range is 100 to 200 ng of DNA per spot.

1. Mark 1 × 1 cm squares on a nylon membrane for each bacterial DNA standard, plus eight squares for λ DNA concentration standards.
2. Dilute λ DNA to 10, 20, 30, 50, 60, 80, and 100 ng/µL concentrations.
3. Dilute microbial standard DNA samples such that 2 µL of DNA solution provides the amount to be spotted on the master filter.
4. DNA is denatured by boiling for 3 min followed by placing on ice for 3 min.
5. Spot 2 µL of each DNA solution onto the nylon membrane.
6. Dry the nylon membrane for 15 min at 80°C in a drying oven.
7. Fix the DNA to the nylon membrane by exposing it to ultraviolet light (365 nm and 7000 µW/cm^2) for 3 min.
8. Wash filters in 6X SSC, and air-dry. Filters can be stored at –20°C indefinitely.

3.1.4. Generation of Probes and Hybridization with Filters

3.1.4.1. Generation of Probe

1. Add 100 ng sample DNA to a 1.5-mL Eppendorf tube; make up to 15 µL with sterile distilled water. Add 5 µL of freshly prepared 0.5 ng/µL λ DNA (*see* **Note 8**).
2. Boil DNA mixture for 3 min, then ice for 3 min. Briefly centrifuge to collect sample.

3. Add 6 µL of PE mix, 2 µL of DNA polymerase I Klenow fragment, and 2 µL of [α^{32}P]dCTP (*see* **Note 9**), then incubate at room temperature for approx 3 h (*see* **Note 10**).

3.1.4.2. Preparation of Filters for Hybridization

1. Place each filter in a polypropylene bag (*see* **Notes 11** and **12**). Add prehybridization mixture to the bag (approx 125 µL per DNA spot), then remove all bubbles and seal.
2. Place sealed bags in a container of water and heat to 68°C in a hybridization oven; incubate rocking for at least 1 h before adding probe.

3.1.4.3. Probe Filter

1. After the probe has incubated for 3 h, boil for 3 min, then place on ice for 3 min.
2. Cut off the corner of the polypropylene bag, add probe to the prehybridized filter, then reseal the bag, removing all bubbles. This step should be done quickly so that the solution does not cool excessively.
3. Return to the 68°C oven and incubate overnight, rocking.
4. Remove filters from bags, wash in 100 mL of 1X SSC for 15 min at room temperature, rocking; then wash a second time in 100 mL of 1X SSC plus 0.2% SDS for 1 h at 68°C, rocking (*see* **Note 13**). Remove filters from wash and air-dry.
5. Expose a phosphoimager plate to the filters for 1 to 3 h, read the results (e.g., using a Fuji Bas1000 Bio-Imaging Analyzer), then quantify the relative intensity of each spot (e.g., using the MacBAS program) (*see* **Notes 14** and **15**).

3.2. Isolation and Purification of DNA From Environmental Samples

3.2.1. Isolation of Bacterial Cells From Soil (see Note 16)

1. This protocol is designed for approx 5 g soil. Dry soil or soil cultures can be used.
2. Place 5 g soil, 20 mL 0.1% $Na_4P_2O_7 \cdot 10H_2O$ (or culture medium, made up to 20 mL total liquid volume with 0.1% $Na_4P_2O_7 \cdot 10H_2O$), and 1 g acid-washed PVPP (*see* **Note 17**) in a 50-mL beaker, and mix on a magnetic stirrer for 20 min.
3. Centrifuge for 10 min at 4°C, 1000*g*; collect supernatant.
4. Wash soil twice more with 10 mL 0.1% $Na_4P_2O_7 \cdot 10H_2O$. Pool collected supernatants.

3.2.2. Cell Lysis and DNA Isolation

1. Centrifuge liquid sample or cells washed from solid support (**Subheading 3.2.1.**) for 20 min at 4°C, 15,000*g*.
2. Resuspend the cell pellet in 5 mL of lysis buffer.
3. Add 250 µL of 300 mg/mL lysozyme then incubate at 37°C for 30–60 min.
4. Add 2.5 mL of 25% SDS per tube, mix by inversion. Lyse cells using three freeze/thaw cycles (30 min at –80°C followed by 30 min at 65°C).
5. Add 0.6 mL of 5 *M* $NaClO_4$ and 3 mL of $CHCl_3$:isoamyl alcohol, then mix on a tube roller for approx 30 min.

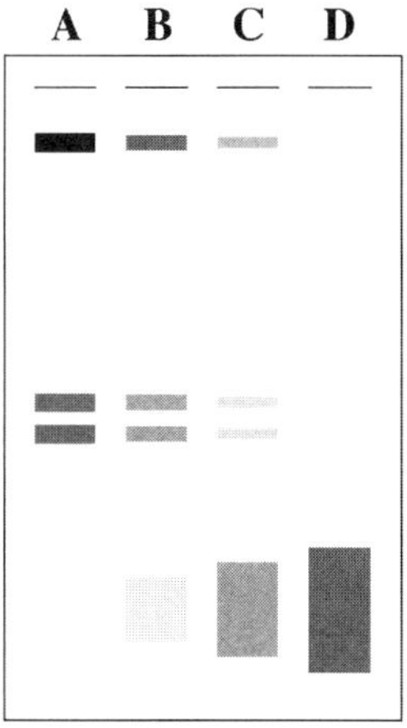

Fig. 2. Schematic of a typical agarose gel showing fractions eluted from spin columns. (A) Fraction 1; (B) fraction 2; (C) fraction 3; (D) fraction 4.

6. Centrifuge at 3500*g* for 10 min at room temperature, then transfer the aqueous phase into 50-mL glass centrifuge tubes.
7. Add 2.5 volumes of 95% ethanol; incubate at –20°C for at least 1 h. Centrifuge at 15,000*g*, 4°C for 20 min. Wash pellet with 1 mL of 70% ethanol and air-dry.
8. Resuspend DNA pellet in TE (*see* **Note 18**).

3.2.3. Remove Humic Acids From Soil DNA (see **Note 19**)

1. Load up to 200 µL of DNA sample onto a spin column (*see* **Notes 20** and **21**).
2. Centrifuge for 4.5 min at 4°C, 1000*g*. Collect eluent in a 1.5-mL Eppendorf tube.
3. Wash the columns 2–3 times, each time by loading 100 µL of TE on the top of the column and centrifuging, collecting each eluent in a fresh Eppendorf tube.
4. Analyze the column eluents by gel electrophoresis. Pool all samples that contain DNA but no humic acids (**Fig. 2**).
5. Add 2.5 vol of ethanol to combined samples, incubate at –20°C for at least 1 h, then centrifuge for 15 min at 4°C, maximum speed in a microfuge. Rinse pellet with 70% ethanol, then air-dry pellet. Resuspend samples in 300 µL TE (*see* **Note 22**).

3.2.4. Final DNA Purification (see **Note 23**)

1. Add 15 µL of 1 mg/mL DNase-free RNase, incubate for 30–60 min at 37°C, then add 5 µL of 14 mg/mL proteinase K and incubate for 30–60 min at 37°C.
2. Add 100 µL of TE-buffered phenol, mix, then centrifuge at maximum speed in a microfuge for 5 min at room temperature.
3. Collect aqueous phase, add approx 750 µL of ice-cold 95% ethanol and precipitate at –20°C for at least 1 h.

4. Centrifuge at maximum speed in a microfuge at 4°C for 15 min, discard supernatant, rinse pellet with 500 µL of 70% ethanol, and air-dry. Resuspend pellet in an appropriate amount of TE (typically around 30 µL for samples isolated from soil; for samples containing more DNA, a larger volume may be required).
5. Quantify DNA (**Subheading 3.1.2.**). Dilute DNA with TE to a concentration that is useful for probe generation (e.g., 10 ng/µL).

3.3. Calculation of f_x

To quantify hybridization of radiolabeled sample DNA to various microbial standards, the hybridization intensity is evaluated using the following equation:

$$f_x = (k_\lambda/k_x)(I_x/c_x)(f_\lambda)(I_\lambda/c_\lambda)^{-1} \qquad (1)$$

- f_x is the weight fraction of genome x present in the DNA mixture used for probing.
- k_λ/k_x is a constant that represents the relative genome complexity, which is unknown for the genomes of various bacterial standard organisms. This constant must be measured for each standard.

1. Mix λ DNA and DNA from a specific standard together such that $f_x = 0.976$ and $f_\lambda = 0.024$.
2. Label DNA and hybridize to the master filter as described in **Subheading 3.1.4.**
3. Calculate k_λ/k_x. This must be done for each standard on the master filter; this constant can be used to calculate the relative f_x for each standard when probed with unknown community DNA samples. The constant k_λ/k_x can be calculated using the following equation:

$$k_\lambda/k_x = (f_x/f_\lambda)(I_\lambda/c_\lambda)(I_x/c_x)^{-1} \qquad (2)$$

4. Repeat this experiment in duplicate; use the average result from the duplicate samples as the value for k_λ/k_x.

- I_x is the relative intensity of standard x on a given master filter, as measured by densitometry or by cpm in a scintillation counter. Background radioactive signal should be subtracted from the total intensity.
- c_x is the amount of standard x DNA spotted on the master filter, in ng.
- f_λ is the fraction of λ DNA present in each labelled probe mixture.
- c_λ is the concentration of λ DNA spotted on the master filter.
- I_λ is the relative intensity of λ DNA on the master filter for each hybridization experiment, as measured by densitometry or by cpm in a scintillation counter. The presence of several concentrations of λ DNA on each master filter allows accurate measurement of this value.

The f_x values calculated with **Eqs. 1** and **2** cannot be corrected for contributions due to cross-hybridization. Cross-hybridization will cause every reported f_x value to be overestimated. This will cause the sum Σf_x of all f_x values to exceed 1. One can report data by setting $\Sigma f_x = 1$. The resulting relative f_x values are reported as %, not as fraction between 0 and 1. A drawback of this is that relative f_x values (%) represent the fractions of standards in the portion of the community represented on the master filter.

4. Notes

1. Solutions can be prepared in advance, autoclaved, and stored at room temperature unless otherwise noted. Solutions stored at 4°C or –20°C are typically not autoclaved.
2. Glass wool used for the spin columns does not have to be acid-washed; however this treatment appears to give higher DNA yields.
3. Before preparing the master filter, it is important to determine which microbial strains of interest do not show significant cross-hybridization. This can be done using DNA filters to assess cross-hybridization between various potential standard genomes. Once standard bacterial strains have been selected, a master filter can be developed. It is important to select bacterial strains that are appropriate to the experimental design and are represented in the community of interest.
4. DNA can be quantified by fluorimetry, A_{260} or visual comparison of fluorescence with ethidium bromide staining and UV light after gel electrophoresis, or after spotting samples onto agarose containing ethidium bromide.
5. Condensation interferes with the accuracy of ethidium bromide plate determinations.
6. Quantification by A_{260} or ethidium bromide plates will work only for pure DNA samples. Humic acids from soil are also detected by these methods *(5)*; when the presence of humic acids is suspected, gel electrophoresis and staining will be a more accurate method of DNA quantification.
7. Used filters can be reused when radioactivity can no longer be detected, or after washing. Washing procedure:
   a. Heat 1X SSC plus 0.2% SDS to boiling.
   b. Pour heated solution over filters. Incubate for 30–60 min at 80°C, rocking.
   c. Allow to cool to room temperature, remove filters, and air-dry.
   d. Repeat procedure until radioactivity is no longer detected.
8. Some RSGP users *(8)* digest DNA with a restriction enzyme (*Sau*3A) before labeling.
9. Other researchers have used ^{35}S *(8)* or a nonisotope labeling kit (DIG DNA Labeling and Detection *[9]*) rather than ^{32}P.
10. While probe is being incubated, master filters can be prepared for hybridization (**Subheading 3.1.4.2.**).
11. Hybridization bags are less cumbersome if they are cut to fit the filter fairly precisely on three sides; there should be extra room on the fourth side as the bag needs to be cut and resealed during the probing process.
12. Rather than using the hybridization bag method, a tube roller and individual hybridization tubes can be used. However, if a large number of samples are being screened, polypropylene bags allow more samples to be processed simultaneously.

13. More than one filter can be washed in the same container; however, some background radioactivity may appear if several filters are present in a single container.
14. If no phosphoimager system is available, DNA hybridization to the master filter can be detected using X-ray film and quantified by densitometry.
15. An alternate method for quantifying sample DNA hybridization to the master filter is to collect each DNA spot and measure the total radioactivity by scintillation counting *(7,8)*; however, this method precludes reuse of the master filters.
16. This protocol is also useful for removing bacterial cells from solid support materials.
17. Acid-washed PVPP is used to remove humic acids; therefore, this compound may be omitted for extracting cells from solids that do not contain these substances.
18. The volume of TE used to resuspend DNA depends on the sample. Typically 300 μL is suitable for soil DNA; larger volumes may be required if the sample still contains substantial amounts of material from soil or if the DNA concentration is high.
19. This step is required only for samples that may contain humic acids.
20. DNA can also be isolated by gel electrophoresis and electroelution, or other means of gel purification. Spin columns are good for processing many samples at once.
21. If the DNA preparation contains suspended solids, a large amount of humic acids, or a large amount of DNA, only 50 to 100 μL should be loaded on the column; the remaining volume should be made up with TE. The column can become plugged, or removal of humic acids can be poor, if it is overloaded.
22. If a sample appears brown after purification on a spin column, it still contains humic acids and must be re-cleaned because they will interfere with the labeling procedure.
23. This step is not necessary if the sample was gel purified, because RNA and protein will have been removed during that process.

References

1. Sambrook, J., Fritsch, E. F., and Maniatis, T. (1989) *Molecular Cloning: A Laboratory Manual,* 2nd ed. Cold Spring Harbor Laboratory Press, Cold Spring Harbor, NY.
2. Voordouw, G., Voordouw, J. K., Jack, T. R., Foght, J., Fedorak, P. M., and Westlake, D. W. S. (1992) Identification of distinct communities of sulfate-reducing bacteria in oil fields by reverse sample genome probing. *Appl. Environ. Microbiol.* **58,** 3542–3552.
3. Voordouw, G., Niviere, V., Ferris, F. G., Fedorak, P. M., and Westlake, D. W. S. (1990) The distribution of hydrogenase genes in *Desulfovibrio* and their use in identification of species from the oil field environment. *Appl. Environ. Microbiol.* **56,** 3748–3754.

4. Holben, W. E., Jansson, J. K., Chelm, B. K., and Tiedje, J. M. (1988) DNA probe method for the detection of specific microorganisms in the soil bacterial community. *Appl. Environ. Microbiol.* **54,** 703–711.
5. Jackson, C. R., Harper, J. P., Willoughby, D., Roden, E. E., and Churchill, P. E. (1997) A simple, efficient method for the separation of humic substances and DNA from environmental samples. *Appl. Environ. Microbiol.* **63,** 4993–4995.
6. Shen, Y., Stehmeier, L. G., and Voordouw, G. (1988) Identification of hydrocarbon-degrading bacteria in soil by reverse sample genome probing. *Appl. Environ. Microbiol.* **64,** 637–645.
7. Voordouw, G., Shen, Y., Harrington, C. S., Telang, A. J., Jack, T. R., and Westlake, D. W. S. (1993) Quantitative reverse sample genome probing of microbial communities and its application to oil field production waters. *Appl. Environ. Microbiol.* **59,** 4101–4114.
8. Bagwell, C. E. and Lovell, C. R. (2000) Persistence of selected *Spartina alterniflora* rhizoplane diazotrophs exposed to natural and manipulated environmental variability. *Appl. Environ. Microbiol.* **66,** 4625–4633.
9. Chao, W.-L., Tien, C.-C., and Chao, C.-C. (1997) Investigation of the effect of different kinds of fertilizers on the compositions of a soil microbial community using the molecular biology technique. *J. of the Chinese Agricultural Chemical Society* **35,** 252–262.

3

T-RFLP Analysis

A Rapid Fingerprinting Method for Studying Diversity, Structure, and Dynamics of Microbial Communities

Werner Liesack and Peter F. Dunfield

1. Introduction

Terminal restriction fragment length polymorphism (T-RFLP) analysis is a method for rapid profiling of mixed populations of an homologous amplicon (i.e., diverse sequences of a single gene). It combines restriction fragment analysis of a PCR-amplified gene marker with automated sequencing gel technology. One primer used in PCR amplification of the marker gene is labeled at the 5' terminus with a fluorescent dye, in order that the terminal restriction fragments (T-RFs) of the digested amplicon can be detected and quantified ***(1–3)***. Detailed evaluations of T-RFLP analysis have shown that, in most cases, both the sizes and relative signal intensities of the individual T-RFs in a sample are highly reproducible. Consequently, T-RFLP analysis is an excellent tool for rapidly comparing microbial communities ***(4–6)***.

Assessment of the diversity, structure, and dynamics of complex microbial communities with T-RFLP has mainly been based on PCR-amplified 16S rRNA genes (16S rDNA). The major advantage of 16S rDNA, a universal phylogenetic gene marker, is that primer systems can be applied in PCR that target a wide range of members of the domain Bacteria ***(5–20)*** or Archaea ***(7,21–31)***. Microbial community patterns have been generated from various environments, including activated sludge ***(14,16)***, bioreactors ***(6,13)***, enrichment cultures ***(9,11)***, marine sediments ***(7,18,20)***, lake sediment ***(28)***, soils ***(5,8,17,19,29,32)***, soil slurries ***(21–23,25,26)***, plant roots ***(12,32)***, rice straw ***(31)***, waters from

From: *Methods in Biotechnology: Environmental Microbiology: Methods and Protocols*
Edited by: J. F. T. Spencer and A. L. Ragout de Spencer

deep gold mines in South Africa *(30)*, the hindgut of soil-feeding termites *(24)*, and the colon of pigs *(15)*. Meaningful applications of the analysis to the fields of forensics *(33)* and food and drinking water quality control *(34,35)* have been reported. T-RFLP analysis has also been used for rapidly screening the diversity of 16S rRNA gene clone libraries *(27,32)*.

PCR assays for genes other than 16S rDNA have also been used to generate T-RFLP community patterns for particular functional groups of bacteria. These assays target genes specific to autotrophic ammonia oxidizers *(36)*, denitrifiers *(7,37)*, methane-oxidizing bacteria *(38)*, methanogens *(39)*, or mercury-resistant bacteria *(40)*.

The procedure for electrophoresis of digested amplicons outlined below is for use with ABI 373 and ABI Prism® 377 automated sequencers (PE Applied Biosystems, Foster City, CA). However, in principle any sequencing apparatus which allows automated detection of at least two different fluorescent dyes in a single lane is applicable to T-RFLP. Simultaneous detection of two fluorescent dyes is necessary because the automated size determination of T-RFs requires an internal lane standard (**Fig. 1**). If other automated sequencers are to be used, the procedure may need minor modifications.

T-RFLP community patterns have been interpreted and compared in many studies using only a qualitative approach (e.g., **refs.** *12,17,21,22,25,30,35*). However, although T-RFLP is a PCR-based technique and PCR-related biases must be considered, comparative numerical analysis of T-RFLP patterns and representation in reduced space can provide meaningful results (e.g., **refs.** *5–7,13,29,37*).

2. Materials

1. Total community DNA.
2. Oligonucleotide primers.
3. PCR equipment.
4. Agarose gel equipment.
5. QIAquick® PCR purification kit (Qiagen, Hilden, Germany).
6. Restriction enzyme(s).
7. Polyacrylamide gel electrophoresis equipment.
8. GeneScan-1000 ROX internal lane standard (ABI).
9. Model ABI 373 or ABI Prism 377 automated sequencer equipped with GeneScan software.
10. Statistical analysis software for comparative data interpretation.

3. Methods

The methods described below outline (1) PCR amplification of target genes and digestion of amplicons, (2) polyacrylamide gel electrophoresis (PAGE) of digested amplicons, (3) assignment of T-RFs, (4) standardization, and (5) data interpretation.

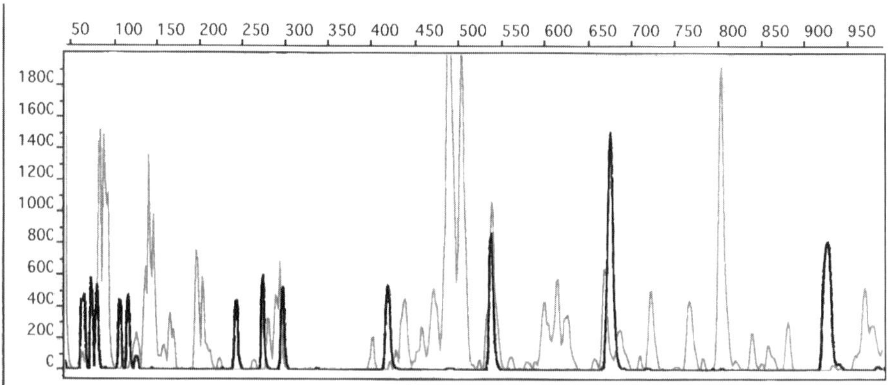

Dye/Sample Peak	Minutes	Size	Peak Height	Peak Area	Data Point
4B, 10	41.90	84.58	1087	9852	419
4B, 11	43.10	89.80	1069	6257	431
4B, 15	55.30	136.54	359	2296	553
4B, 16	55.90	138.76	473	3007	559
4B, 17	56.80	142.08	974	7327	568
4B, 18	58.20	147.24	712	7153	582
4B, 23	72.00	197.64	556	5770	720
4B, 24	74.00	204.88	432	3537	740
4B, 29	98.00	290.84	347	3817	980
4B, 30	99.30	295.66	505	3824	993
4B, 35	139.70	440.28	331	4917	1397
4B, 39	152.80	489.89	2021	42136	1528
4B, 40	156.80	505.09	1501	24424	1568
4B, 42	165.90	539.80	767	12650	1659
4B, 44	180.60	600.80	314	4560	1806
4B, 46	184.10	615.97	420	6164	1841
4B, 50	196.20	670.48	474	5709	1962
4B, 54	207.30	724.07	364	4705	2073
4B, 55	216.00	768.62	318	4641	2160
4B, 57	222.70	804.55	1375	16712	2227
4R, 5 •	37.90	67.00	358	1704	379
4R, 6 •	39.80	75.00	418	1726	398
4R, 7 •	41.10	81.00	389	1747	411
4R, 8 •	47.60	108.00	340	1823	476
4R, 9 •	50.30	118.00	342	2192	503
4R, 10 •	84.90	244.00	319	3134	849
4R, 11 •	93.70	275.00	436	3344	937
4R, 12 •	100.20	299.00	383	3679	1002
4R, 14 •	134.60	421.00	394	3455	1346
4R, 15 •	165.70	539.00	620	6367	1657
4R, 16 •	197.60	677.00	1079	14599	1976
4R, 17 •	243.80	928.00	585	8865	2438

Fig. 1. Output format of T-RFLP data. The pattern was generated from the 5' region of PCR-amplified 16S rDNA, using PAGE of the digested amplicon in the GeneScan mode of an ABI 373 automated sequencer. The T-RFLP fingerprint pattern of a bacterial community from soil (gray) is shown in relation to the GeneScan-1000 ROX size standard (black). The table includes the (1) fluorescent dye used (gray = FAM, black = ROX) and T-RF number, (2) retention time in minutes, (3) fragment size in base pairs, (4) peak height, (5) peak area, and (6) number of data points. The sizes of DNA fragments in the GeneScan-1000 standard are 29, 33, 37, 64, 67, 75, 81, 108, 118, 244, 275, 299, 421, 539, 674, 677, and 928 bp.

3.1. PCR of Target Gene and Restriction of Amplicons

1. After extraction of total community DNA (*see* **Note 1**), amplify by PCR the target gene of interest. PCR assays used in T-RFLP analysis are described in **Subheadings 3.1.1.** and **3.1.2.** One of the two oligonucleotide primers used in PCR must be labeled at the 5' terminus with a fluorescent dye (*see* **Note 2**).
2. Verify amplicon by agarose gel electrophoresis of an aliquot.
3. Purify the amplicon using the QIAquick® PCR purification kit (Qiagen) according to the manufacturer´s instructions (*see* **Note 3**).
4. Digest the amplicon using the appropriate restriction endonuclease as indicated in **Subheadings 3.1.1.** and **3.1.2.**

3.1.1. PCR Amplification of 16S rDNA

Various oligonucleotide primer combinations have been used for PCR amplification of 16S rDNA from members of the domain Bacteria (PCR assays a to d) or Archaea (PCR assays e, f). These are listed below as target stretch according to the nomenclature of *Escherichia coli* 16S rRNA; oligonucleotide PCR primers 5'-3' (* = labeled primer) (*see* **Note 4**); PCR program; restriction enzyme; and reference:

a. Target stretch: 8 to 536; *27f (AGAGTTTGATCCTGGCTCAG) and 519r (GWATTACCGCGGCKGCTG); initial denaturation (5 min at 94°C), 35 cycles of denaturation (30 s at 95°C), primer annealing (60 s at 54°C), and extension (2 min at 72°C); combination of *Hha*I and *Hae*III ***(19)***.
b. Target stretch: 8 to 1406; *27f (AGAGTTTGATCCTGGCTCAG) and 1392r (ACGGGCGGTGTGTRC); initial denaturation (2 min at 94°C), 30 cycles of denaturation (45 s at 94°C), primer annealing (60 s at 48°C), and extension (2 min at 72°C), plus a final extension (8 min at 72°C); *Msp*I ***(5)***.
c. Target stretch: 8 to 1541; *27f (AGAGTTTGATCCTGGCTCAG) and 1525r (AAGGAGGTGWTCCARCC); initial denaturation (5 min at 95°C), 30 cycles of denaturation (30 s at 94°C), primer annealing (30 s at 55°C), and extension (90 s at 72°C), plus a final extension (10 min at 72°C); *Mnl*I ***(12)***.
d. Target stretch: 43 to 1406; *63f (CAGGCCTAACACATGCAAGTC) and 1389r (ACGGGCGGTGTGTACAAG); initial denaturation (2 min at 94°C), 30 cycles of denaturation (60 s at 94°C), primer annealing (60 s at 55°C), and extension (2 min at 72°C), plus a final extension (10 min at 72°C); *Alu*I or *Hha*I ***(4)***.
e. Target stretch: 109 to 931; Arch109f (ACKGCTCAGTAACACGT) and *Arch912r (CTCCCCCGCCAATTCCTTTA); initial denaturation (5 min at 94°C), 28 cycles of denaturation (60 s at 94°C), primer annealing (60 s at 52°C), and extension (90 s at 72°C), plus a final extension (6 min at 72°C); *Taq*I ***(25)***.
f. Target stretch: 7 to 976; Arch21f (TTCCGGTTGATCCYGCCGGA) and *Arch958r (YCCGGCGTTGAMTCCAATT); initial denaturation (3 min at 94°C), 30 cycles of denaturation (60 s at 94°C), primer annealing (60 s at 55°C),

and extension (60 s at 72°C), plus a final extension (7 min at 72°C); *Hha*I, *Rsa*I, or *Hae*III ***(27)***, or a combination of *Hha*I and *Hae*III ***(30)***.

3.1.2. PCR Amplification of Functional Gene Markers

Various genes other than 16S rDNA have been used to generate T-RFLP community patterns for particular functional groups of bacteria. These are listed below as target gene; enzyme; functional group of bacteria; oligonucleotide PCR primers 5'–3' (* = labeled primer) (*see* **Note 4**); PCR program; restriction enzyme; and reference:

a. *amoA;* ammonia monooxygenase; autotrophic ammonia-oxidizing bacteria; **amoA*-1F (GGGGTTTCTACTGGTGGT) and *amoA*-2R (CCCCTCKGSAAA-GCCTTCTTC); initial denaturation (5 min at 94°C); 30–35 cycles of denaturation (60 s at 94°C), annealing (90 s at 60°C), and extension (90 s at 72°C), plus a final extension (10 min at 72°C); *Taq*I, plus *Cfo*I or *Alu*I for finer resolution ***(36)***.
b. *nirS;* nitrite reductase; denitrifiers; *nirS1F (CCTAYTGGCCGCCRCART) and nirS6R (CGTTGAACTTRCCGGT); initial denaturation (5 min at 95°C), 30 "touchdown" cycles of denaturation (30 s at 95°C), primer annealing (40 s at 56–51°C for 10 cycles, 54°C for 25 cycles), and extension (40 s at 72°C), plus a final extension (7 min at 72°C); *Hha*I, *Msp*I, or *Taq*I ***(7)***.
c. *nosZ;* nitrous oxide reductase; denitrifiers; *Nos661F (CGGCTGGGGGCTGA CCAA) and Nos1773R (ATRTCGATCARCTGBTCGTT); initial denaturation (5 min at 94°C), 35 cycles of denaturation (30 s at 95°C), annealing (30 s at 55°C), and extension (90 s at 72°C), plus a final extension (10 min at 72°C); *Hin*PI ***(37)***.
d. *pmoA;* particulate methane monooxygenase; methane-oxidizing bacteria; *A189 (GGNGACTGGGACTTCTGG) and A682 (GAASGCNGAGAAGAASGC); initial denaturation (2 min at 94°C), 30 "touchdown" cycles of denaturation (45 s at 94°C), primer annealing (60 s at 62–52°C for 20 cycles, 55°C for 10 cycles), and extension (2 min at 72°C), plus a final extension (6 min at 72°C); *Msp*I ***(38)***.
e. *mcrA;* methyl coenzyme M reductase; methanogenic archaea; *MCRf (TAYGAYCARATHTGGYT) and MCRr (ACRTTCATNGCRTARTT); initial denaturation (3 min at 94°C), 30–35 cycles of denaturation (45 s at 94°C), annealing (45 s at 50°C) and extension (90 s at 72°C), plus a final extension (5 min at 72°C); *Sau96*I ***(39)***.
f. *merR, merRTΔ;* mercury resistance genes; mercury-resistant bacteria; *FluRX (ATAAAGCACGCTAAGGCRTA) and either PX (TTCTTGACWGTGA TCGGGCA) or MARB (GTCAAYGTGGAGACVATCCG); initial denaturation (4 min at 95°C), 30–35 cycles of denaturation (60 s at 94°C), primer annealing (60 s at 55°C), and extension (4 min at 72°C), plus a final extension (10 min at 72°C); *Fok*I ***(40)***.

3.2. PAGE

1. Prepare a mixture of digested amplicon, deionized formamide, loading buffer (ABI), and internal GeneScan-1000 (ROX) size standard, denature it at 94°C for 5 min, and immediately chill on ice until loading onto the gel. Note that the maximum loading capacity of the wells produced by the shark-tooth combs is 2 to 3 µL.
2. Electrophorize the digested mixture in the GeneScan mode with the following conditions:
   Model ABI 373 ***(5,21,22)***:
   12-cm 6% (w/v) polyacrylamide gel containing 8.3 *M* urea and 1X TBE buffer (89 m*M* Tris-borate, 2 m*M* EDTA); electrophoresis for 6 h with the following settings: 2500 V, 40 mA, and 27 W.
   Model ABI Prism 377 ***(26)***:
   36-cm 5% (w/v) polyacrylamide gel containing 6 *M* urea and 1X TBE buffer; electrophoresis for 14 h with the following settings: 2500 V, 40 mA, and 30 W.

3.3. Assignment of T-RFs

GeneScan software contains several size-calling algorithms to calculate T-RF size in relation to the internal GeneScan-1000 (ROX) standard. The appropriate algorithm should be determined empirically for particular running conditions and equipment. The errors in assignment are generally less than 1 bp (**Fig. 1**), but manual rounding up or down may be necessary (i.e., assignment of fragments based on predicted T-RF sizes from clone libraries; *see* **Notes 5** and **6**).

3.4. Standardization

Peaks less than 35 bp in size are discarded to avoid detection of primers. All other peaks above a baseline noise level (usually peak height between 25 and 100 fluorescence units) are kept. The choice of baseline noise will affect the number of peaks detected. A higher threshold will decrease the number of small, irreproducible peaks obtained, but may also discard some reproducible peaks (*see* **Note 7**).

The proportional abundance (A_i) of each T-RF is calculated as:

$$A_i = n_i/N$$

in which n_i represents the peak area (or height) of a T-RF (i) and N is the sum of all peak areas (or heights) in a given T-RFLP pattern. Peak area is preferable to peak height when it can be calculated. However, in profiles with many closely spaced peaks, area becomes difficult to integrate properly and peak height is more reliable.

There may be considerable variability in the detection limit across samples because of variables such as differing amounts of product loaded into the gel. A correction method has been suggested by Dunbar et al. ***(41)***, in which the profile with the smallest total peak area (or height) is used to standardize results as follows:

$$A_{i(corr)} = (N_{std} / N) \times A_i$$

where $A_{i(corr)}$ is the corrected area of each T-RF and N_{std} is the lowest N of all samples. If small peaks fall below the minimum detection limit after the adjustment, they are eliminated and the correction performed iteratively until the total areas of the profile and standard are nearly equal. Essentially, this correction sets the detection limit of all samples to that of the worst (least PCR product). This may entail a loss of information, but most of the loss will be small peaks which are difficult to reproduce in any case. Whether this standardization is useful or not is system-specific. If a quantitative extract of environmental DNA can be obtained and profiles are very reproducible, as is often the case *(4–6)*, the correction is not necessary. However, where this is difficult to assess, this correction may improve reproducibility and facilitate subsequent comparative analyses.

3.5. Data Interpretation

3.5.1. Single T-RFs

Biases involved in environmental DNA extraction and in primer annealing to different templates during PCR (e.g., **refs.** ***42,43***) mean that certain DNA sequences (or T-RFs) are preferentially retrieved from a sample. Therefore, one particular T-RF cannot be compared to a different T-RF in a single profile (i.e., $A_1 > A_2$ does not mean that Population 1 > Population 2). However, it is generally assumed that a T-RF can be compared to itself over different samples (i.e., if A_1 in sample X > A_1 in sample Y, then population 1 probably comprises a greater proportion of the total population in sample X than in sample Y). Although not necessarily true (*see* **Note 8**), this generalization is the logic behind the presentation of stacked bar diagrams to compare all T-RFs across samples (e.g., ***7,24,29,31***). Whether the variation of T-RF abundances across samples is significant can be statistically tested with multivariate analysis of variance ***(5,29)***.

3.5.2. Community Analysis

Because of the severe method-inherent limitations in comparing single T-RFs (*see* **Note 8**), a greater value of T-RFLP analysis is in characterizing and comparing entire communities. Provided one is aware of method-inherent limitations when drawing conclusions, the same statistical analyses can be applied to T-RFLP patterns that are applied to any other community-profiling method. We give here only the briefest overview of common analyses; for details, the reader is referred to original literature and basic texts on analytical methods—for example, the text of Legendre and Legendre ***(44)***.

Most simply, diversity indices such as the Shannon-Weaver index [H = $-\Sigma (A_i) \log (A_i)$] and Simpson's index [$D = 1 - \Sigma(A_i)^2$] can be calculated and compared across samples. Comparison of samples can also be made using a similarity coefficient. This can incorporate both presence/absence and relative abundance data, as the Morisita coefficient ***(6)***. Alternatively, to avoid all but the most severe problems arising from extraction and PCR biases, the coefficient can be based on presence/absence data only, as the Sørenson coefficient or the Jaccard coefficient ***(7,8,18,41)***. The Ribosomal Database Project (http://www.cme.msu.edu/RDP/html/analyses.html) contains an online utility for calculating the Jaccard coefficient of different T-RFLP patterns ***(45)***. When choosing a similarity coefficient, one should be aware of the double-negative problem ***(44)***, that the presence of a T-RF in two communities should be given more weight than the absence of a T-RF. Both the Jaccard and Sørensen coefficients are recommended for this reason.

Cluster analysis based on a matrix of similarity coefficients can be used to visualize the relationships among many samples ***(6–8,18,41)***. As a complement to cluster analysis, a variety of multivariate techniques involving ordination in reduced space can also be used to analyze entire community patterns. For example, principal component analysis has been applied ***(6,13,37)***. However, because this technique does not weight against double negatives, other types of factor analysis such as correspondence analysis ***(7)*** are to be preferred in most cases ***(44)***. In order to formally test the effects of particular environmental variables (plot treatments) on community patterns, Lukow et al. ***(5)*** employed a multivariate analysis of variance with canonical analysis.

4. Notes

1. Total community RNA has also been used as the starting material for T-RFLP analysis. Community patterns were generated via RT-PCR of bacterial 16S rRNA ***(17)***. Detailed protocols for the extraction and purification of total community DNA and RNA from various environments ***(46)***, and excellent reviews addressing various procedural and technical aspects of the extraction of total community DNA and RNA from soils and sediments ***(47–51)***, have been published. For more information, the reader is also referred to the environmental studies cited here.
2. T-RFLP studies usually employ the 5' region of 16S rDNA because it provides a greater discrimination (i.e., an increased number of T-RFs) than does the 3' region. This is a consequence of length heterogeneities within the V1, V2, and V3 regions at the 5' region of the 16S rRNA gene ***(4)***. In a few cases, the forward and reverse primers have been labeled with different fluorescent dyes and used simultaneously in PCR ***(4,27,30)***. In this way it is possible to generate T-RFLP

datasets for the 5'- and 3'- T-RFs of a single amplicon simultaneously in one GeneScan run. As predetermined by the fluorescence detection facility of the ABI 373 and ABI Prism 377 automated sequencers, dyes used to label PCR primers in T-RFLP analysis are the fluorescein derivatives carboxy-fluorescein (5'6-FAM), 5-hexachlorofluorescein (HEX), 6-carboxy-4', 5'-dichloro-2', 7'-dimethoxyfluorescein (5'6-JOE), and 5-tetrachlorofluorescein (TET). For the simultaneous use of fluorescently labeled forward and reverse primers, the combinations FAM/HEX ***(4)***, FAM/JOE ***(27)***, or HEX/TET ***(30)*** have been employed.

3. The use of the QIAquick® PCR purification kit is recommended, but purification kits from other manufacturers can also be used (e.g., Prep-A-Gene kit [Bio-Rad, Munich, Germany]).
4. Degenerated nucleotide positions are indicated using the following standard degeneracy symbols:
   K = G:T // M = A:C // R = A:G // S = G:C // W = A:T // Y = C:T // B = G:C:T // H = A:C:T // V = A:G:C // N = A:G:C:T // all 1:1.
5. Unlike the related community-fingerprinting techniques of denaturing gradient gel electrophoresis ***(52–54)*** and single-strand-conformation polymorphism analysis ***(55,56)***, it is not yet possible in T-RFLP to correlate individual phylotypes (T-RFs) with phylogenetic information obtained either by probing or recovery of sequence information. However, one advantage of T-RFLP compared to these other methods is its automated quantification of both size and relative abundance of individual T-RFs. This enables a more objective comparison of community fingerprint patterns. In addition, the T-RFLP technique is easier to handle and more sensitive ***(5)***.

   A Web-based research tool (designated TAP T-RFLP) is available at the Ribosomal Database Project Web site (http://www.cme.msu.edu/RDP/html/analyses.html) ***(57)***. Here one can perform *in silico* restriction digestions of the entire 16S rDNA sequence database and derive T-RF sizes, measured in base pairs, from the 5' terminus of the user-specified primer to the 3' terminus of the restriction endonuclease target site. The output can be sorted and viewed either phylogenetically or by size. However, some T-RFs might be common to many phylogenetically diverse microorganisms (this is especially true for T-RFs generated from the 3' terminus of the 16S rRNA gene). As a consequence, an approach combining T-RFLP analysis and generation of clone libraries is recommended to correlate individual T-RFs with phylogenetic information. Individual clones should be analyzed by sequencing and determination of T-RF sizes, and these compared to the entire T-RFLP community pattern ***(17,38)***.
6. It has been observed that multiple peaks are often generated in T-RFLP analysis of individual nucleotide sequences (e.g., from pure cultures or from cloned DNA fragments). The generation of extra "pseudo T-RFs" from a particular sequence can seriously complicate T-RFLP analysis. In an environmental sample the generation of pseudo T-RFs may lead to an overestimation of microbial diversity, and impair the mapping of taxonomically meaningful OTUs to single peaks.

Thus, as already outlined, if the correlation of individual T-RFs with phylogenetic information is a major goal, a combined approach of T-RFLP analysis and generation of clone libraries is recommended.

There are several explanations for the generation of multiple T-RFs from a single sequence. The problem has usually been attributed to incomplete digestion of PCR product, which is easily correctable by optimizing the digestion procedure. However, other causes of the phenomenon are not so easily corrected. If the two restriction sites closest to the fluorescently labeled terminus are located within a 15-bp stretch, the restriction endonucleases may be unable to digest the first restriction site quantitatively (unpublished data). Pseudo T-RFs also arise through the generation of single-stranded DNA during PCR. Secondary structure formation in this single-stranded DNA can protect some restriction sites from attack ***(58)***. The use of single-strand-specific mung bean nuclease before T-RFLP digestion removes these pseudo T-RFs; however, one must be aware that this further complicates the "semiquantitative" nature of T-RFLP in microbial community analysis by selectively removing a fraction of the total PCR product (see below).

7. Despite great care taken to exactly reproduce a T-RF profile, small, irreproducible peaks are occasionally observed ***(4,5,41)***. This can occur even when the replication is only at the level of using different aliquots of a single restriction digest ***(4,41)***. The problem of irreproducible small peaks is usually handled by discarding peaks smaller than 100 fluorescence units in height or <1% of total peak area. To more accurately identify irreproducible peaks it is necessary to perform replicate T-RFLP analyses of a sample, at best beginning with separate DNA extractions in order to encompass the error occurring at every procedural step. However, major peaks in T-RF profiles are generally very reproducible, even when generated from multiple extracts of a single sample, and multiple extractions are therefore usually not necessary ***(4–6)***.
8. There are several obvious pitfalls in any analysis of single T-RFs. Comparison is based on proportions rather than absolute numbers, and therefore all T-RFs are cross-correlated. A change in the *absolute* abundance of one template in the DNA extract will cause the *relative* abundances of all others to change. Thus the relative abundance of a T-RF can change even if the absolute abundance of its DNA template is constant. A more insidious problem arising from method-inherent biases is that the relative abundance of a T-RF can change even when the relative abundance of the respective microbial population, or the relative abundance of the respective DNA template, remains the same. To illustrate this, imagine a pattern containing three T-RFs numbered 1–3. The extraction and PCR biases are such that the respective templates are preferentially extracted or amplified in the order 1>2>3. Suppose that some members of Population 3 in the environment are replaced with an equal number of Population 1. This will cause the relative proportion of T-RF 2 to decrease in the T-RFLP profile, despite the fact that neither the absolute nor the relative abundance of Population 2 has changed. Ecologically, such a situation could result from the replacement of one species with another species that is functionally similar but phylogenetically different.

Acknowledgments

This work was supported by grants from the Bundesministerium für Bildung, Wissenschaft, Forschung und Technologie awarded to W.L. (contracts 0311955 and 0312319) and from the Deutsche Forschungsgemeinschaft awarded to P. D. (DU377/1-1)

References

1. Liu, W. T., Marsh, T. L., Cheng, H., and Forney, L. J. (1997) Characterization of microbial diversity by determining terminal restriction fragment length polymorphisms of genes encoding 16S rRNA. *Appl. Environ. Microbiol.* **63,** 4516–4522.
2. Clement, B. G., Kehl, L. E., DeBord, K. L., and Kitts, C. L. (1998) Terminal restriction fragment patterns (TRFPs), a rapid, PCR-based method for the comparison of complex bacterial communities. *J. Microbiol. Methods* **31,** 135–142.
3. Marsh, T. L. (1999) Terminal restriction fragment length polymorphism (T-RFLP): an emerging method for characterizing diversity among homologous populations of amplification products. *Curr. Opin. Microbiol.* **2,** 323–327.
4. Osborn, A. M., Moore, E. R. B., and Timmis, K. N. (2000) An evaluation of terminal-restriction fragment length polymorphism (T-RFLP) analysis for the study of microbial community structure and dynamics. *Environ. Microbiol.* **2,** 39–50.
5. Lukow, T., Dunfield, P. F., and Liesack, W. (2000) Use of the T-RFLP analysis to assess spatial and temporal changes in the bacterial community structure within an agricultural soil planted with transgenic and non-transgenic potato plants. *FEMS Microb. Ecol.* **32,** 241–247.
6. Dollhopf, S. L., Hashsham, S. A., and Tiedje, J. M. (2001) Interpreting 16S rDNA T-RFLP data: application of self-organizing maps and principal component analysis to describe community dynamics and convergence. *Microb. Ecol.* **42,** 495–505.
7. Braker, G., Ayala-del-Rio, H. L., Devol, A. H., Fesefeldt, A., and Tiedje, J. M. (2001) Community structure of denitrifiers, *Bacteria,* and *Archaea* along redox gradients in pacific northwest marine sediments by terminal restriction fragment length polymorphism analysis of amplified nitrite reductase (*nirS*) and 16S rRNA genes. *Appl. Environ. Microbiol.* **67,** 1893–1901.
8. Buckley, D. H. and Schmidt, T. M. (2001) The structure of microbial communities in soil and the lasting impact of cultivation. *Microb. Ecol.* **42,** 11–21.
9. Covert, J. S. and Moran, M. A. (2001) Molecular characterization of estuarine bacterial communities that use high- and low-molecular weight fractions of dissolved organic carbon. *Aquat. Microb. Ecol.* **25,** 127–139.
10. Dunbar, J., Ticknor, L. O., and Kuske, C. R. (2000) Assessment of microbial diversity in four southwestern United States soils by 16S rRNA gene terminal restriction fragment analysis. *Appl. Environ. Microbiol.* **66,** 2943–2950.
11. Flynn, S. J., Loffler, F. E., and Tiedje, J. M. (2000) Microbial community changes associated with a shift from reductive dechlorination of PCE to reductive dechlorination of cis-DCE and VC. *Environ. Sci. Technol.* **34,** 1056–1061.
12. Garland, J. L., Cook, K. L., Adams, J. L., and Kerkhof, L. (2001) Culturability as an indicator of succession in microbial communities. *Microb. Ecol.* **42,** 150–158.

13. Guieysse, B., Wikström, P., Forsman, M., and Mattiasson, B. (2001) Biomonitoring of continuous microbial community adaptation towards more efficient phenol-degradation in a fed-batch bioreactor. *Appl. Microbiol. Biotechnol.* **56,** 780–787.
14. Hiraishi, A., Iwasaki, M., and Shinjo, H. (2000) Terminal restriction pattern analysis of 16S rRNA genes for the characterization of bacterial communities of activated sludge. *J. Bioscience & Bioengineering* **90,** 148–156.
15. Leser, T. D., Lindecrona, R. H., Jensen, T. K., Jensen, B. B., and Møller, K. (2000) Changes in bacterial community structure in the colon of pigs fed different experimental diets and after infection with *Brachyspira hyodysenteriae. Appl. Environ. Microbiol.* **66,** 3290–3296.
16. Liu, W. T., Marsh, T. L., and Forney, L. J. (1998) Determination of the microbial diversity of anaerobic-aerobic activated sludge by a novel molecular biological technique. Water *Sci. Technol.* **37,** 417–422.
17. Lüdemann, H., Arth, I., and Liesack, W. (2000) Spatial changes in the bacterial community structure along a vertical oxygen gradient in flooded paddy soil cores. *Appl. Environ. Microbiol.* **66,** 754–762.
18. Moeseneder, M. M., Arrieta, J. M., Muyzer, G., Winter, C., and Herndl, G. J. (1999) Optimization of terminal-restriction fragment length polymorphism analysis for complex marine bacterioplankton communities and comparison with denaturing gradient gel electrophoresis. *Appl. Environ. Microbiol.* **65,** 3518–3525.
19. Sessitsch, A., Weilharter, A., Gerzabek, M. H., Kirchmann, H., and Kandeler, E. (2001) Microbial population structures in soil particle size fractions of a long-term fertilizer field experiment. *Appl. Environ. Microbiol.* **67,** 4215–4224.
20. Urakawa, H., Yoshida, T., Nishimura, M., and Ohwada, K. (2000) Characterization of depth-related population variation in microbial communities of a coastal marine sediment using 16S rDNA-based approaches and quinone profiling. *Environ. Microbiol.* **2,** 542–554.
21. Chin, K. J., Lukow, T., Stubner, S., and Conrad, R. (1999) Structure and function of the methanogenic archaeal community in stable cellulose-degrading enrichment cultures at two different temperatures (15 and 30°C). *FEMS Microbiol. Ecol.* **30,** 313–326.
22. Chin, K.-J., Lukow, T., and Conrad, R. (1999) Effect of temperature on structure and function of the methanogenic archaeal community in an anoxic rice field soil. *Appl. Environ. Microbiol.* **65,** 2341–2349.
23. Fey, A., Chin, K.-J., and Conrad, R. (2001) Thermophilic methanogens in rice field soil. *Environ. Microbiol.* **3,** 295–303.
24. Friedrich, M. W., Schmitt-Wagner, D., Lueders, T., and Brune, A. (2001) Axial differences in community structure of Crenarchaeota and Euryarchaeota in the highly compartmentalized gut of the soil-feeding termite *Cubitermes orthognathus. Appl. Environ. Microbiol.* **67,** 4880–4890.
25. Lueders, T. and Friedrich, M. (2000) Archaeal population dynamics during sequential reduction processes in rice field soil. *Appl. Environ. Microbiol.* **66,** 2732–2742.

26. Lueders, T. and Friedrich, M. (2002) Effects of amendment with ferrihydrite and gypsum on the structure and activity of methanogenic populations in rice field soil. *Appl. Environ. Microbiol.* **68,** 2484–2494.
27. Moeseneder , M. M., Winter, C., Arrieta, J. M., and Herndl, G. J. (2001) Terminal-restriction fragment length polymorphism (T-RFLP) screening of a marine archaeal clone library to determine the different phylotypes. *J. Microbiol. Methods* **44,** 159–172.
28. Nüsslein, B., Chin, K. J., Eckert, W., and Conrad, R. (2001) Evidence for anaerobic syntrophic acetate oxidation during methane production in the profundal sediment of subtropical Lake Kinneret (Israel). *Environ. Microbiol.* **3,** 460–470.
29. Ramakrishnan, B., Lueders, T., Dunfield, P. F., Conrad, R., and Friedrich, M. W. (2001) Archaeal community structures in rice soil from different geographical regions before and after initiation of methane production. *FEMS Microbiol. Ecol.* **37,** 175–186.
30. Takai, K., Moser, D. P., DeFlaun, M., Onstott, T. C., and Fredrickson, J. K. (2001) Archaeal diversity in waters from deep South African gold mines. *Appl. Environ. Microbiol.* **67,** 5750–5760.
31. Weber, S., Lueders, T., Friedrich, M. W., and Conrad, R. (2001) Methanogenic populations involved in the degradation of rice straw in anoxic paddy soil. *FEMS Microbiol. Ecol.* **38,** 11–20.
32. Derakshani, M., Lukow, T., and Liesack, W. (2001) Novel bacterial lineages at the (sub)division level as detected by signature nucleotide-targeted recovery of 16S rRNA genes from bulk soil and rice roots of flooded rice microcosms. *Appl. Environ. Microbiol.* **67,** 623–631.
33. Horswell, J., Cordiner, S. J., Maas, E. W., Martin, T. M., Sutherland, B. W., Speir, T. W., et al. (2002) Forensic comparison of soils by bacterial community DNA profiling. *J. Forensic Sci.* **47,** 350–353.
34. Giraffa, G. and Neviani, E. (2001) DNA-based, culture-independent strategies for evaluating microbial communities in food-associated ecosystems. *Int. J. Food Microbiol.* **67,** 19–34.
35. Regan, J. M., Harrington, G. W., and Noguera, D. R. (2002) Ammonia- and nitrite-oxidizing bacterial communities in a pilot-scale chloraminated drinking water distribution system. *Appl. Environ. Microbiol.* **68,** 73–81.
36. Horz, H.-P., Rotthauwe, J. H., Lukow, T., and Liesack, W. (2000) Identification of major subgroups of ammonia-oxidizing bacteria in environmental samples by T-RFLP analysis of *amoA* PCR products. *J. Microbiol. Methods* **39,** 197–204.
37. Scala, D. J. and Kerkhof, L. J. (2000) Horizontal heterogeneity of denitrifying bacterial communities in marine sediments by terminal restriction fragment length polymorphism analysis. *Appl. Environ. Microbiol.* **66,** 1980–1986.
38. Horz, H.-P., Yimga, M. T., and Liesack W. (2001) Detection of methanotroph diversity on roots of submerged rice plants by molecular retrieval of *pmoA, mmoX, mxaF,* and 16S rRNA and ribosomal DNA, including *pmoA*-based terminal restriction fragment length polymorphism profiling. *Appl. Environ. Microbiol.* **67,** 4177–4185.

39. Lueders, T., Chin, K.-J., Conrad, R., and Friedrich, M. (2001) Molecular analyses of methyl-coenzyme M reductase alpha-subunit (*mcrA*) genes in rice field soil and enrichment cultures reveal the methanogenic phenotype of a novel archaeal lineage. *Environ. Microbiol.* **3,** 194–204.
40. Bruce, K. D. (1997) Analysis of *mer* gene subclasses within bacterial communities in soils and sediments resolved by fluorescent-PCR-restriction fragment length polymorphism profiling. *Appl. Environ. Microbiol.* **63,** 4914–4919.
41. Dunbar, J., Ticknor, L. O., and Kuske, C. R. (2001) Phylogenetic specificity and reproducibility and new method for analysis of terminal restriction fragment profiles of 16S rRNA genes from bacterial communities. *Appl. Environ. Microbiol.* **67,** 190–197.
42. Suzuki, M. T. and Giovannoni, S. J. (1996) Bias caused by template annealing in the amplification of mixtures of 16S rRNA genes by PCR. *Appl. Environ. Microbiol.* **62,** 625–630.
43. v. Wintzingerode, F., Göbel, U. B., and Stackebrandt, E. (1997) Determination of microbial diversity in environmental samples—pitfalls of PCR-based rRNA analysis. *FEMS Microbiol. Rev.* **21,** 213–229.
44. Legendre, P. and Legendre, L. (1998). *Numerical Ecology,* second English edition. Elsevier, Amsterdam, The Netherlands.
45. Maidak, B. L., Cole, J. R., Lilburn, T. G., Parker, C. T., Saxman, P. R., Stredwick, J. M., et al. (2000) The RDP (Ribosomal Database Project) continues. *Nucl. Acids Res.* **28,** 173–174.
46. Akkermans, A. D. L., van Elsas, J. D., and de Bruijn, F. J., eds. (1995) *Molecular Microbial Ecology Manual.* Kluwer Academic Publishers, Dordrecht, The Netherlands.
47. Hurt, R. A., Qiu, X. Y., Wu, L. Y., Roh, Y., Palumbo, A. V., Tiedje, J. M., et al. (2001) Simultaneous recovery of RNA and DNA from soils and sediments. *Appl. Environ. Microbiol.* **67,** 4495–4503.
48. Miller, D. N., Bryant, J. E., Madsen, E. L., and Ghiorse, W. C. (1999) Evaluation and optimization of DNA extraction and purification procedures for soil and sediment samples. *Appl. Environ. Microbiol.* **65,** 4715–4724.
49. Krsek, M. and Wellington, E. M. H. (1999) Comparison of different methods for the isolation and purification of total community DNA from soil. *J. Microbiol. Methods* **39,** 1–16.
50. Griffiths, R. I., Whiteley, A. S., O'Donnell, A. G., and Bailey, M. J. (2000) Rapid method for coextraction of DNA and RNA from natural environments for analysis of ribosomal DNA- and rRNA-based microbial community composition. *Appl. Environ. Microbiol.* **66,** 5488–5491.
51. Frostegård, A., Courtois, S., Ramisse, V., Clerc, S., Bernillon, D., LeGall, F., et al. (1999) Quantification of bias related to the extraction of DNA directly from soils. *Appl. Environ. Microbiol.* **65,** 5409–5420.
52. Kowalchuk, G. A., Naoumenko, Z. S., Derikx, P. J. L., Felske, A., Stephen, J. R., and Arkhipchenko, I. A. (1999) Molecular analysis of ammonia-oxidizing bacte-

ria of the β subdivision of the class *Proteobacteria* in compost and composted materials. *Appl. Environ. Microbiol.* **65,** 396–403.

53. Weber, S., Stubner, S., and Conrad, R. (2001) Bacterial populations colonizing and degrading rice straw in anoxic paddy soil. *Appl. Environ. Microbiol.* **67,** 1318–1327.
54. Liu, W. T., Linning, K. D., Nakamura, K., Mino, T., Matsuo, T., and Forney, L. J. (2000) Microbial community changes in biological phosphate-removal systems on altering sludge phosphorus content. *Microbiology* **146,** 1099–1107.
55. Schwieger, F. and Tebbe, C. C. (1998) A new approach to utilize PCR-single-strand-conformation polymorphism for 16S rRNA gene-based microbial community analysis. *Appl. Environ. Microbiol.* **64,** 4870–4876.
56. Peters, S., Koschinsky, S., Schwieger, F., and Tebbe, C. C. (2000) Succession of microbial communities during hot composting as detected by PCR-single-strand-conformation polymorphism-based genetic profiles of small-subunit rRNA genes. *Appl. Environ. Microbiol.* **66,** 930–936.
57. Marsh, T. L., Saxman, P., Cole, J., and Tiedje, J. (2000) Terminal restriction fragment length polymorphism analysis program, a web-based research tool for microbial community analysis. *Appl. Environ. Microbiol.* **66,** 3616–3620.
58. Egert, M. and Friedrich, M. W. (2003) Formation of pseudo-terminal restriction fragments, a PCR-related bias affecting terminal restriction fragment length polymorphism analysis of microbial community structure. *Appl. Environ. Microbiol.* **69,** 2555–2562.

4

Assessing Bacterial DNA Content in Aquatic Systems

Garrett W. Pernèy, Betsy R. Robertson, and D. K. Button

1. Introduction

These protocols were developed for the quantitative analysis of DNA in aquatic bacteria and have been used to characterize both marine and freshwater samples and cultures *(1)*. Apparent DNA content is a valuable tool when used in conjunction with forward light scatter for biomass *(2–4)* to characterize heterotrophic bacterioplankton, organisms which are too small for observation by light microscopy and difficult to cultivate *(5)*.

Quantitative analysis of flow cytometric data is obtained through use of internal standards. These are used to determine flow rates for population density calculations and relative fluorescence indexing, which makes comparative analysis of signal intensities within and across sample sets possible. The principle of relative fluorescence indexing is based on the constant proportionality of signal intensities of two particles, while signal intensities for each particle may change from sample to sample. This proportionality allows particles of known size, fluorescence, and concentration, such as latex microspheres, to be used as an index. The relative fluorescence intensity of the internal-standard microspheres is calibrated to the relative fluorescence intensity of a DAPI (4',6-diamidino-2-phenylindole) stained organism having a known DNA content and AT/GC ratio. The calibrated internal standard is then used as the relative fluorescence index for determining the DNA content of the organisms being analyzed.

DAPI, a multi-AT-based DNA stain, was used because its binding properties have been well studied *(6)*, it appears to be a highly specific DNA stain, and its fluorescence decay is slower than that of Hoecht 33258. Reports of

From: *Methods in Biotechnology: Environmental Microbiology: Methods and Protocols*
Edited by: J. F. T. Spencer and A. L. Ragout de Spencer

nonspecificity *(7)* are thought to be because of overstaining and/or insufficient permeabilization. Without adequate permeabilization, the stain becomes species and culture specific. However, with ample permeabilization and when staining conditions are held constant, sensitivity is sufficient for the analysis of the smallest bacteria, with a high degree of repeatability and without significant interspecies variability due to differences in cell envelope permeability *(8)*. It should also be noted that since the AT/CG ratio varies between organisms, only an approximate or "apparent" DNA content value is obtained, but this error appears to be small for the bulk of aquatic bacteria *(9)*.

The effect of temperature on membrane permeability was investigated in the development of these protocols (data not published). These data indicate that 10°C is sufficient to reach near-plateau levels of DAPI fluorescence in preserved organisms treated with Triton X-100.

Preservation of the organisms with formaldehyde increases the light-scatter intensity of bacteria due to additional mass resulting from aldehyde–protein crosslinking, and changes the refractive index of the organism *(2)*. Formaldehyde treatment also permeabilizes the cell envelope and is essential because DAPI penetration, even with the addition of Triton X-100, is marginally adequate in lithotrophs *(1)*. If there is doubt about a particular species, examine the kinetics of staining over an increase in permeabilization conditions. While conditions specified do not attain perfect plateau in signal, and thus require careful control of staining time and temperature, lithotrophs remain mostly unstained without Triton X-100. Fluorescence increases rapidly during the first few minutes, then slows. The concentration of DAPI is adjusted so that the rapid staining period requires at least 10 min to ensure that fluorescence is the result of high-affinity AT triplet sites. By extending staining time to 1 h, the exact staining time remains important but becomes less critical. Formaldehyde-treated and refrigerated *Escherichia coli* (ATCC 33849) and *Cycloclasticus oligotrophus* cells were indistinguishable after several months of storage when compared to cells that were freshly preserved from either freshwater or saltwater media, on the basis of dry mass *(2,8)* or apparent DNA content. All samples, both reference and unknown, should be preserved in the same manner to improve quantitative values.

Consideration must also be given to media chemistry. Owing to salt effects on the DAPI–DNA binding constants, the fluorescence intensity changes with the ion content of the medium. This is particularly true for aquatic bacteria, given the differences between freshwater and marine environments. For this reason, the internal standard must be calibrated to the DNA standard organism, in a medium of the same ionic concentration as that of the unknown samples to be analyzed.

Instrument sensitivity is also a factor in the measurement of the ultrasmall aquatic bacteria. Maintaining the system at peak performance is necessary for their detection. Measurements in our laboratory are obtained with a modified Ortho Cytofluorograf IIs equipped with a 5-W argon laser. Computerized operation is accomplished with Cicero system and Cyclops software (Cytomation, Inc.). The laser is tuned to multi-line ultraviolet operating at 100 mW output. An adjustable beam-shaping and focusing lens is used to narrow the beam width, increasing illumination intensity and decreasing illumination of the sheath fluid, resulting in a reduction of background noise. The signal-to-noise ratio is further improved by filtering the sheath fluid through a 0.1 µm filter.

Light scatter in the forward direction is segregated from the laser beam by a vertical 1.5-mm beam-blocker bar, then reflected by a 424-nm long-pass dichroic filter through a 310 to 370 nm bandpass filter, and focused onto a shielded fiber-optic cable leading to a photomultiplier detector (Hamamatsu #R1104). Orthogonal or 90° light scatter, used to evaluate the frequency of 0.96-µm internal-standard microspheres for determining flow rates and populations, is isolated with a similar optical filter set to that of the forward direction and focused onto an unshielded fiber-optic cable connected to a second photomultiplier detector. Blue fluorescence from DAPI–DNA complex and the 0.60-µm internal-standard microspheres is collected orthogonally by passing through a 424-nm long-pass dichroic filter and a 450 to 490 nm bandpass filter to a third unshielded fiber-optic cable and photomultiplier detector unit **(Fig. 1)**.

Logarithmic amplifiers with a dynamic range of 3.5 decades are calibrated to establish the range of linearity between signal input and numeric response. Data acquisition, analysis, and storage utilize PC-based software (Cytomation, Inc.). Acquisition is triggered by DAPI–DNA blue fluorescence to eliminate the forward light scatter signals from nonfluorescent debris.

Conversion of 256 channels resolved by the logarithmic amplifiers to $10^{3.5}$ linear channels is accommodated by the software, and mean linear values for the forward light scatter and DNA-bound DAPI fluorescence intensity of each particle are analyzed and assigned to the appropriate channel, and recorded. Detected signals are electronically plotted, and resulting subpopulations are evaluated for fluorescence intensity. Ratiometric comparison of the mean DAPI–DNA fluorescence intensity of a subpopulation to that of the 0.60-µm internal-standard microspheres fluorescence normalizes among samples, accounts for instrument drift, and gives apparent DNA content of the subpopulation.

DNA content is determined from DAPI–DNA fluorescence intensity and standardized to the relative fluorescence intensity signal of *E. coli* (ATCC, cat.

no. 33849) with a known genome size of 4.7 Mbp, or 5.17 fg ***(10,11)***, and a GC content of 50 mol%. To account for the effects of medium salinity on DAPI fluorescence intensity, which amounts to a 10–30% reduction with increases in salt concentration, standards are prepared in the analyzed medium using preserved freshly stained *E. coli.* Linear regression analysis of modal values of integral numbers of chromosome copies gives the fluorescence intensity associated with single chromosome copy, which is then related to the fluorescence intensity of the internal-standard 0.60-µm microspheres. Cellular DNA content for other strains or subpopulations within a culture are obtained ratiometrically from the mean DAPI–DNA fluorescence intensity of the subpopulation and the mean fluorescence intensity of the 0.60-µm internal-standard microspheres, with a correction for AT bias of DAPI based on their G + C contents *(8)*.

Cultivation of the DNA standard, *E. coli,* under conditions producing polyploidy subpopulations within a single culture is needed to calibrate the fluorescence signal produced by the 0.60-µm internal-standard latex microspheres. Polyploidy subpopulations are produced in culture by adding rifampicin (Rifamycin AMP) to a rapidly growing, log-phase culture ***(12)***. Once preserved in 0.5% (W/V) formaldehyde, this culture is used to calibrate the 0.60-µm internal-standard microspheres and as a control for each day's runs, allowing adjustments for differences in stain preparations and instrumentation.

2. Materials

2.1. Escherichia coli *DNA Standard*

1. *E. coli* DH1 (ATCC, cat. no. 33849).
2. Glass-distilled water.
3. NH_4Cl.
4. KCl.
5. Na_2HPO_4.
6. KH_2PO_4.
7. NaOH.
8. Nutrient broth (Gibco, cat. no. 0003-01). Autoclave and store at 4°C.
9. Dimethylsulfoxide (DMSO).
10. Rifampicin (Sigma, cat. no. R 3501). Caution (*see* **Note 1**).
11. M9 minimal culture medium (*see* **Note 2**).
12. 37% Formaldehyde as commercially supplied.

2.2. Internal-Standards Calibration

1. 0.96 µm Microspheres: Poly Science's 1.00 µ Fluoresbrite™ YG Microspheres, Calibration Grade #18860 (*see* **Notes 3** and **4**).

2. 0.60 μm Microspheres at a concentration of 1×10^8: Poly Science's Flow Check High Intensity Alignment Grade Particles 0.50 μm (YG) #23516 (*see* **Notes 3**, **5**, and **6**).
3. Internal-standard stock solution; a mixture of 0.96 μm and 0.60 μm microspheres, with the concentration of 0.96 μm microspheres at 1×10^7/mL and the concentration of 0.60 μm microspheres 1×10^7/mL or slightly less in freshly filtered glass distilled water (*see* **Notes 4** and **6**).
4. Internal-standard working solution; mixed fresh daily. 0.96 μm microspheres at a concentration of 1×10^6/mL and 0.60 μm microspheres at a concentration of 1×10^6/mL or slightly less (**item 3**) in freshly filtered glass-distilled water (*see* **Note 6**).
5. 4',6-Diamidino-2-phenylindole (DAPI). Aqueous stock solution of 0.5 mg/mL DAPI, stored frozen and protected from light (*see* **Note 7**).
6. Triton X-100: 5% (v/v) aqueous solution of Triton X-100, stored frozen (*see* **Note 8**).
7. DAPI/Triton X-100 staining solution: 50 μL DAPI stock solution (**item 4**) into 1 mL Triton X-100 stock solution (**item 5**), mixed and filtered through a 0.2 μm filter immediately before use daily, in the dark.
8. *E. coli* DNA standard cell suspension at $\leq 1 \times 10^7$/mL. Stock of rifampicin-treated cells, preserved with formaldehyde to a final concentration of 0.5% and refrigerated at 4°C.
9. Sheath fluid: glass-distilled water or basal medium filtered through a 0.1 μm filter.

2.3. Sample Analysis

1. A 5-W argon laser tuned to UV emission (351.1 and 363.8 nm) at 100 mW power output and equipped with the focusing lens and filter train as described in **Fig. 1**.
2. Internal-standard stock mixture of 0.60-μm and 0.96-μm microspheres, with the concentration of 0.96 μm microspheres at 1.0×10^6/mL (**Subheading 2.2., item 3**).
3. Filtered basal medium: M9 for fresh water, SAS for salt water. Filtered through a 0.2 μm filter immediately before use.
4. Filtered glass-distilled H_2O: filtered through a 0.2 μm filter immediately before use.
5. DAPI/Triton X-100 staining solution: filter and freshly prepared (**Subheading 2.2., item 6**).
6. *E. coli* DNA standard. A stock of rifampicin-treated cells, preserved and stored at 4°C in the dark (**Subheading 2.2., item 7**).

3. Methods

3.1. E. coli *DNA Standard*

1. Inoculate 50–100 mL of 50%-strength nutrient broth with *E. coli* DH1 (ATCC, cat. no. 33849) from frozen glycerol stock to produce a healthy inoculum for the experiment. Incubate at room temperature on a shaker table to provide good aeration.
2. When growth (turbidity) is observed, determine the population and biomass of the culture by Coulter counter or spectrophotometer.

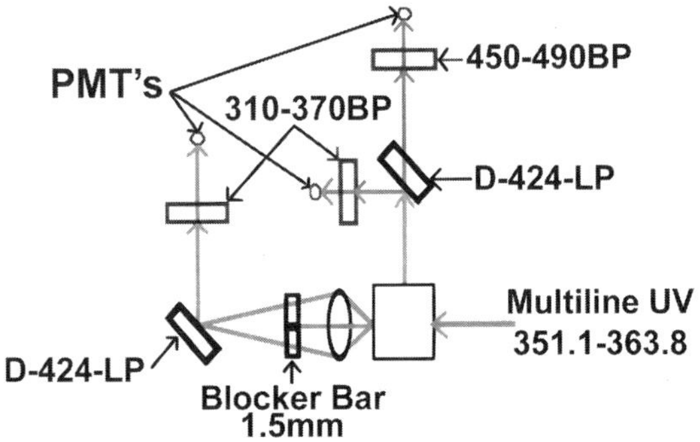

Fig. 1. Optical schematic. Laser beam is modified by cells in the sample chamber at lower right. Light scattered in the forward direction is focused around the blocking bar to one of three photomultipliers (PMTs). Side scatter is deflected by the long pass (LP) dichroic (D). The DAPI signal is purified by a band pass filter (BP), all at wavelengths shown.

3. Use the culture to inoculate three cultures at an initial biomass of 0.01, 0.1, and 1 mg wet/L.
4. Incubate the cultures overnight at room temperature on a shaker table.
5. In the morning, analyze each culture two or three times over a period of 3–4 h by Coulter counter, spectrophotometer, or epifluorescence microscopy to estimate the growth rate.
6. Select the best culture to treat with rifampicin according to the following criteria:
   a. Culture is in log phase.
   b. Biomass at least 10 times the biomass at t = 0 h.
   c. Culture is unlikely to reach stationary phase before you use it to inoculate the rifampicin treated medium.
7. Rifampicin stock solution: Dissolve rifampicin in DMSO to make the most concentrated stock solution possible, minimizing the amount of DMSO added to the culture. Add enough to 10 mL of M9 minimal culture medium to give a final concentration of 150 μg/mL (*see* **Notes 1** and **9**).
8. Inoculate the rifampicin-treated medium with *E. coli* at a biomass approx 20 mg (wet)/L or approx 1×10^7 cells/mL. It is always wise to make replicate cultures.
9. Inoculate rifampicin-free controls.
10. Measure the growth rate of all the cultures and determine when growth in the rifampicin-treated culture has ceased.
11. Allow enough time for DNA replication to be completed, 4–6 h after the biomass has ceased to increase.
12. Harvest the cells by centrifugation (10 min at 10,000*g*), resuspend them in filtered (0.2 μm) M9 medium, and preserve with formaldehyde to a final concentration of 0.5% (w/v).
13. Check for polyploidy subpopulations by flow cytometry. Samples should be preserved with formaldehyde and refrigerated at 4°C for at least 12 h before evalua-

tion. The selected standard culture should be kept refrigerated at 4°C, to be used as stock DNA standard (*see* **Notes 10–12**).

3.2. Internal-Standard Calibration

1. Set up and align the flow cytometer as described in **Subheading 3.3.**
2. Controls: system equilibration and negative controls should be run as described in **Subheading 3.3.2.**
3. "EC" *E. coli* DNA standard (rifampicin treated) cultures: transfer 0.6 mL of a formaldehyde-fixed *E. coli* DNA standard culture diluted to approx 1×10^5/mL cellular concentration with filtered (0.2 µm) basal M9 medium into a sterile, 1.5-mL microfuge tube. Treat and analyze as a sample (**Subheading 3.3.1.**).
4. Adjust gains for blue fluorescence and forward scatter so that all polyploidy populations are on scale.
5. Repeat "C" resetting the gains as described in **Subheading 3.3.**
6. Repeat *E. coli* DNA standard "EC" in order to establish the fluorescence value of n1 cells when the gains are in the range of those used during sample analysis.

The apparent DNA content of organisms being analyzed is calculated from their DAPI–DNA fluorescence as compared to standard cells in the presence of fluorescent microspheres used as an internal standard.

The relationship above makes it possible to equate the fluorescence intensity of the internal standard to that of an unknown, but only if the internal-standard fluorescence intensity has been accurately calibrated to the fluorescence intensity of a known sample. The fluorescence intensity per 5.17 fg DNA chromosome in the *E. coli* standard is determined by regression analysis of the mean fluorescence intensities of the resolved subpopulations containing multiple chromosome copies ***(13)***. The greater the number of integral genome subpopulations within the culture, the greater the accuracy of the genome relative fluorescence index resulting from the slope of n1 through nX. Correlation coefficients of 0.999 are typical, but slopes vary among stain preparations and conditions ***(4)***. Therefore, the *E. coli* DNA standard should be included in each day's sample set (**Subheading 3.3.2.**).

3.3. Flow Cytometry

1. Alignment of flow cytometer: Using 0.96 µm microspheres with linear scaling and pulse integration/area mode, optimize for forward scatter, orthogonal or 90° fluorescence, and 90° scatter, the filter train described in **Fig. 1**.
2. Data collection: Use logarithmic amplification, pulse integration/area mode. Blue fluorescence from DAPI/DNA is collected at 90° to the beam through a 424 nm long-pass dichroic filter and a 450–490-nm bandpass filter. Fluorescence intensity is determined with a calibrated dynamic-range logarithmic amplifier with acquisition triggered by fluorescence. Gains are set with reference to the 0.60-µm-diameter internal-standard microspheres, and formaldehyde-preserved *E. coli*

DNA standard. The gains should be set so that the 0.60-µm microspheres are in the upper 15% to 25% of the logarithmic amplifier's range for forward scatter and as high as possible for blue fluorescence, but low enough that 1n *E. coli* DNA standard remains on scale in both forward scatter and blue fluorescence. Setting the gains in this way should place the 0.96-µm microspheres in the upper 10% or just above full scale. Adjust the gain for the 90° scatter detector so only the 0.96-µm microspheres of region 9 are displayed in histogram 4.

3. Run each sample for 5–8 min at an event rate no greater than 50/s for a total of 6000 to 10,000 events. Record listmode data.
4. Histograms:

   Histogram 1—Bivariant: Forward Scatter vs Blue Fluorescence with regions for bacteria (region 1), 0.60-µm microspheres (region 8), and 0.96-µm microspheres (region 9)

   Histogram 2—Univalent: Forward Scatter gated inside region 1 of histogram 1

   Histogram 3—Univalent: Blue Fluorescence gated inside region 1 of histogram 1

   Histogram 4—Univalent: 90° Scatter gated inside region 9 of histogram 1

3.3.1. Sample Preparation

1. Fix samples in formaldehyde at a final concentration of 0.5% (w/v) for at least 12 h before analysis.
2. Filter sample through a 1.0-µm filter.
3. Transfer 0.6 mL of formaldehyde-fixed and filtered sample to a sterile, 1.5-mL microfuge tube. Label tubes (*see* **Note 12**).
4. Add 12 µL of staining solution to the sample to give 0.5 µg DAPI/mL, mix sample well and incubate 1 h ± 5 min at 10°C in the dark. Label tubes with time of stain addition.
5. Add 6 µL of daily internal-standard solution (**Subheading 2.2., item 4**) microspheres, for a final concentration of 0.96-µm microspheres of 1×10^5/mL, vortex, and analyze.

3.3.2. Sample Array

1. "R" Rinse and reference: 0.6 mL of basal medium stained as a sample with internal-standard microspheres added to 10 times concentration of that used in sample analysis, or 60 µL of daily internal-standard solution (**Subheading 2.2., item 4**) for a final concentration of 6×10^5/mL. Adjust gains in logarithmic scale during system equilibration. Gains should be adjusted as described in **Subheading 3.3.** (*see* **Note 13**).
2. "C" Negative control: 0.6 mL of basal medium stained and treated as a sample with internal-standard microspheres added to a final concentration of 1×10^5/mL (6.0 µL of daily internal-standard solution), the same as that used for sample analysis (*see* **Note 14**).
3. Numbered samples: 0.6 mL of sample as described in **Subheading 3.3.2.**

4. "EC" *E. coli* DNA standard: Formaldehyde-fixed *E. coli* DNA standard culture diluted to a population of 1×10^5/mL with filtered (0.2 μm) basal M9 medium (*see* **Note 10**).

3.3.3. Data Analysis

1. Calculations for apparent DNA content. For bacteria with unknown GC/AT ratio, the GC content of the standard *E. coli* is used as an approximation.
2. DNA content:

$$(I_u/I_s) \times (200/I_{n1}) \times 5.17 = \text{fg DNA per cell}$$

Where I_u = mean fluorescence intensity of sample
I_s = mean fluorescence intensity of the 0.60-μm microspheres
I_{n1} = mean fluorescence intensity of *E. coli* (single chromosome) when corrected to mean fluorescence intensity = 200 for internal-standard 0.60-μm microspheres.
5.17 = fg DNA per genome of *E. coli* ***(10,11)***.

For simplicity, all calculations are based on the fluorescence intensity of the 0.60-μm microspheres of 200.

3. Correction for AT bias of DAPI ***(6)***. This correction is based on the statistical probability of finding an AT triplet in a random sequence.

$$P = (1 - A)\, A^n/1 - A^n$$

Where P = Probability of binding
A = Decimal proportion of AT content
n = Number of consecutive AT pairs needed for binding (DAPI: $n = 3$)

For *E. coli*, GC content is 50 mol% ***(8)*** or 51.7% ***(8,13)***

$0.0714 = (1 - 0.5)\, 0.5^3/1 - 0.5^3$

If AT = 0.517, $P = 0.0657$

If GC content of a bacterium is 40 mol% and the apparent DNA content is 5 fg per cell, then $P = 0.110$ and DNA content corrected for GC content is:

$$\text{DNA} = (P_{\text{E. coli}}/P_{\text{Bact.}})\, \text{DNA}_{\text{app}};\ \text{DNA} = (0.0714 / 0.110)\, 5 = 3.2 \text{ fg per cell}$$

4. Notes

1. Rifampicin (Rifamicin AMP) poses a serious safety hazard. Please read the Material Safety Data Sheet available from the product distributor.
2. M9 medium (minimal culture medium for freshwater organisms). In 500 mL glass-distilled water: NH_4Cl, 0.5 g; KCl, 0.75g; Na_2HPO_4, 2.855 g; KH_2PO_4, 1.25 g. Adjust the pH to 7.5 with 1 *M* NaOH, autoclave, and store at ≤ 10°C. The above is not a complete medium. For more information, *see* **ref. *14***.

   SAS medium (basal saltwater medium): NaCl, 30.0 g/L; $MgCl_2$-$6H_2O$, 1.0 g/L; Na_2SO_4 (anhydrous), 4.0 g/L; KCl, 0.70 g/L; $CaCl_2$–$2H_2O$, 0.15 g/L; NH_4Cl, 0.50 g/L; $NaHCO_3$, 0.20 g/L; KBr, 0.10 g/L; $SrCl_2$–$6H_2O$, 0.04 g/L; H_3BO_3, 0.025

g/L; MOPS buffer, 2.09 g/L. Autoclave. Adjust pH to 7.9 with 10 *N* NaOH. The above is not a complete medium. Phosphate and trace metals are autoclaved separately and added cold. For more information, *see* **ref. 5**.

3. Poly Science's microspheres are not exactly 1.0 µm or 0.50 µm in size. Each production lot varies in size and fluorescence.
4. Microsphere selection is very important. The microspheres used should be as uniform as possible and should not produce signal in the range of that produced by the target bacteria. The concentration of these microspheres is also important, as they are used to determine the sample population concentration. Verify the concentration of microspheres by Coulter counter.
5. These microspheres are the internal relative fluorescence standard, which is calibrated to the *E. coli* DNA standard.
6. Store microsphere stock suspensions with a concentration of 1×10^8/mL of 0.96 µm and approx 1×10^7/mL of 0.60 µm in the dark in the refrigerator. Working internal-standard solutions can be made by diluting the internal-standard stock solution at a ratio of 1:10 in freshly filtered glass-distilled water, giving a final concentration of 0.96-µm and 0.60-µm microspheres of 1×10^7/mL. This will enable the addition of 6.0 µL of working internal-standard solution to a 0.6-mL sample yielding an internal-standard concentration of 1×10^5/mL for the ratiometric determination of bacterial population.
7. 500 µL aliquots of stock solution can be made up and used as needed. DAPI stock solution should be protected from light. Stock solutions can be stored for up to 3 mo at –4°C.
8. 1-mL Aliquots of Triton X-100 can be made up and used as needed. Stock solutions can be stored for up to 3 mo at –4°C.
9. Mix 0.150 g Rifamicin AMP to 1.0 mL DMSO. This can then be added to cultures at a rate of 1.0 µL to 10 mL of culture.
10. Stained with DAPI and analyzed by flow cytometry, polyploidy subpopulations with integral numbers of genome should be visible. Rifampicin inhibits the initiation of replication. In rapidly growing *E. coli*, more then one node for replication initiation can be formed on the chromosome *(8)*. When rifampicin is introduced, replication of each node will continue until complete, but initiation of further replication will not occur. In the method described above, before the addition of rifampicin, there should be cells in which the single chromosome has no nodes for replication. They will retain their single chromosome (1n). Those cells containing a single chromosome, with replication initiated at a single node, will complete that replication after the addition of rifampicin, producing cells with two genome copies (2n), and so on.
11. Alternative E. coli DNA standard prep: (1) In 50 mL of 50% nutrient broth start a fresh inoculum culture of DH 1; place in a 37°C shaker incubator. (2) After 20 h growth, use 10 to 20 µL of this culture to inoculate 30 to 60 mL, respectively, of 50% nutrient broth. Place this new subculture into a 37°C shaker incubator. (3) After 2.5 h in the incubator, remove a 10-mL subculture and add Rifampicin to a

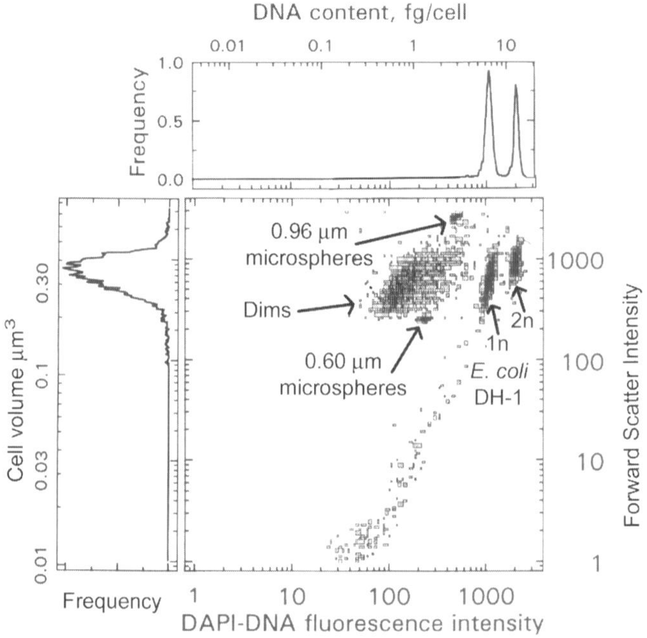

Fig. 2. Cytogram of *E. coli* showing cells containing one and two chromosomes along with cell volume and DNA content.

final concentration of 150 µg/mL. Label and place back into the shaker incubator. (4) Repeat step 3 every 30 min for 60-mL culture or every hour for 30-mL culture. (5) Incubate all subcultures overnight at 37°C in a shaker incubator. (6) Harvest cells from each subculture by centrifugation at 10,000 rpm for 10 min. Draw off superanate and resuspend cell pellet with filtered (0.2 µm) basal M9 medium and fix with formaldehyde. Samples should be preserved with formaldehyde for at least 12 h before evaluation by flow cytometry as described in **Subheading 3.1.** (7) Evaluate each subculture for polyploidy by flow cytometry as described in **Subheading 3.1.**

12. Laboratory cultures should be diluted to approx 1×10^6 cells/mL with filtered (0.2 µm) basal medium. Aquatic samples are usually dilute enough for direct analysis by flow cytometry after preservation.
13. "R," the Rinse and reference, is used to equilibrate the sample stream to DAPI and to give a reference in gain adjustment for that day's sample analysis.
14. "C" is a negative control and is used to evaluate the ambient noise level in the target range of the organisms to be analyzed.

References

1. Button, D. K. and Robertson, R. B. (2000) Effect of nutrient kinetics and cytoarchitecture on bacterioplankton size. *Limnol. Oceanogr.* **45,** 499–505.

2. Robertson, B. R., Button, D. K., and Koch, A. L (1998) Determination of the biomasses of small bacteria at low concentrations in a mixture of species with forward light scatter measurements by flow cytometry. *Appl. Environ. Microbiol.* **64,** 3900–3909.
3. Robertson, B. R. and Button, D. K. (1999) Determination of bacterial biomass from flow cytometric measurements of forward light scatter intensity. *Current Protocols in Cytometry* **11(9),** 1–10.
4. Robertson, B. R. and Button, D. K. (1989) Characterizing aquatic bacteria according to population, cell size, and apparent DNA content by flow cytometry. *Cytometry* **10,** 70–76.
5. Schut, F., DeVries, E. J., Gottschol, J. C., Robertson, B. R., Harder, W., Prins, R. A., et al. (1993) Isolation of typical marine bacteria by dilution culture: growth, maintenance, and characteristics of isolates under laboratory conditions. *Appl. Environ. Microbiol.* **59,** 2150–2160.
6. Wilson, W. D., Tanious, F. A., Barton, H. J., Jones, R. L., Fox, K., Wydra, R. L., et al. (1990) DNA sequence dependent binding modes of 4',6-diamidino-2-phenylindole(DAPI). *Biochemistry* **29,** 8452–8461.
7. Monger, B. C. and Landry, M. R. (1993) *Flow Cytometric Analysis of Marine Bacteria* with Hoechst 33342.
8. Button, D. K. and Robertson, B. R. (2001) Determination of DNA content of aquatic bacteria by flow cytometry. *Appl. Environ. Microbiol.* **67**, 1636–1645.
9. Krawiec, S. and Riley, M. (1990) Organization of the bacterial chromosome. *Microbiol. Rev.* **54,** 504–539.
10. Kohara, Y., Akiyame, K., and Isono, K. (1987) The physical map of the whole *E.coli* chromosome: application of a new strategy for rapid analysis and sorting of a large genomic library. *Cell* **50,** 495–508.
11. Rudd, K. E., Miller, W., Ostell, J., and Benson, D. A. (1990) Alignment of *Escherichia coli* K12 DNA sequences to a genomic restriction map. *Nucleic Acids Res.* **18,** 313–321.
12. Cooper, S. (1991) *Bacterial Growth and Division: Biochemistry and Regulation of Prokaryotic and Eukaryotic Division Cycles.* Academic Press, Inc., New York, NY.
13. Gillis, M., De Ley, J., and De Cleene, M. (1970) The determination of molecular weight of bacterial genome DNA from rematuration rates. *Eur. J. Biochem.* **12,** 143–153.
14. Miller, H. (1987) Practical aspects of preparing phage and plasmid DNA: growth, maintenance, and storage of bacteria and bacteriophage. *Methods Enzymol.* **152,** 145–170.

5

Multiplexed Identification and Quantification of Analyte DNAs in Environmental Samples Using Microspheres and Flow Cytometry

Mary Lowe, Alex Spiro, Anne O. Summers, and Joy Wireman

1. Introduction

Complex mixtures of nucleic acids occur in numerous systems, including environmental samples (e.g., groundwater, sediment), skin, feces, and blood. Often, it is desirable to be able to identify and measure the amount of a particular analyte DNA in a mixture. When there are multiple analyte DNAs of interest, multiplexing techniques can speed up the analysis.

This chapter describes protocols associated with a microsphere-based method for multiplexed detection and quantification of analyte DNAs in PCR products obtained from environmental DNA extracts. Very often the concentrations of target DNAs are low. The basic principles have been described in *(1,2)*. The method involves the following elements: microscopic polystyrene beads bearing carboxyl groups on the surface, two or three fluorophores for multiplexed detection, and flow cytometry instrumentation. One or two classification fluorophores (e.g., red and IR dyes) are impregnated within the beads in varying discrete amounts, thereby creating distinct bead types, each with a unique spectral code. The DNA hybridization assay is conducted on the surface of the beads, and another fluorophore is attached to the DNA amplicon as the reporter. The reporter fluorophore has a distinct fluorescence (e.g., green or orange), which can be detected by a flow cytometer separately from the classification fluorescence. In flow cytometry, the beads are directed single file into a thin fluid column, where they are interrogated one at a time by a laser.

From: *Methods in Biotechnology: Environmental Microbiology: Methods and Protocols*
Edited by: J. F. T. Spencer and A. L. Ragout de Spencer

An oligonucleotide ("capture probe"), designed to be complementary to a particular target sequence, is attached to the surface of a unique bead type, creating a "bead probe." The analyte consists of a mixture of DNA amplicons, some of which are targeted by distinct capture probes. By mixing different microsphere types in a single hybridization reaction and exposing them to the same analyte, direct hybrid capture occurs between matching capture probes and amplicons. The target DNAs are labeled with a reporter dye either prior to capture (direct labeling) or after capture (indirect labeling). Using flow cytometry, multiplexed detection is accomplished through simultaneous measurements of the red, IR, and reporter emission intensities, and the forward (optional) and side scatter. Detection times are typically a few seconds to a few minutes per hybridization reaction.

A number of multiplexed, flow cytometric DNA assays have been reported in the literature, mostly for single nucleotide polymorphism analysis for medical applications *(3–13)*. In our work, the stress is on quantitative analyses of mixtures of analyte amplicons from environmental samples. The overall procedure is as follows:

1. Collection of environmental samples.
2. Extraction of DNA from environmental samples.
3. PCR amplification of environmental DNA and purification of PCR product.
4. Exonuclease digestion to prepare ss-PCR products.
5. Attachment of capture probes to beads.
6. Hybridization and fluorescence labeling of PCR products to bead probes.
7. Flow cytometry detection of the beads with labeled hybrids.
8. Raw data processing.
9. Determination of the amounts of individual target amplicons in the PCR product.

This chapter describes **steps 3–9** from the point of view of developing a new assay. We assume that the reader has a method for collecting samples and extracting the DNA. We use the Bio 101 SPIN Kit for Soil for sediment and groundwater samples, and a similar bead-beater-phenol extraction method for fecal specimens. Also we assume the reader has PCR primers and a protocol to amplify community DNA, and has designed capture probes specific for the desired target molecules. Most of our work has used universal bacterial primers flanking various regions of the 16S rRNA gene, and we will use this example to illustrate the basic procedures. Although we have worked with various reporter dyes and assays, the main focus in this chapter is on indirect labeling with streptavidin-R-phycoerythrin (**Fig. 1**) which can be detected on the Luminex 100 and Becton Dickinson FACSCalibur flow cytometers (or equivalent). We also discuss standards that we use to compare results at different time points and between different instruments. Basic procedures for quantifying tar-

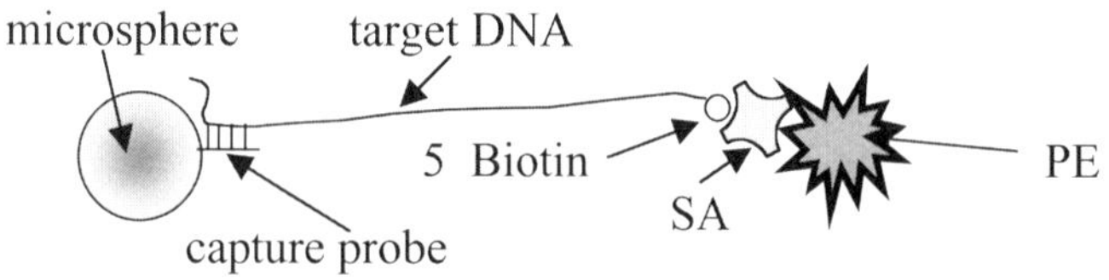

Fig. 1. Direct hybrid capture and fluorescence labeling on microspheres. Capture probes are attached at the 5' end to the surface of spectral-coded fluorescent polystyrene microspheres. The beads are then exposed to an analyte consisting of a mixture of single-stranded biotinylated amplicons. Direct hybrid capture occurs between matching capture probes and targets. After capture, the target amplicons are labeled with streptavidin-R-phycoerythrin (SA-PE).

get amplicons are described. In all protocols, we concentrate on commercially available products that can be readily used in other laboratories.

We also wish to note several other developments related to flow cytometry detection of DNA on bead surfaces. There is the potential to clone captured amplicons, which may expedite the search for new genes in environmental samples. A single bead probe can be used in a hybridization reaction ***(14)***, or multiple bead probes can be separated with a cell sorter. Multiplexed detection of RNAs for gene expression studies has been reported ***(15)***. Improved capture of target DNA was accomplished with peptide nucleic acid (PNA) capture probes attached to beads ***(16)***. Chandler et al. have used these beads in an automated, renewable hybridization column ***(17)***, which has been interfaced to a Luminex 100 (personal communication).

2. Materials

The typical work area is located in a hood, and contains a vortexer, microcentrifuge, two dry heat baths with conical wells to accommodate 1.5-mL microfuge tubes, one water bath, and a timer. For all steps, only molecular-biology-grade water (Gibco) is used.

1. Oligonucleotide capture probes, synthesized with a 5'-unilinker (Oligos Etc. or Operon), 200 μ*M* in water, store at –20°C.
2. Fluorescent, carboxylated, polystyrene beads (5.5–5.6 μ in diameter), store at 4°C (*see* **Note 1**):
   Red/infrared beads (Luminex Corp.), assorted (*see* **Note 2**).
   Red or infrared beads (Molecular Probes), assorted.
   Plain beads (Bangs Laboratories).
3. PCR reagents and PCR purification:
   Community DNA in water or TE (10 m*M* Tris-HCl, pH 8.0, 1 m*M* EDTA).
   Biotinylated, phosphorothioated forward PCR primer in water.
   Reverse PCR primer in water.
   dNTP.

Taq polymerase and 10X buffer (*see* **Note 3**).
5 mg/mL Bovine serum albumin (*see* **Note 4**).
QIAquick PCR purification kit (Qiagen).

4. Gel reagents:
   Low DNA Mass Ladder (Life Technologies, cat. no. 10068-013).
   Agarose gel equipment, TAE buffer, loading dye.
   Digital imaging and quantification system.
5. Attachment:
   EDC (1-ethyl-3-[3-dimethyaminopropyl] carbodiimidehydrochloride; Pierce) (*see* **Note 5**).
   MES, 0.1 *M,* pH 4.5 (2-[*N*-morpholino] ethanesulfonic acid, Sigma) (*see* **Note 6**).
   0.02% Tween-20 (Pierce).
   0.1% SDS (sodium dodecyl sulfate) (i.e., 1 mg SDS in 1 mL water).
   Coulter counter or hemacytometer.
6. T7 gene 6 exonuclease (U.S. Biochemical Corp.) and 5X exonuclease buffer (200 m*M* Tris-HCl pH 7.5, 100 m*M* $MgCl_2$, and 250 m*M* NaCl), store at –20°C.
7. Hybridization and secondary labeling reagents:
   1.5X TMAC buffer (*see* **Note 7**), store at room temperature.
   Streptavidin-R-phycoerythrin (Molecular Probes), 1 mg/mL, store at 4°C. Do not agitate.
8. Flow cytometry detection:
   Sheath fluid (Luminex Corp. or Becton-Dickinson FACsFlow).
   Ethanol, 70%.
   Flow cytometer (e.g., Luminex 100 or Becton Dickinson FACSCaliber).
   FCS Express v. 1.065 (De Novo Software) data analysis software, if using Luminex.
   CellQuest (Becton Dickinson) data analysis software, if using BD.
   Spreadsheet software (e.g., Microsoft Excel) (*see* **Note 8**).
9. Commercial bead standards:
   LinearFlow Orange Flow Cytometry Intensity Calibration Kit, 6 µ (Molecular Probes, cat. no. L-14815).
   Quantum R-PE MESF Medium Level Kit, 7.8 µ (Bangs Laboratories, cat. no. 827A).
   LinearFlow Green Flow Cytometry Intensity Calibration Kit, 6 µ (Molecular Probes, cat. no. L-14824).

3. Methods

3.1. Oligonucleotides

3.1.1. PCR Primers

Typically we use modified, universal PCR primers to amplify community DNA. For example, for the 16S rRNA gene, universal and all bacteria sequences

Table 1
16S rRNA Gene PCR Primers

Primer	Sequence	Target	Reference
8Fbs	5'[biotin]*a*g*a*g*tttgatcmtggctcag	All bacteria	***18***
338Fbs2	5'[biotin]*t*c*c*t*a cgg gag gca gc	All bacteria	***19***
533R	5'ttaccgcggctgctggcac	Universal	***18***
907R	5'ccgtcaattcmtttragttt	Universal	***18***
1392R	5'acgggcggtgtgtrc	Universal	***20***
1492R	5'ggttaccttgttacgactt	Universal	***18***

are shown in **Table 1**. Combinations used with the bead method are 8Fbs/533R, 8Fbs/907R, 338Fbs2/907R, 338Fbs2/1392R, 338Fbs2/1492R (*see* **Note 9**).

To prepare ss-PCR products at a later step (**Subheading 3.4.**), the forward PCR primer is modified. As an example, for an indirect labeling assay using SA-PE, the 338Fbs2 primer is synthesized with a 5' biotin modification, 12-carbon linker, five phosphorothioate bonds (*), and reverse-phase high-performance liquid chromatography purification (Synthegen, Inc.).

The reverse primer is unmodified with no special instructions for synthesis.

3.1.2. Capture Probes

To determine the capture probe sequences, we use literature sources and various software, including the Wisconsin package (Genetics Computer Group, Inc., Madison, WI); ARB (http://www.arb-home.de); Oligo (Molecular Biology Insights, Inc., Cascade, CO); and Probemer (http://probemer.cs.loyola.edu). The capture probes are designed to target specific sequences and to minimize the length of homologous regions with nontarget sequences to < 8 nt. We have not included degeneracies in the capture probes. The melting temperatures of stem-loops and homodimers are less than the hybridization temperature (46°C). We have tried oligos ranging from 16 to 31 nt in length, and have not found a consistent pattern in performance based on length.

The capture probes are synthesized with a 5'-amino modification called "unilinker" (Operon Technologies, Inc. or Oligos Etc.). They are reconstituted in water to a concentration of 200 μ*M*.

3.1.3. Helper Oligos

Variation in the magnitude of the signal is consistently observed among different capture probes for the same amount of pure strain target amplicons. We assume that much of this variation is due to secondary structure in the target DNA, which can block probe target sites and hybridization between capture probe and target. Fuchs et al. *(21)* describe a method of using helper oligos to increase hybridization signals. Helper oligos are short oligonucleotides that bind the target amplicon adjacent to the probe target site. The helper oligos are included in the hybridization reaction in amounts equivalent to the molarity of the amplicon, and they compete with the probe target site for annealing to adjacent regions and thereby prevent secondary structure formation between these adjacent sequences and the target site. Using this method, Fuchs observed signal increases of 4- to 25-fold in fluorescence *in situ* hybridization (FISH) with helper oligos for different regions of *E. coli* 16S rDNA.

We have used alignments of target sequences for genus-specific capture probes to determine whether regions adjacent to target sites are conserved and, thus, are suitable locales for genus-specific helper oligos. If the regions adjacent to target regions are not conserved, species-specific helpers were designed so that the temperature of dissociation (Td) of the helper was ≤5°C above that of the capture probe.

Hybridization reactions using different combinations of control amplicon(s), capture probe(s), and helpers, singly and multiplexed, showed variable signal increases as high as 13.5-fold (**Table 2**). A given capture probe signal was enhanced by its specific helper used alone or in a mixture of other helpers, but its signal was not enhanced by nonspecific helpers.

3.2. Attachment of Oligonucleotides to Carboxylated Microspheres

Oligos with a 5'-amino modification can be covalently attached to carboxylated beads using a procedure adapted from Luminex Corp (personal communication). For quantitative analyses, we prefer to work with small quantities of beads that are used within 1.5 wk of attachment. This is due to aging of the bead probes, in which there is degradation in the DNA hybridization signal over time and an increase in noise. The bead probes can be used over a longer period if the signals are large or if precise quantification is not needed.

3.2.1. General Information

All centrifugations are done at >10,500*g* for 60 s.

The wash steps are important in this assay for good reproducibility and detection of low signals. Washing involves the following: Add a buffer to the beads in the microfuge tube, vortex the mixture vigorously for a few seconds, and spin. If enough beads are present, there should be a visible pellet or a smear of beads on one side of the tube. To remove the supernatant, take a 200 μL pipet tip and slide it along the wall with no beads. If bubbles are present, they need to be removed first. Remove as much of the supernatant as possible without aspirating the beads. You may need to respin if the pellet becomes loose.

Table 2
Effects of Helper Oligos on Various Hybridization Reactions

Capture probe	Standard strain ATCC no.	Genus-specific capture probe sequence	Fold signal increase Amplicon[a] 8/533	Fold signal increase Amplicon[a] 388/1392
CC482	*Clostridia clostridiforme* 29084	5' gcttcttagtcaggtaccgt	1.0 (0.1)	0.9
LAA1023	*Lactobacillus johnsonii* 332	5' ctcttaggtttgcactggatgt	n/a	4.9 (0.8)
BAC1195	*Bacteroides vulgatus* 8482	5' taagggccgtgctgatttgac	n/a	2.8 (1.1)
BAC303	*Bacteroides vulgatus* 8482	5' ccaatgtgggggacctt	13.5	n/a

[a]Numbers refer to positions in the *E. coli* 16S rDNA sequence.

3.2.2. Procedure for Attachment

This takes about 1.5–3 h, depending on the number of bead types.

1. Select a set of bead types. We usually include a plain bead to determine the effect of the internal dyes upon the background and hybridization signals.
2. For each bead type, pipet 1.25×10^6 carboxylated beads from the stock into a 1.5 mL microfuge tube and add water to a convenient volume (*see* **Note 10**). Wash the beads.
3. Add 25 µL 0.1 *M* MES pH 4.5 to each tube. Vortex to resuspend the beads.
4. Add 0.5 nmol capture probe (e.g., 2.5 µL at 200 µ*M*) to each tube. Vortex.
5. Make fresh EDC solution at approx 10 mg/mL. For example, weigh 8 mg EDC and add 800 µL water. Vortex. EDC dissolves easily.
6. Add 1.25 µL fresh EDC solution to each tube. Vortex immediately. Cover with foil to keep the tubes dark, and incubate at room temperature for 30 min. Discard the EDC solution.
7. Repeat **steps 4** and **5** by making fresh EDC solution, adding 1.25 µL to each tube, and incubating for 30 min.
8. During the incubations, prepare 0.02% Tween by combining 2 µL 10% Tween with 1 mL water. You will need 500 µL for each tube of beads.
9. Also prepare 0.1% SDS by combining 100 µL 1% SDS with 900 µL water for a total volume of 1 mL. You will need 500 µL for each tube of beads.
10. After the incubations, add 500 µL 0.02% Tween to each tube of beads. Wash.
11. Add 250 µL 0.1% SDS to each tube of beads. Wash.
12. Again add 250 µL 0.1% SDS to each tube of beads. Wash.
13. Resuspend the beads in 25 µL 0.1 *M* MES pH 4.5 to obtain a nominal final concentration of 5×10^4 beads/µL. Store beads at 4°C in the dark.
14. Important as the assay becomes more developed: Determine the bead concentration in each tube using a Coulter counter (1:5000 dilution) or a hemacytometer.

3.3. PCR, Purification, and Gel-Based Quantification of Product

The PCR reaction mix should contain primers capable of amplifying a variety of sequences. The researcher should follow the protocol which works most effectively for producing the desired amplicons. For example, we use 16S rDNA universal primers (shown in **Table 1**) and the protocol described in **Note 11.** The PCR product is stored at 4°C and is never frozen.

In preparation for determining the background levels of the beads, it is best to prepare a PCR reaction mix without DNA template ("PCR mix").

Because the labeled primers can adhere nonspecifically to the surface of the bead, resulting in higher background signals, we often purify the PCR product to remove the unincorporated primers before introducing it to the beads. The PCR mix is also purified in exactly the same way as the PCR product. After testing four kits according to the manufacturer's instructions, we found that the QIAquick PCR purification kit works the best for this bead assay. The final elution from the spin column is done with 50 μL water. The PCR product can also be concentrated in this way (*see* **Note 12**).

The concentrations of purified or unpurified PCR products are determined on an agarose gel stained with ethidium bromide and a titration of Low DNA Mass Ladder. All reagents are vortexed briefly before each use. With a digital imaging system, the bands closest in length to the amplicon are used to prepare a standard curve. Knowing the volume of the PCR product and the length of the sequence, we convert the number of nanograms into fmol/μL. For example, a 570 bp PCR product has a molecular weight of $2 \times 570 \times 325$ g/mol = 3.71×10^5 g/mol. If 60 ng of amplicons are dissolved in 4 μL diluent, the concentration of the amplicons is 60×10^{-9}g/4 μL/(3.71×10^5 g/mol) = 40.5 fmol/μL.

3.4. Exonuclease Digestion, Hybridization, and Labeling

The most common DNA hybridization procedure that we use involves direct hybrid capture of single-stranded (ss), biotinylated target molecules by the capture probes on the bead surface. The use of ss-PCR products greatly improves the hybridization efficiency *(1)*. After capture, the target molecules are labeled with streptavidin-R-phycoerythrin (SA-PE). With good laboratory technique and careful data processing, several hundred amols of specific target amplicons in a PCR product can be detected. But generally we recommend that the researcher use 50–200 fmol in the preliminary studies.

In the preparation of ss-PCR products, T7 gene 6 exonuclease is added to the ds-PCR product. The 5'-to-3' hydrolytic activity of exonuclease is inhibited by the phosphorothioate bonds on the forward PCR primer. The strand without the modifications is digested *(22)*. The reaction is stopped by heating to 95°C.

In the protocol below, there are three general aspects which should be noted. First, careful washing, described in **Subheading 3.2.**, is critical for good reproducibility and improved lower detection limit. The washes eliminate excess PCR primers and SA-PE, both of which can adhere nonspecifically to the bead surface. Second, all of the pipeting should be done quickly. The tubes should be left open for as short a time as possible. Third, the incubation time for labeling should be as uniform as possible.

3.4.1. Preparation

1. Biotinylated PCR products, purified or unpurified. Vortex briefly before use.
2. Set of bead probes.
3. 1.5X TMAC, pH 8.5. Warm to dissolve crystals.
4. 5X Exonuclease buffer. Thaw.
5. Equilibrate heat block 1 to 37°C for exonuclease digestion.
6. Equilibrate heat block 2 to 95°C.
7. After exonuclease digestion, increase block 1 to 46°C.
8. Set centrifuge to approx 13,000*g*, 90 s.

3.4.2. Protocol for Exonuclease Digestion of PCR Product Prepared From a Phosphorothioated Primer

1. Prepare reaction mix with a total volume in each tube of 17 µL: 13.26 µL PCR/diluent + 3.4 µL 5X exo buffer + 0.34 µL exonuclease (*see* **Note 13**). Vortex briefly. Spin briefly or tap the tubes to collect the liquid at the bottom.
2. Incubate at 37°C for 45 min.

3.4.3. Protocol for DNA Hybridization

1. Resuspend beads in a single tube containing 1.5X TMAC. Calculate the total quantity assuming 34 µL per tube and approx 5000 beads of each type (*see* **Note 14**) per hybridization reaction. Include enough for a no-DNA control (PCR mix; *see* **Note 13**). Vortex. Keep at 46°C. (If helper oligos are used, they may be added at this stage.)
2. Place exonuclease/ss-amplicon mixture in 95°C heat block.
3. Incubate at 95°C for 10 min. Quickly spin to pull down condensation, and reheat at 95°C for 1–2 min.
4. With the ss-amplicons (target) still in the heat block at 95°C, add 34 µL of the bead/TMAC mixture to each tube with target DNA. Vortex immediately and place in 46°C heat block. The total volume in each tube is now 51 µL.
5. Incubate at 46°C for 2 h. (Optional: Vortex briefly after 1 h.)

*3.4.4. Protocol for SA-PE Labeling (see **Note 15**)*

1. Prepare 1X TMAC assuming 1.8 mL for each hybridization reaction. Keep 1X TMAC at 46°C.

2. Prepare 1:50 SA-PE solution in 1X TMAC. Pipet SA-PE from the middle of the stock. Do not shake the stock. Calculate the total quantity assuming 11.76 µL of 1X TMAC and 0.24 µL of SA-PE (1 mg/mL stock) for each hybridization reaction. Vortex. Cover with foil. Let it sit at room temperature.
3. Make sure the flow cytometer is warmed up.
4. Intermediate wash: Add 500 µL of 1X TMAC to first hybridization reaction with beads/DNA, vortex, load centrifuge. Repeat for all hybridization reactions. Spin. For the first hybridization reaction, look at the pellet, aspirate the fluid, and store the hybridization reaction at 46°C. Repeat for the other hybridization reactions.
5. Add 12 µL SA-PE solution to the first hybridization reaction. Vortex immediately and place it back in the 46°C heat block. Start the timer for 10 min. Continue to the other hybridization reactions, noting the amount of time to complete the whole process.
6. Wash 1: Add 500 µL of 1X TMAC to the first tube. Vortex, load centrifuge. Repeat for the other hybridization reactions. This process should take as long as **step 5**. Spin. Look at the pellet. Aspirate the fluid from each tube.
7. Wash 2: Add 500 µL of 1X TMAC to all tubes. Vortex, centrifuge. Aspirate the fluid from each tube.
8. Resuspend the beads by adding 46°C 1X TMAC (70 µL for Luminex detection, 200 µL for BD detection) to each tube. Vortex. Put the tubes back at 46°C to prevent crystalization. Detect on flow cytometer (*see* **Notes 16** and **17**).

3.5. Flow Cytometry Detection

We have conducted the bead assay on two flow cytometers with different laser excitation and detection filters: the Luminex 100 and the Becton Dickinson FACSCaliber. Recommendations for operating each cytometer are described below.

3.5.1. Luminex 100 Flow Cytometer

This instrument uses an excitation wavelength of 532 nm and an emission wavelength of 575 ± 10 nm for the reporter signal (RP1 channel). Classification emission of beads is excited at 635 nm and is detected at two spectral ranges: 658 ± 10 nm (CL1 channel) and ≥ 720 nm (CL2 channel). The software version is 1.7. Appropriate beads and detection channels are shown in **Table 3.**

When developing a quantitative bead assay for DNA, we prefer to adjust the reporter PMT voltage and create a dot plot of classification color vs reporter signal showing all of the bead types used in the assay. This is not the standard mode of operation for the instrument, but has the advantages that (1) we can see the quality of many bead distributions at the same time; (2) we can use plain (unstained) beads to determine the effects of the internal dyes on background and hybridization; and (3) we can use bead standards and fluorescent microspheres from different companies because we are not restricted to the

Table 3
Detection Channels for Various Commercial Beads

	Beads			
Instrument	Bangs plain beads[c]	Luminex Red/IR beads	Molecular Probes red beads[c]	Molecular Probes IR beads[c]
Becton Dickinson	FL3 FL4 FL3 and FL4[a]	FL3 and FL4[a]	FL3	FL3[b]
Luminex 100	CL1 CL2 CL1 and CL2[a]	CL1 CL2 CL1 and CL2[a]	CL1 CL2 CL1 and CL2[a]	CL2 CL1 and CL2[a]

[a]Both detection channels need to be set simultaneously.
[b]Authors acknowledge Yu-Zhong Zhang at Molecular Probes, Inc. for the data.
[c]These beads will not fall on the Luminex bead map.

Luminex bead map. With proper selection of classification colors, we often work with approx 12 bead types because they can all be displayed without overlap on the dot plot. We must, however, analyze the raw data files after acquisition because the Luminex software cannot calculate numbers associated with non-Luminex beads (*see* **Note 18**).

For general operational details, the researcher should follow the manufacturer's instructions for warming up, calibration, washing, and gating the beads based on side scatter (doublet discriminator mode) to select the monomers. We normally choose the following:

Number of events is set to 1000 × N, where N is the number of bead types
Volume sample = 30 μL
Flow rate = fast (default)
RP1 PMT = 700 V
Parameters for the other detectors (DD APD, CL1 APD, CL2 APD, and threshold) are set automatically when the instrument is calibrated according to the Luminex instructions.
dot plot abscissa: reporter
dot plot ordinate: CL1 or CL2, depending on choice of beads

Before running the samples, we run 1X TMAC to check if there are residual beads in the flow cytometer. For each sample, the instrument produces a separate run file containing data for each hybridization reaction.

The manufacturer can provide instructions for saving the session containing the run files, ending the session, and washing the instrument.

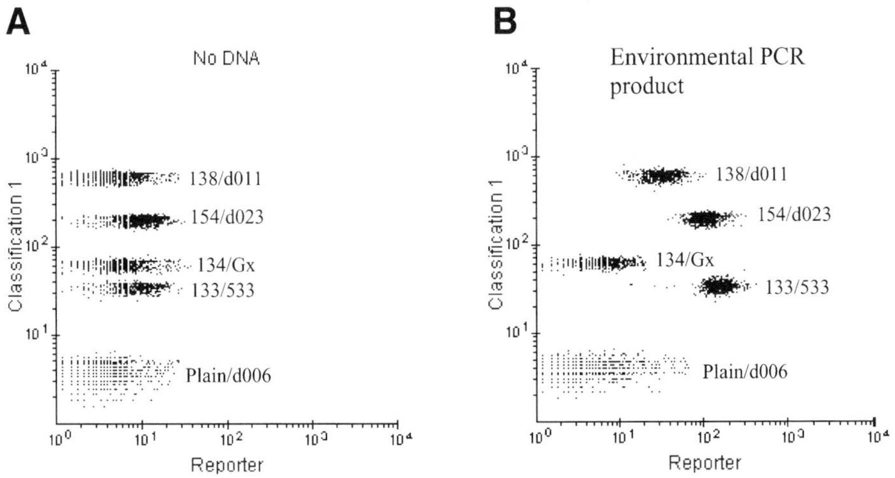

Fig. 2. Multiplexed fluorescence detection of the bead-probes with the Luminex 100. (Reprinted from **ref.** ***2***). The graphs were obtained with FCS Express. The dot plots show the orange reporter signal and the red bead classification intensities. **(A)** Negative control (PCR mix only). **(B)** Analyte containing 16S rDNA from contaminated groundwater. Designation XXX/YYY means that the YYY capture probe was attached to bead type XXX.

For all of our data, we postprocess the run files by setting gates on the following: (1) the monomer population based on DD; (2) each bead type based on CL1 and CL2; (3) the bead population with the reporter signal indicative of DNA hybridization (*see* **Note 19**) to determine the mean and statistical characteristics of the distribution. For fast postprocessing, we use in-house software. Slow, careful data analyses can be accomplished with FCS Express software. An example of the dot plots is shown in **Fig. 2.**

3.5.2. Becton-Dickinson FACSCalibur

Two lasers are available: 488 nm and 635 nm. The emission filters are: 530 nm ± 15 nm at FL1, 585 nm ± 21 nm at FL2, ≥670 nm at FL3, and 661 nm ± 8 nm at FL4. Appropriate beads and detection channels are shown in **Table 3.**

Voltage settings: FSC = E00 (no signal amplification), SSC = 350, FL1 = 628, FL2 = 610, FL3 = 300, and FL4 = 500. The FL1 and FL2 channels are used to detect the reporter dyes fluorescein and SA-PE, respectively.

Thresholds for FSC (forward scatter), FL1, FL2 and FL3 were set at the default of 52. The SSC (side scatter) threshold, selected as the primary parameter, was set at 70.

Flow rate was set on high (= 60 μL ± 6 per min).
Sheath fluid was FACsFlow.

Between samples, 1X TMAC (45°C) is run for 30 s to remove residual beads from the flow system. No beads are detectable after this washing procedure. After use, the FACSCaliber is cleaned according to the manufacturer's specifications.

Becton-Dickinson CellQuest software is used to collect and analyze data. An example of multiplexed detection is shown in **Fig. 3** for five types of Molecular Probes red beads with captured DNA labeled with SA-PE. We plot FSC (*x* axis, log scale) vs SSC (*y* axis, linear scale) as a dot plot and draw a gate around the monomer population of beads **(Fig. 3A)** (*see* **Note 20**). We usually count approx 10,000 events in this gate, which takes 1–2 min. Owing to bead loss or aggregation, it is not always possible to count 10,000 monomer events. However at this flow rate, the sample is exhausted after 3 min.

To distinguish the various red beads, the monomer population is plotted as an SSC (*y*-axis, linear scale) vs FL3 (*x* axis, log scale) dot plot; a gate is drawn around each bead cluster **(Fig. 3B)**. We then plot each cluster as a reporter signal histogram (**Fig. 3C,D**) and place a region marker over the peak, excluding the outer 10% on each side of the peak.

3.6. Data Analysis and Quantification

3.6.1. Bead Standards

Fluorescent bead standards are used to calibrate the flow cytometer and to compare results between different labs, between measurements within one lab at different times, and at different instrument settings. We have used three main types of fluorescent bead standards: beads with internal dyes, beads with surface labeling calibrated in MESF units, and beads with a known number of fluorescent nucleic acid sequences (FNAS) on the bead. The latter enabled us to determine fundamental characteristics of the hybridization and labeling, and the quality of commercial standards (**ref.** ***1***, also unpublished data). However, FNAS standards are not readily implemented in other labs. Therefore, for general use, we will concentrate on procedures using commercially available standards.

1. *LinearFlow Flow Cytometry Intensity Calibration Kits (Molecular Probes).* The kits are designed for intensity calibration of different detection channels of flow cytometers (green, orange, and so on). Each kit contains fluorescent beads stained at several relative fluorescence intensity levels with a dye impregnated throughout the bead volume. As reference standards, these beads are used to evaluate the signal intensities of the samples in relative fluorescence intensity units and to check instrument reproducibility. The calibration procedure is described in the BD manual. These beads are not necessarily spectral-matched to specific fluorescent labels in the DNA assay and cannot be used directly to compare the results from different types of flow cytometers.

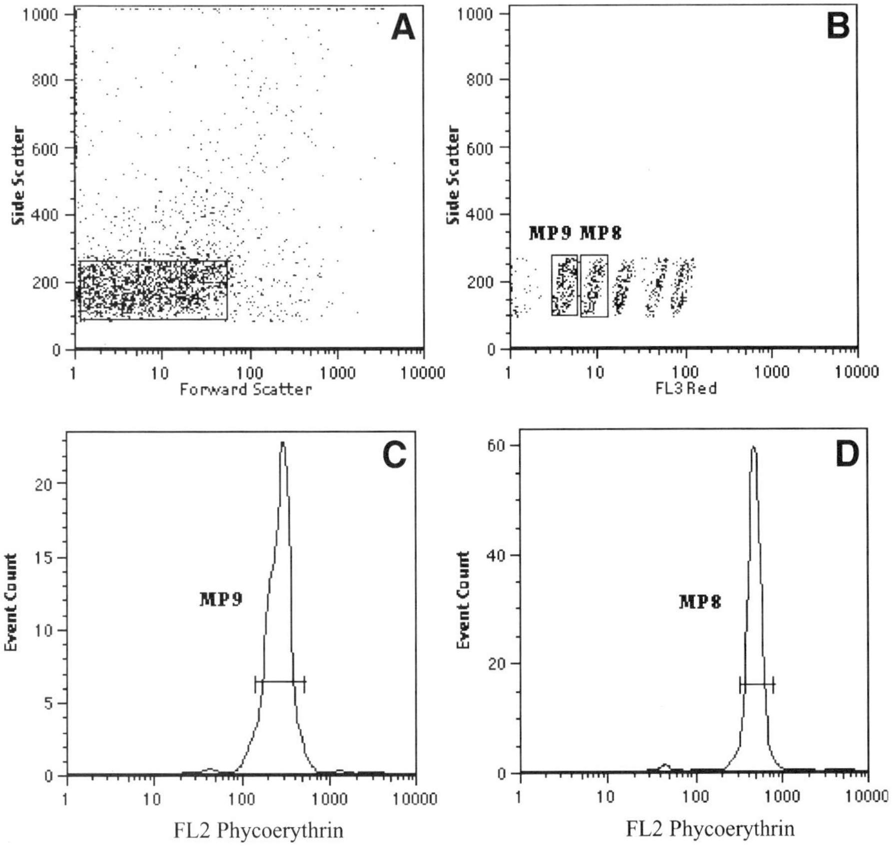

Fig. 3. Multiplexed fluorescence detection of bead-probes with Becton-Dickinson FACSCaliber. **(A)** Dot plot of side scatter vs forward scatter. A gate is drawn around the monomer bead population. **(B)** The monomers are plotted as a side scatter/FL3 dot plot where gates are drawn around distinct red bead populations. **(C)** For each bead type, a histogram is plotted for the reporter signal (FL2 phycoerythrin), and marker regions are placed over each peak. Molecular Probes red bead 9 is shown. **(D)** The reporter histogram for MP red bead 8.

2. *Quantum Fluorescence Kits (Bangs Labs).* These are fluorescence reference standards based on surface-labeled beads calibrated in molecules of equivalent soluble fluorochrome (MESF) units. These beads are matched to the spectral properties of the specific fluorescent labels. Each kit contains beads with several different levels of intensities. For quantitative analysis of an assay based on SA-PE labeling, we use Quantum R-PE MESF Medium Level Kit (cat. no. 827A, 500–50,000 MESF range) which is spectral-matched to a solution of R-phycoerythrin. Since PE molecules have a wide absorption spectrum, the R-PE MESF standard can be used with any cytometer equipped with a laser emitting between

Table 4
Flow Cytometry Bead Standards

DNA label	Instrument	Excitation wavelength	Detection channel	MP calibration kit[a]	Bangs Labs reference standard[a]
Fluorescein	BD	488 nm	FL1	LinearFlow Green	
PE	BD	488 nm	FL2	LinearFlow Orange	Quantum™ R-PE MESF Medium Level Kit
PE	Luminex 100	532 nm	Reporter	LinearFlow Orange	Quantum™ R-PE MESF Medium Level Kit
TAMRA	Luminex 100	532 nm	Reporter	LinearFlow Orange	
HEX	Luminex 100	532 nm	Reporter	LinearFlow Orange	

[a]*See* **Subheading 2**.

480 nm and 570 nm and a detection system between 570 nm and 590 nm. The procedure for converting reporter intensities into MESF units is described in the BD and Quantum MESF kit manuals.

Table 4 indicates the commercially available kits that are recommended for different labels and instruments based on our experience.

The Quantum MESF Standard is more expensive and less stable than MP beads (*see* **Note 21**). After working with different lots of Quantum MESF beads we found a significant variation of the standard from lot to lot. Therefore, for routine work, we recommend working with MP beads in the following way:

1. To monitor instrument performance and to compare the data over time, periodically calibrate your instrument in relative fluorescence intensities.
2. To compare the data between instruments of the same type in different labs (for example, Luminex to Luminex, or BD to BD) calibrate the instruments in relative fluorescence intensities.
3. To compare the data among different instruments (e.g., BD to Luminex) MP beads can be calibrated in MESF units using the Quantum MESF standard. This calibration involves three steps: (1) calibrate the instrument with Quantum MESF standard as described in the kit instruction, (2) run the MP bead kit with the same instrument settings used for the Quantum kit, (3) quantify each fluorescence level of MP beads in MESF units as described in the Quantum instructions under "Fluorescence quantitation of an unknown sample." The MP beads can be routinely run on this particular instrument for at least 1 yr to convert the intensity of the

samples to MESF units. For correct data comparison among different instruments, the MP beads must be calibrated on each instrument using the same lot number of the Quantum kit.

3.6.2. Evaluating Attachment and Hybridization Using Bead Standards

For evaluating the attachment to the bead surface, an oligo can be synthesized with 5'-unilinker, 3'-fluorescein modifications and attached to plain beads (5.5–5.6 μm in diameter) (*see* **Note 22**). After attachment, suspend the beads in 1X TMAC, and detect with a BD flow cytometer. Calibrate the instrument in relative fluorescence intensities (RFI) with the MP LinearFlow Green calibration kit. A mean signal from properly attached beads is approx 0.35% of RFI. This corresponds to 3×10^5 to 5×10^5 capture probes per bead on average. Alternatively, for evaluating attachment using SA-PE labeling, an oligo can be synthesized with 5'-unilinker, 3'-biotin modifications. After attachment, these beads can be labeled and detected with a Luminex or BD flow cytometer.

For preliminary hybridization tests, it is helpful to use a 5' biotinylated target oligo that is complementary to a capture probe sequence on the beads. After hybridization (30–60 min incubation) and SA-PE labeling, detect the beads on a BD or Luminex flow cytometer. Calibrate the instrument in R-PE MESF units. With proper attachment and hybridization, the mean signal is approx 10^5 R-PE MESF units (*see* **Note 23**).

For ss-PCR products ≤1 kb, the maximum hybridization signal is approx 10-fold less than signals measured from labeled oligos.

3.6.3. Quantifying Abundance

There are two procedures that we have adopted for quantifying the concentration of target amplicons in the analyte. Both are based on standard additions of control strain amplicons to an environmental PCR product. This approach is necessary because the behavior of the concentration curve in the environmental sample matrix may be different from that of an analyte consisting of one type of amplicon.

3.6.3.1. Method 1: Multiple Standard Additions With Dilution to a Constant Volume

The first quantification procedure maintains a constant volume of the environmental PCR product in a constant total analyte volume *(**23**)*. For our protocols, the total analyte volume is 13.26 μL, which includes the environmental PCR product and the addition, diluted in PCR mix. In general, the amount of the standard spikes must be small enough to not perturb the sample matrix. Also, the concentration of the target in the analyte must be small enough so that all measurements remain in the linear dynamic range. Dilution of the environmental PCR product may be necessary.

Table 5
Fluorescence Response From Standard Additions to an Environmental PCR Product

Hybridization reaction	Amount of spike C_i (fmol)	Amount of environ. PCR product (fmol)	Total signal (MESF)	Difference between total signal and PCR mix, F_i (MESF)
0	0	100	480	$F_0 = 230 = (480–250)$
1	5	100	750	$F_1 = 500$
2	10	100	1090	$F_2 = 840$
3	15	100	1250	$F_3 = 1000$
4	30	100	1750	$F_4 = 1500$
5	PCR mix	0	250	0

As an example, consider the data in **Table 5** for five hybridization reactions, four of which are spiked with amplicons from Species A. An additional tube contains PCR mix to determine the background signal.

The standard addition plot F_i vs C_i is shown in **Fig. 4** for hybridization reactions 0–3. Since the graph appears nearly linear, the slope S can be calculated using any standard fitting program: $S = 53$ MESF/fmols. The concentration C_e of the target in the environmental PCR product can be found from $F_0 \approx SC_e$. For the above data, $C_e \approx 4.3$ fmol. This is the amount of Species A amplicons in 100 fmol of environmental PCR product. Therefore, the relative abundance is about 4%.

3.6.3.2. METHOD 2: MODEL CURVE

Another method for determining the abundance of target amplicons in an environmental PCR product is based on a model describing DNA hybridizations *(2)*:

$$P + T \rightleftarrows PT$$

where P is the capture probe on the bead surface, T is the target amplicon, and PT is the hybrid on the surface. Under equilibrium conditions, the fluorescence response of the hybrid-bearing beads may be described by a saturation curve:

$$F = I_{max}K_aT/[1 + K_a T] \quad (1)$$

where F is the fluorescence response of a particular target (with background subtracted), T is the concentration of target amplicons in the hybridization reaction, K_a is the association constant, and I_{max} is the maximum response corresponding to complete saturation ($K_aT >> 1$). This equation assumes that the

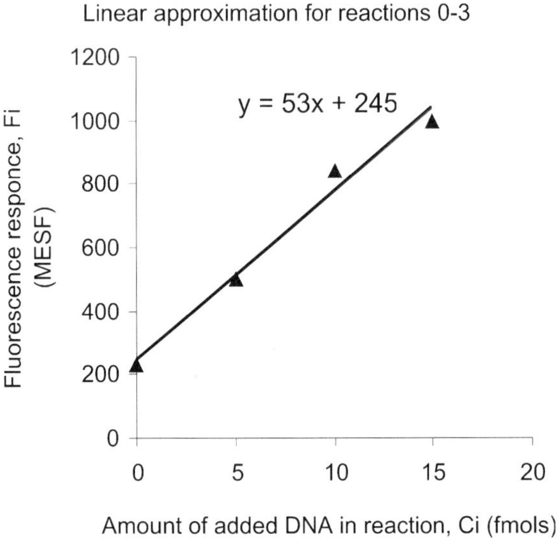

Fig. 4. Multiple standard additions with dilution to constant volume. When the concentration of target amplicons is small, a linear fit can be used to determine the unknown concentration of the target in the environmental PCR product.

concentration of unbound target amplicons remains essentially constant during the reaction, as shown in our measurements *(1,2)*.

In practice, when dealing with the same volume ($V = 51$ µL, *see* **Subheading 3.4.**) for all hybridization reactions, it is convenient to convert molar concentrations T in **Eq. 1** to amounts C in fmol, to change the association constant K_a to the parameter $k_a = K_a/V$ in $(\text{fmol})^{-1}$, and to introduce the parameter $S = I_{max}k_a$. Then for the hybridization reaction involving an unknown amount C_e of target amplicons in the environmental PCR product and a known amount C_i of the standard spike, we have:

$$\frac{S(C_e + C_i)}{[1 + k_a(C_e + C_i)]} \tag{2}$$

Three-point solution. At minimum, three hybridization reactions are needed to determine C_e: two standard additions and one no-addition. Then **Eq. 2** can be solved analytically (*see* **Note 24**). For example, consider the subset of data from **Table 5** which are shown in **Table 6:**

To obtain C_e, use the following formulas:

$$C_e = \alpha\, C_1\, C_2 / [C_2(1 - \alpha) - C_1]$$

where $\alpha = (F_0/F_1)[(F_2 - F_1)/(F_2 - F_0)]$

Inserting the data from **Table 6,**

Table 6
Data for Three-Point Solution to Model Equation

Amount of spike (Ci), fmol	Fluorescence response, MESF (Fi)
$C_0 = 0$	$F_0 = 230$
$C_1 = 15$	$F_1 = 1000$
$C_2 = 30$	$F_2 = 1500$

$$\alpha = 230\ (1500 - 1000)\ /\ [1000\ (1500 - 230)] = 0.0906$$

$$C_e = 0.091\ (15 \times 30)\ /\ [30 \times (1 - 0.091) - 15] = 3.32 \text{ fmol}$$

Nonlinear fit. Additional understanding about the hybridization efficiency may be obtained from a standard addition plot of the data in **Table 5.**

A nonlinear fit of **Eq. 2** to the data, shown in **Fig. 5**, yields values for C_e, S, and k_a. In this example, the fitting parameters are C_e = 3.32 fmol, S = 72.8 MESF/fmol, and k_a = 0.0179/fmol. Note that for $k_a(C_e + C_i) << 1$, the model equation becomes linear with slope S, and C_e may be found approximately according to Method 1 (multiple standard additions with dilution to constant volume).

4. Notes

1. DNA hybridization tests using a variety of bead types show variation in the hybridization signal. Generally, beads loaded with low or middle levels of internal dyes perform better than those with high levels of dyes.
2. Qiagen and Biorad beads have the same dyes as Luminex beads, and require the same cytometer settings.
3. We have used Amplitaq with GeneAmp 10X PCR buffer (Perkin-Elmer), which contains 100 m*M* Tris-HCl, pH 8.3; 500 m*M* KCl; 15 m*M* $MgCl_2$; and 0.01% w/v gelatin. We have also used *Taq* DNA Polymerase in Storage Buffer B with Thermophilic DNA Polymerase 10X buffer, Magnesium Free (Promega), which contains 100 m*M* Tris-HCl pH 9.0; 500 m*M* KCl; and 1% Triton X-100. $MgCl_2$ is added to a final concentration of 1.5 m*M*.
4. BSA can increase PCR product yields. We have not, however, compared DNA hybridizations on bead surfaces with and without BSA in the PCR product.
5. In a dry N_2 glove box, aliquot approx 15 mg EDC into each of a set of 1.5-mL microfuge tubes. Store the tubes and stock bottle at –20°C indefinitely. Before attachment (**Subheading 3.2.**), pull two tubes out of the freezer.
6. To prepare 0.1 *M* MES pH 4.5, combine 1.062 g MES with 50 mL water in a sterile bottle. Adjust the pH by adding 1.4 *M* KOH drop by drop while stirring (approx 100 µL). Store at 4°C.
7. 1.5X TMAC buffer: 1X TMAC buffer contains 3 M tetramethylammonium chloride (Sigma); 0.1% SDS; 50 m*M* Tris-HCl, pH 8.0 (Amresco); 4 m*M* EDTA pH

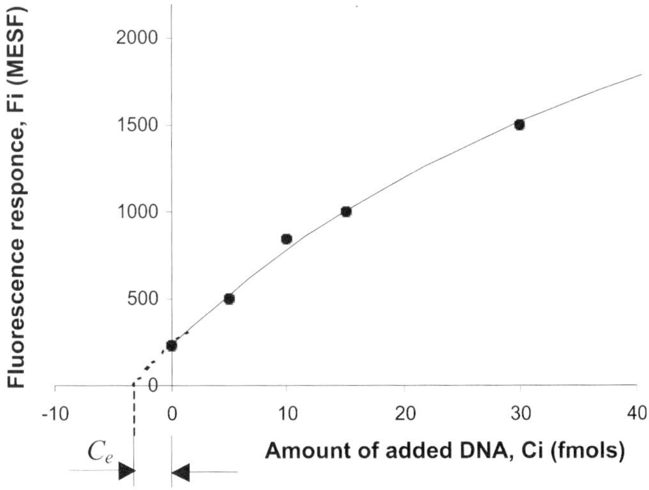

Fig. 5. Standard additions with a nonlinear fit based on a model describing DNA hybridizations in this assay. This method is more general and can handle saturation data more effectively than Method 1.

8.0). The pH is about 8.5. To prepare 1.5X TMAC buffer, combine the following to make 200.0 mL total volume:

180.0 mL	5 *M* TMAC
1.5 mL	20% SDS
15.0 mL	1 *M* Tris-HCl, pH 8.0
2.4 mL	500 m*M* EDTA, pH 8.0
1.1 mL	Water
200.0 mL	Total volume

When SDS is added to TMAC, crystals form. Before using the buffer, the bottle should be heated gently on a hotplate to dissolve most of the crystals. We have noticed that crystal formation varies with different lots of TMAC. All procedures involving TMAC are conducted under a hood.

8. When developing an assay, we recommend postprocessing the run files for analysis of the bead populations. In our laboratory, we use in-house software for rapid analysis. Commercial packages like FCS Express or CellQuest are slow but effective.
9. For PCR primers 8Fbs and 338Fbs2, we have had them synthesized without the phosphorothioate bond between the biotin and the 5' position of the terminal nucleotide. This also appears to result in a successful hybridization, but we have not made a thorough comparison of having or not having a phosphorothioate bond at that position. For a direct labeling assay using fluorescein, an example of a forward primer is 5' [fluorescein-ON] T*C*C*T*A Cgg gAg gCA gC (Operon Technologies, Inc.), where a phosphodiester bond exists between the fluorescein and the 5' position of the terminal nucleotide.

10. We typically use 100 μL.
11. In our studies, the PCR reaction mix has a total volume of 100 μL and contains 65 μL water, 10 μL 10X buffer without $MgCl_2$, 6 μL $MgCl_2$ at 25 m*M*, 5 μL forward primer at 10 μ*M*, 5 μL reverse primer at 10 μ*M*, 8 μL dNTP at 10 m*M*, 0.5 μL *Taq* at 5 U/μL, and 0.5–1 μL DNA template. Bovine serum albumin can be included to increase product yields. The PCR reactions are loaded into a thermal cycler after it has reached 94°C. Then follows 3 min denaturation at 94°C, and 25 cycles of denaturation (30 s at 94°C), annealing (45 s at 55°C), extension (8 min at 72°C), and storage at 4°C. The PCR products are never frozen.
12. Preliminary tests show that 1–2 fmol primers remain in 13 μL of a product that has been purified by the Qiagen kit. This represents more than a 1000-fold decrease in the primer concentration after purification. We elute with water.
13. When the PCR product is not purified, the diluent must be PCR mix and the background tube should contain PCR mix. When the PCR product is purified, the diluent is preferably purified PCR mix, but water can be used. The background tube must be processed following all steps for DNA samples as described in **Subheadings 3.4.3.** and **3.4.4.**
14. When the number of bead probes of a given type were varied, our studies and calculations showed that fewer beads result in higher signals. In practice, however, working with small numbers of beads can be difficult because the pellet cannot be seen easily. When multiplexing 10–15 bead types, we use 3000–5000 beads of each type. However, for smaller levels of multiplexing (e.g., <5 bead types), we use approx 10,000 beads of each type, and have gone as high as 25,000.
15. To achieve more uniform results, several steps should be noted. Maintaining the same incubation time for labeling across all hybridization reactions is important. When using nonpurified PCR products, a careful intermediate wash is important for removing excess biotinylated primers. For removing unbound SA-PE, the first of the final washes is important.
16. After hybridizing, we have noticed increased formation of bead aggregates (e.g., dimers, trimers, and higher multimers) for certain commercial beads. 95% of Bangs unstained beads, Luminex, and Molecular Probes beads remain monomers throughout the hybridization procedure. However, for Bangs Quantum Plex beads, as much as 50% of the events counted by the cytometer can be multimers. Reducing the concentration of tetramethyl ammonium chloride in 1X TMAC buffer reduced aggregation but lowered the reporter signal. Using 2*M* tetramethyl ammonium chloride rather than 3 *M* reduced multimers to less than 20% while decreasing the reporter signal by less than 15%. One molar tetramethyl ammonium chloride reduced multimers to less than 10% but the reporter signal was reduced by more than 50%.
17. We conducted preliminary tests on alternate wash buffers, and found that 2X SSC may be used in the final washes and final resuspension of bead probes labeled with SA-PE. The doublet discriminator gate on the Luminex 100 must be reset. When fluorescein is used as the reporter, resuspending the beads in high pH 1X TMAC buffer, prepared with 2 *M* Tris, optimizes the signal.

18. We typically operate the Luminex 100 in the "Advanced mode" with the "Simplex" option. The Luminex Corp. can provide details on how to set it up. In this mode, it is possible to calculate statistics for one bead population of any type (Luminex or non-Luminex beads). To obtain numbers for all bead types, Luminex software can be used with Luminex beads in the multiplexed analysis mode, subject to the limitation described in **Note 19.** Alternatively, FCS Express or in-house software can be used to postprocess the data for all bead populations of any type.
19. With Luminex software, the mean and median are calculated for the entire dataset. The software does not set markers around the correct bead peak in the reporter signal histogram. When the signal is low, this approach is undesirable.
20. The FSC/SSC dot plots for beads suspended in 1X TMAC buffer show a lot of extraneous events. We suspect some of them to be bubbles. When saline is used to resuspend the beads, the gates must be set differently.
21. Quantum MESF Standard 827A costs $235 for 5 mL (5 tubes, 1 mL each) at 2×10^6 beads/mL with a guaranteed storage time of seven months. The MP LinearFlow Green Calibration Kit (cat. no. L-14815) costs $182 for 6 mL (6 tubes, 1 mL each) at 2×10^7 beads/mL, 1 yr guarantee.
22. We used oligos 5'-(UL)AT GGT CTT CTG GTT GCC CCC-FL and (UL)AT GGT CTT CTG GTT GCC CCC-B (Oligos Etc.), where UL = unilinker, FL = fluorescein, and B = biotin, for evaluating attachment and SA-PE labeling.
23. This number was obtained on a Luminex 100 cytometer calibrated with Bangs R-PE MESF Medium Level Kit, lot no. A-052200.
24. For the three-point solution, the three tubes must form a "good" dataset, which follows a saturation curve.

Acknowledgments

We wish to thank Julie Nelson, Steve Hilliard, and Gina Hamlin Green for assisting us at the flow cytometry facilities at the University of Georgia and the Johns Hopkins University. We also thank the Luminex Corporation staff and Yu-Zhong Zhang at Molecular Probes for bead samples and advice on protocols. This work was supported by DOE NABIR grant DE-FG02-01ER63264, NSF MRI Award BES-0116765, and the University Exploratory Research Program sponsored by Procter & Gamble.

References

1. Spiro, A., Lowe, M., and Brown, D. (2000) A bead-based method for multiplexed identification and quantitation of DNA sequences using flow cytometry. *Appl. Environ. Microbiol.* **66,** 4258–4265.
2. Spiro, A. and Lowe, M. (2002) Quantitation of DNA sequences in environmental PCR products by a multiplexed, bead-based method. *Appl. Environ. Microbiol.* **68,** 1010–1013.

3. Fulton, R. J., McDade, R. L., Smith, P. L., Kienker, L. J., and Kettman, J. R. (1997) Advanced multiplexed analysis with the FlowMetrix system. *Clin. Chem.* **43,** 1749–1756.
4. Smith, P. L., WalkerPeach, C. R., Fulton, R. J., and DuBois, D. B. (1998) A rapid, sensitive, multiplexed assay for detection of viral nucleic acids using the FlowMetrix system. *Clin. Chem.* **44,** 2054–2056.
5. Dunbar, S. A. and Jacobson, J. W. (2000) Application of the Luminex LabMAP in rapid screening for mutations in the cystic fibrosis transmembrane conductance regulator gene. *Clin. Chem.* **46,** 1498–1500.
6. Iannone, M. A., Taylor, J. D., Chen, J., Li, M.-S., Rivers, P., Slentz-Kesler, K. A., et al. (2000) Multiplexed single nucleotide polymorphism genotyping by oligonucleotide ligation and flow cytometry. *Cytometry* **39,** 131–140.
7. Taylor, J. D., Briley, D., Nguyen, Q., Long, K., Iannone, M. A., Li, M. S., et al. (2001) Flow cytometric platform for high-throughput single nucleotide polymorphism analysis. *BioTechniques* **30,** 661–669.
8. Ye, F., Li, M. S., Taylor, J. D., Nguyen, Q., Colton, H. M., Casey, W. M., et al. (2001) Fluorescent microsphere-based readout technology for multiplexed human single nucleotide polymorphism analysis and bacterial identification. *Hum. Mutat.* **17,** 305–316.
9. Armstrong, B., Stewart, M., and Mazumder, A. (2000) Suspension arrays for high throughput, multiplexed single nucleotide polymorphism genotyping. *Cytometry* **40,** 102–108.
10. Colinas, R. J., Bellisario, R., and Pass, K. A. (2000) Multiplexed genotyping of beta-globin variants from PCR-amplified newborn blood spot DNA by hybridization with allele-specific oligodeoxynucleotides coupled to an array of fluorescent microspheres. *Clin. Chem.* **46,** 996–998.
11. Cai, H., White, P. S., Torney, D., Deshpande, A., Wang, Z., Marrone, B., et al. (2000) Flow cytometry-based minisequencing: a new platform for high-throughput single-nucleotide polymorphism scoring. *Genomics* **66,** 135–143.
12. Nolan, J. P. and Mandy, F. F. (2001) Suspension array technology: new tools for gene and protein analysis. *Cellular and Molecular Biology* **47,** 1241–1256.
13. Nolan, J. P. and Sklar, L. A. (2002) Suspension array technology: evolution of the flat-array paradigm. *Trends in Biotechnology* **20,** 9–12.
14. Jacobsen, C. S. (1995) Microscale detection of specific bacterial DNA in soil with a magnetic capture-hybridization and PCR amplification assay. *Appl. Environ. Microbiol.* **61,** 3347–3352.
15. Yang, L., Tran, D. K., and Wang, X. (2001) BADGE, Beads Array for the Detection of Gene Expression, a high-throughput diagnostic bioassay. *Genome Research* **11,** 1888–1898.
16. Chandler, D. P., Stults, J. R., Anderson, K. K., Cebula, S., Schuck, B. L., and Brockman, F. J. (2000) Affinity capture and recovery of DNA at femtomolar concentrations with peptide nucleic acid probes. *Analytical Biochem.* **283,** 241–249.
17. Bruckner-Lea, C. J., Stottlemyre, M. S., Holman, D. A., Grate, J. W., Brockman, F. J., and Chandler, D. P. (2000) Rotating rod renewable microcolumns for automated, solid-phase DNA hybridization studies. *Anal. Chem.* **72,** 4135–4141.

18. Dojka, M. A., Hugenholtz, P., Haack, S. K., and Pace, N. R. (1998) Microbial diversity in a hydrocarbon- and chlorinated-solvent-contaminated aquifer undergoing intrinsic bioremediation. *Appl. Environ. Microbiol.* **64,** 3869–3877.
19. Ravenschlag, K., Sahm, K., Knoblauch, C., Jørgensen, B. B., and Amann, R. (2000) Community structure, cellular rRNA content, and activity of sulfate-reducing bacteria in marine arctic sediments. *Appl. Environ. Microbiol.* **66,** 3592–3602.
20. Lin, C. and Stahl, D. A. (1995) Taxon-specific probes for the cellulolytic genus *Fibrobacter* reveal abundant and novel equine-associated populations. *Appl. Envir. Microbiol.* **61,** 1348–1351.
21. Fuchs, B. M., Glockner, F. O., Wulf, J., and Amann, R. (2000) Unlabeled helper oligonucleotides increase the *in situ* accessibility to 16S rRNA of fluorescently labeled oligonucleotide probes. *Appl. Environ. Microbiol.* **66,** 3603–3607.
22. Nikiforov, T. T., Rendle, R. B., Kotewicz, M. L., and Rogers, Y.-H. (1994) The use of phosphorothioate primers and exonuclease hydrolysis for the preparation of single-stranded PCR products and their detection by solid-phase hybridization. *PCR Methods Appl.* **3,** 285–291.
23. Ramette, R. W. (1981) *Chemical Equilibrium and Analysis.* Addison-Wesley, Reading, MA. Also, a useful website is http://oncampus.richmond.edu/~rdominey/301/SAM_General.html.

6

Molecular Characterization of Microbial Communities From Marine Environments

Nicolás G. Barengo, Juliana M. Benito, and Carlos M. Abate

1. Introduction

Over 70% of the Earth's surface is covered by saltwater, and most of the biomass and biogeochemical activity occurring therein can be attributed to picoplankton *(1)*. Marine prokaryotes are considered to be the major primary producers and heterotrophic consumers in these systems *(2)*. Ferguson et al. showed that more than 99.9% of the natural bacterioplankton community in seawater could not be cultured on marine agar 2216 medium *(3)*.

However, the molecular approach to the marine environment has revolutionized the concept of the microbial ecology of the oceans. Owing to the lack of culturability of the vast majority of microbes in the sea, the application of such techniques has resulted in detection of forms never before isolated *(4)*.

Methods that rely on direct amplification and analysis of 16S rRNA gene (rDNA) sequences are rapidly replacing cultivation as a way to assess the structure of microbial communities. A full description of community structure includes the number of species present (diversity), the number of individual cells of each species (evenness), as well as the physiological role of each species in the environment and its interaction(s) with other species *(5)*.

As the limitations of culture methods became clear, many different molecular techniques for evaluating microbial communities were developed. Terminal restriction fragment length polymorphism (T-RFLP) is one of the newest tools for evaluating microbial communities; it provides investigators with a large amount of easily analyzed data to compare microbial communities and assess community dynamics in a rapid and reproducible way *(6)*. Most investi-

From: *Methods in Biotechnology: Environmental Microbiology: Methods and Protocols*
Edited by: J. F. T. Spencer and A. L. Ragout de Spencer

gators report using an automated fragment analysis program that calculates TRF length (bp) by comparing TRF peak retention time to a DNA size standard. These programs integrate the electropherograms and give the TRF peak height and area. The patterns of TRF peaks can then be numerically compared between samples using a variety of multivariate statistical methods. In addition, individual TRF peaks in a pattern can be identified by comparison to a clone library or by predictions from an existing database of sequences *(6)*.

The objective of this chapter is to provide new protocols for studying the microbial diversity of seawater samples. We hope that readers will evaluate these protocols, using them as guidelines and improving upon them for their particular application.

2. Materials

2.1. Sampling

1. Acid-washed Niskin bottles, thoroughly prerinsed with seawater.
2. White polycarbonate filters (47 mm diameter; 0.2 µm pore size; type GTTP2500; Millipore) (*see* **Note 1**).
3. Filtration system.

2.2. DNA Isolation, Purification, and Visualization

1. Lysis buffer: 20 m*M* EDTA; 400 m*M* NaCl; 750 m*M* sucrose; Tris-HCl 50 m*M*, pH 9.0; and lysozyme 2 mg/mL (*see* **Note 2**).
2. 10 mg/mL Proteinase K in distilled water.
3. 20% SDS (w/v).
4. 3 *M* Sodium acetate, pH 4.8.
5. Phenol-chloroform-isoamyl alcohol (25:24:1) *(7)*.
6. Chloroform-isoamyl alcohol 24:1 (v/v).
7. Isopropanol and 70% ethanol.
8. RNAase A solution (stock 10 mg/mL) in distilled water *(7)*.
9. 1X TAE buffer: 0.04 *M* Tris-acetate, 0.001 *M* EDTA, pH 8.0 *(7)*.
10. 0.8% Agarose in 1X TAE buffer.
11. Ethidium bromide staining solution: 1 µg/mL in 1X TAE buffer *(7)*.
12. Molecular weight markers: K562 High Molecular Weight (Promega).
13. Loading buffer (Gibco BRL).

2.3. PCR Amplifications and Purification of PCR Products

1. Primers (final concentration 0.5 µ*M*): for eubacterial Domain 27F, 5'-AGAGTTTGATCMTGGCTCAG-3; 1492R, 5'-GGTTACCTTGTTACGACTT-3' *(8)* (*see* **Note 3**). For checking the rDNA 16S Library: SP6 and T7 promoter primers (Biodynamics).
2. *Taq* polymerase and 10X STR buffer (Promega).
3. Thermal cycler (e.g., Gene Amp PCR System 9700, Applied Biosystems, CA).

4. Reagents and apparatus for an agarose gel electrophoresis.
5. Molecular weight marker: 1 kb Ladder (Promega).
6. Wizard® PCR Preps DNA Purification System (Promega).
7. Sterile-distilled water.

2.4. rDNA 16S Library

1. pGEM®-T Easy Vector System (Promega).
2. *Escherichia coli* DH5 α competent cells *(7)*.
3. Cell-Porator® Electroporation System and Cell-Porator® Voltage Booster (Gibco BRL).
4. LB agar plates: 20 g/L agar; 10 g/L tryptone; 5 g/L yeast extract; 10 g/L NaCl; pH 7.0 supplemented with ampicillin (final concentration 100 µg/mL), nalidixic acid (10 µg/mL final concentration), IPTG, and X-gal (both at final concentrations of 40 g/mL).
5. SOB medium: 20 g/L tryptone; 5 g/L yeast extract; 18 g/L glucose; 0.584 g/L NaCl; 0.186 g/L KCl; pH 7.0.
6. SOC medium: add to 98 mL of SOB medium 1 mL of a 2 *M* stock solution of glucose, and 1 mL of a 2 *M* stock solution of Mg^{++} ($MgCl_2 \cdot 6H_2O$, 20.33 g; $MgSO_4 \cdot 7H_2O$, 24.65 g).
7. LB liquid medium supplemented with ampicillin (100 µg/mL final concentration) and nalidixic acid (10 µg/mL final concentration) (Promega).
8. Reagents for sequencing and DNA sequencer.

2.5. Reagents and Software Programs for T-RFLP and Sequence Analysis

1. Restriction endonucleases *Hae*III, *Hha*I, and *Rsa*I (Gibco BRL) (*see* **Note 4**).
2. DNA fragment length standard (TAMRA GS 2500; ABI).
3. DNA sequencer (e.g., 373 ABI Stretch).
4. GeneScan® Analysis software.
5. Programs available at http://rdp.cme.msu.edu/.
6. BLAST tools at http://www.ncbi.nlm.nih.gov/.

3. Methods

3.1. Collection of Samples

1. Collect the seawater samples at the selected depth in acid-washed and seawater-prerinsed Niskin bottles.
2. Filter the seawater through the polycarbonate filters.
3. Store the filters at –20°C until processing.

3.2. DNA Isolation, Purification, and Visualization (see Note 5)

1. Cut one of the filters obtained in **Subheading 3.1.** in small pieces and submerge them in 1 mL of lysis buffer.
2. Vortex 1 min and incubate at 37°C for 30 min.

3. Add SDS (final concentration 1% w/v) and proteinase K (100 μg/mL final concentration).
4. Vortex 10 s and incubate at 55°C for 2 h.
5. Discard the filter pieces and add an equal volume of phenol-chloroform-isoamyl alcohol and mix by inverting the tube several times.
6. Centrifuge (10,000*g* for 15 min). Transfer the aqueous phase (upper) to a new tube and repeat phenol extraction once.
7. Transfer the upper aqueous phase to a clean tube and add an equal volume of chloroform-isoamyl alcohol. Again mix well and centrifuge (10,000*g* for 5 min). Repeat this extraction two times.
8. Transfer the aqueous phase to a new tube and precipitate the DNA by adding 1/10 vol of 3 *M* sodium acetate and 0.6 to 1 vol of 2-propanol. Mix gently and incubate at –20°C from 1 h to overnight.
9. Centrifuge (10,000*g* for 15 min). Discard the supernatant and wash DNA with 500 L of 70% ethanol to remove residual salts and isopropanol. Centrifuge (10,000*g* for 5 min), carefully discard the ethanol, and dry until ethanol has been removed.
10. Resuspend DNA in 20–30 μL of double-distilled sterile water and 0.1–0.2 μL RNAase A. Allow to dissolve at 37°C at least 3 h (*see* **Note 6**).
11. Visualization of the extracted DNA is done by running 5 μL of the sample in a 0.8% agarose gel electrophoresis using 1X TAE electrophoresis buffer and high-molecular-weight marker. Electrophoresis is carried out at 10 V/cm for 1 h at room temperature; gels should be stained with ethidium bromide solution and observed under UV light.

3.3. PCR Conditions

3.3.1. PCR Conditions of 16S rDNA for Library Construction and T-RFLP Analysis

Prepare the reaction mixture to a final volume of 25 μL (*see* **Note 7**):

1. 100 ng isolated DNA.
2. 2.5 μL 10X STR buffer.
3. 0.2 μL *Taq* polymerase (1 U).
4. 0.2 μL of each primer: 27F and 1492R (0.5 μ*M* final concentration) (*see* **Note 3**).
5. Double-distilled sterile water to 25 μL (*see* **Note 8**).

Amplification is performed with an initial denaturation at 94°C for 5 min; 25–30 cycles of 94°C for 1 min, 55°C for 1 min, and 72°C for 2 min; and a final extension at 72°C for 10 min (*see* **Note 9**).

3.3.2. PCR Conditions for Screening of 16S rDNA Gene Library

Extract the plasmids from the library clones using Wizard Plus SV Minipreps DNA Purification Systems (Promega) following the manufacturer's instructions. Prepare the reaction mixture to a final volume of 25 μL:

1. 100 ng Isolated plasmidic DNA.
2. 2.5 µL 10X STR buffer.
3. 0.2 µL *Taq* polymerase (1 U).
4. 0.5 µL of each primer: SP6 and T7 (0.5 µ*M* final concentration).
5. Double-distilled sterile water to 25 µL (*see* **Note 8**).

Amplification is performed with an initial denaturation at 94°C for 5 min; 35 cycles of 94°C for 30 s, 52°C for 1 min, and 72°C for 3 min; and a final extension at 72°C for 10 min (*see* **Notes 9** and **10**).

3.4. T-RFLP

1. After visualization and purification, aliquots (15 µL) of each sample are separately digested with restriction endonucleases *Hae*III, *Hha*I, and *Rsa*I (Gibco BRL) according to the supplier's instructions.
2. To protect the internal standard from cleavage, the restriction endonucleases are deactivated by heating the reaction mixture at 65°C for 30 min.
3. Aliquots (2 µL) of the digest are mixed with 2 µL of deionized formamide, 0.5 µL of loading buffer (Applied Biosystems Instruments [ABI], Foster City, CA), and 0.5 µL of a DNA fragment length standard (TAMRA GS 2500; ABI).
4. After denaturing the DNA at 94°C for 5 min and immediate chilling on ice, aliquots (2.5 µL) are loaded onto a 36-cm long 6% denaturing polyacrylamide gel of an automated DNA sequencer (373 ABI Stretch). Electrophoresis is run for 14 h with limits of 1.680 V and 40 mA.
5. After electrophoresis, the lengths of fluorescent labeled terminal restriction fragments (T-RFs) are analyzed by comparison with the internal standard using GeneScan software (ABI).

3.4.1. Analysis of T-RFLPs

For each sample, peaks over a threshold of 50 U above background fluorescence are analyzed by manually aligning fragments to the size standard. To avoid detection of primers and uncertainties of size determination, terminal fragments smaller than 50 bp and larger than 800 bp are excluded from the analysis. Communities are characterized by the numbers of peaks and the heights of the peaks. The relative abundance of T-RFs within the sections is determined by calculating the ratio between the peak height of each peak and the total peak height of all peaks within one sample. Ratios are converted to percentages, and the results are displayed as histograms.

3.5. Construction of 16S rDNA Library

1. After visualization, purify the PCR products using the Wizard PCR Preps DNA Purification System (Promega) according to the manufacturer's instructions.
2. Ligate the purified 1.5 kb fragments into pGEM-T Easy vector following the supplier's instructions.

3. Mix carefully 20 µL of competent cells with 2 µL of the ligation product in a sterile microcentrifuge tube.
4. Use a micropipet to suspend each cell-DNA sample between the electrode bosses of a disposable microelectroporation chamber, and perform the electroporation procedure as described in the Cell-Porator manual.
5. After electroporation add 1 mL of SOC medium to the disposable microelectroporation chamber and incubate the cells at 37°C for 1 h, to allow the bacteria to recover and to express the antibiotic resistance marker encoded by the plasmid.
6. Transfer the cell suspension to a new sterile microtube and plate 100 µL on LB agar plates supplemented with ampicillin, nalidixic acid, IPTG, and X-gal.
7. Incubate at 37°C for 12–16 h until blue and white colonies develop. If desired, store the plates at 4°C for several hours. This allows the blue color to develop fully.
8. Pick each white colony into LB liquid medium supplemented with ampicillin and nalidixic acid and incubate at 37°C overnight.
9. Screen the isolated colonies as indicated in **Subheading 3.3.2.** to identify bacterial colonies that contain recombinant plasmids (*see* **Note 11**).
10. Once the colonies are selected, sequencing of the inserts is done (*see* **Note 10**).
11. Retrieved sequences are analyzed using BLAST from http://www.ncbi.nlm.nih.gov and programs from http://rdp.cme.msu.edu.

4. Notes

1. DNA extracts from the particulate or dissolved pools have to be concentrated. For the microbial pool (usually > 0.2 µm) a variety of approaches have been taken for cell concentration. The most effective and simplest technique is showed by Fuhrman et al. ***(9)***, which is filtering large volumes of seawater through membrane filters. Volumes to be filtered may vary, but to get the greater part of a microbial community it should be more than 10 L. Concerning the filters, the sample can be filtered first through an 8 µm pore-size filter to separate nonbacterial particles, including protists.
2. The lysis buffer without lysozyme has to be aseptically prepared and stored at room temperature for long periods of time. The lysozyme is added just before using.
3. For T-RFLP, primer 27F is 5'-labeled with a fluorescent molecule, e.g., 27F-FAM (phosphoramidite fluorochrome 5-carboxyfluorescein).

   The primer pair cited in this chapter is used for analyzing bacterial communities. Nevertheless it is possible to use primer pairs to amplified specific groups within the domain Eubacteria. Also there are several reports in which the investigators use primers to specifically amplify genes from the domain Archaebacteria as well as particular groups within this domain. This methodology can be also applied to the study of the diversity of metabolic genes. For a more detailed review, *see* **ref. *6***.
4. Selection of restriction endonucleases for T-RFLP will depend on each investigator's previous work or preference. If there is not any previous experi-

ence, one can test a set of enzymes and then select the best based on the major number of peaks obtained.

5. Gloves have to be used to minimize the risk of DNAase contamination. In the case of PCR preparation, gloves should be powder free because powder inhibits DNA polymerases.

   Depending on the size of a sample, pooling of replicate extractions is recommended to limit random bias, although systematic biases will persist.

   DNA samples can be stored for several weeks at 4°C. Aliquots of samples can be preserved at –20°C, taking into account that repeated freeze and thaw of the samples can damage DNA.
6. DNA concentration should be determined by reading the absorbance of the samples at 260 nm in a UV spectrophotometer, or using gel analyzer software. It is recommended to standardize the concentrations of the DNA at 100 ng/L with the aim of making a comparative analysis of the samples.
7. Although it is clear that systematic PCR bias cannot be controlled, many investigators in the TRF literature pooled multiple PCR reactions from a single sample to ensure random PCR artifacts are minimized. In our personal experience, we work with triplicates of each sample, which are pooled after PCR amplification.
8. It is convenient to prepare a master mix with all the reaction components except the DNA, considering total number of samples including a positive control plus an additional control tube that will not include template DNA (negative control).
9. Visualization of the PCR products is done by running aliquots (5 μL) of the products on a 1% agarose gel using a molecular-weight marker, e.g., 1 kb Ladder (Promega). Electrophoresis is run at 10 V/cm for 1 h. Bands are visualized after staining the gel with ethidium bromide. If there are not any unspecific amplification bands, PCR products can be purified directly from the solution. But if unspecific amplification is observed, the entire PCR product has to be run on an agarose gel and the 1.5 kb band purified from the gel. In both cases, the purification is carried out using the Wizard PCR Preps DNA Purification System (Promega).
10. Positive clones are those that show amplification products of approx 1650 bp. To avoid sequencing the same fragment two or more times, it is convenient to treat each insert with two tetrameric restriction endonucleases and then compare the different patterns obtained. Based on the restriction profiles, the clones are grouped and one representative of each group is sequenced.
11. The 16S rDNA gene library can be screened by a variety of methods. Colony hybridization procedures using rRNA gene-specific oligonucleotide probes of defined phylogenetic resolution may be used. Alternatively, plasmid minipreps and restriction digests are used to confirm the presence of cloned DNA of the correct size, or else colony PCR (using, e.g., sequencing primers with priming sites that flank the insert DNA) can be used as a rapid screening procedure to detect cloned PCR products, and can also rapidly provide template DNA suitable for sequencing.

References

1. Fenchel, T. (1988) Marine plankton food chains. *Annu. Rev. Ecol. System.* **19,** 19–38.
2. Fuhrman, J. A., McCallum, K., and Davis, A. A. (1993) Phylogenetic diversity of subsurface marine microbial communities from the Atlantic and Pacific oceans. *Appl. Environ. Microbiol.* **59,** 1294–1302.
3. Ferguson, R. L., Buckley E. N., and Palumbo, A. V. (1984) Response of marine bacterioplankton to differential filtration and confinement. *Appl. Environ. Microbiol.* **47,** 49–55.
4. Giovannoni, S. J., Britschgi, T. B., Moyer, C. L., and Field, K. G. (1990a) Genetic diversity in Sargasso Sea bacterioplankton. *Nature* **345,** 60–63.
5. Tiedje, J. M., Auming-Brempong, S., Nusslein, K., Marsh, T. L., and Flynn, S. J. (1999) Opening the black box of soil microbial diversity. *Applied Soil Ecology* **13,** 109–122.
6. Kitts, C. L. (2001) Terminal restriction fragment patterns: a tool for comparing microbial communities and assessing community dynamics. *Curr. Issues Intest. Microbiol.* **2,** 17–25.
7. Sambrook, J., Fristch, E. F., and Maniatis, T., ed. (1989) *Molecular Cloning: A Laboratory Manual, 2nd ed.* Cold Spring Harbor Laboratory Press, Cold Spring Harbor, NY.
8. Lane, D. J. (1991) 16S/23S rRNA sequencing, in *Nucleic Acid Techniques in Bacterial Systematics* (Stackebrandt, E. and Goodfellow, M., eds.), John Wiley & Sons, New York, NY.
9. Fuhrman, J. A., McCallum, K., and Davis, A. A. (1992) Novel measure archae bacterial group from marine plankton. *Nature* **356,** 148–149.

7

Yeasts

Ecology in Northwest Argentina

Lucía I. C. de Figueroa, M. Alejandra Martínez, and John F. T. Spencer

1. Introduction

Yeast ecology, biogeography, and biodiversity are important and interesting topics of research. The ecology of yeasts is valuable especially for its practical value and as the fundamental basis of their evolution. New species of yeast are formed due to selection pressures exerted by the environment, and observation of the similarities and differences among the yeasts found in a particular environment can assist the investigator to see species formation as it happens ***(1,2)***.

Because ecology is the study of any aspect of organisms in the context of their environment, all investigations of yeast morphology, physiology, or genetics, be they descriptive or predictive, may be spiced with an ecological flavor. Yeast ecology has, therefore, progressed largely in parallel with the development of studies of yeast taxonomy and physiology. Also, it is concerned with the manner in which yeasts and yeast-like organisms propagate in nature; the kind of substrates that support specific yeasts; competition among yeast species, between yeasts and other microorganisms, and with larger organisms, especially insects; the effect of the chemical composition of substrates on the microflora; and the vectors responsible for introducing yeasts to suitable habitats. Ecology often addresses the study of life from many of the same perspectives as evolutionary biology, but on a different time scale. Whereas evolutionary responses involve changes in gene frequencies over generations, ecological adaptation may occur repeatedly during the life of an organism, and does not usually involve changes in the genotype of that organism. Ecological development examines the interaction between the genetic profiles of organisms as members of communities with their natural ecosystems.

From: *Methods in Biotechnology: Environmental Microbiology: Methods and Protocols*
Edited by: J. F. T. Spencer and A. L. Ragout de Spencer

This concept had a lot of influence in yeast ecology, leading to the study of the interdependence of, e.g., cacti, yeasts, and drosophilids ***(3)***. Yeasts form communities of species. Each community may be defined by its habitat, which is the place where an assemblage of yeasts lives, and by the niches of its component species. The niche, or ecosystem, consists of the attributes that make a yeast capable of sharing the habitat with other members of the community ***(4–6)***. The rationale for studying plants, insects, and soils is that photosynthesis and the soil furnish the necessary nutrients for the growth of plants, and plants in turn can support the growth of yeasts. Insects are by far the most important vectors in the distribution of yeasts in nature, whereas soil is the ultimate repository for organic and inorganic materials and constitutes a suitable medium for storage and even development of certain species of yeast. Bacteria, fungi, and insects, including fly larvae and boring beetles, degrade dead and dying plant material to forms which can be metabolized by yeasts. Yeasts do not photosynthesize or fix nitrogen, and therefore depend on the other organisms to convert their substrates to available forms. They have the ability to ferment sugars and also the ability to respire sugars and other organic compounds. The utilization of different substrates by yeasts is highly selective and is often related to niche specificity, because most are metabolic products of specific plant species or products of bacteria or fungi that produce them from precursors in plants ***(2,7–10)***. The greatest populations of yeasts are usually associated with concentrations of assimilable sugars and other carbon sources. Leaf surfaces and plant exudates commonly support large numbers of this microorganism. Flowers and decaying fruits and other plant material support a wide spectrum of species in high numbers. Insects represent rich sources of yeasts, especially wood borers and species of *Drosophila.* The majority of yeasts are obligate aerobes and therefore require oxygen for growth and reproduction. Because of this, yeasts are usually found in the upper layers of soils and sediments ***(11)***.

The association between yeasts and insects is well known. Citrus fruits are known to be attacked by numerous insect pests and scavengers, *Drosophila* species in particular being vectors for transmission of a number of yeast species from one food source to another. De Camargo and Phaff ***(12)*** and Vacek et al. ***(13)*** found a number of yeast species in fermenting tomatoes and decaying citrus fruits (oranges, grapefruits, mandarins, and so on) and *Drosophila* species associated with them, the yeasts including *Pichia kluyveri, Pichia fermentans, Torulopsis stellata, Hanseniospora uvarum, Kloekera apiculata,* and a few isolates of other species. Yeasts were associated with various species of *Drosophila,* vectors carrying the yeasts.

To the best of our knowledge, in the Tucumán region, Northwest Argentina, the South American fruit fly (*Anastrepha fraterculus* Wied.) and the Mediterranean fruit fly (*Ceratistis capilata* Wied.), which are endemic to the region ***(14)***, have not been implicated in the infection of citrus with yeasts. However, these flies may well be vectors carrying the yeasts from fruit to fruit during the growing and ripening season. It is not known whether the yeasts participate directly in the spoilage process, but it has been observed that they are present in considerable numbers in the spoiled fruit in the Northwest Argentina region. There are occasional instances of spoilage by filamentous fungi (probably *Aspergillus* or *Penicillium* species), but the majority of infected material yielded yeasts and/or yeast-like fungi, probably *Geotrichum* sp. ***(2,15)***.

Yeast microflora associated with trees of several types has been investigated by numerous workers. These yeasts are mostly associated with exudates of sap (slime fluxes) and are part of a varied population of yeasts, bacteria, worms, larvae of numerous insects (such as *Drosophila* spp.), and so on ***(8)***. However there are few references to yeasts associated with leguminous plants. Yeast species in exudates of leguminous trees may have a selective advantage favoring their development in these habitats ***(2,16)***. There is a wide variety of this kind of tree in South America, occurring over a wide geographic range. Leguminous trees that form slime fluxes are relatively rare, but there are several species of algarrobo (*Prosopis* spp., family Leguminosae, subfamily Mimosoidea) that are common in Northwestern Argentina and form slime fluxes readily ***(17)***.

The yeast–*Drosophila*–cactus–*Cactoblastis cactorum* system is well known ***(3)***. The first invaders of damaged cactus tissue are probably pectolytic bacteria. These attract *Drosophila* species, which are vectors for the different cactophilic yeasts. Yeasts invade rotting tissue of numerous species of giant cacti, where they metabolize sugars and related compounds released by bacterial action. The *Drosophila* feed on the yeasts and lay their eggs in the rotting tissue, the cactus eventually dies, the *Drosophila* move on to other rooting cacti, carrying yeasts with them, and the cycle is repeated ***(2)***. In Northwest Argentina numerous cactophilic yeasts have been isolated, especially from rotting giant cacti ***(18)***. In this region are cacti of the families Pachycereinae and Stenocereinae (**Table 1**), and several strains have also been isolated from agricultural residues (corn silage and viticulture residues) in order to select xylitol-producing yeasts ***(19)***.

The study of the distribution of yeasts in nature requires the isolation, identification, and enumeration of the yeast species present in a given microenvironment. Once a habitat has been identified and suitable methods designed, a collection scheme should be formulated to yield sufficient numbers of samples. Lachance and Starmer *(7)* have determined empirically that a minimum of 15 samples is required to obtain an accurate reflection of community composition for cactus necroses, insects, tree exudates, or flowers. The number varies

Table 1
Specific and Common Names of Giant Cacti Associated With Yeasts in Northwest Argentina

Family	Specific name	Common name
Pachycereinae	*Lophocereus schottii*	Senita
Pachycereinae	*Carnegia gigantea*	Saguaro
Pachycereinae	*Pachycereus pringlei*	Cardon
Stenocereinae	*Stenocereus thurberi*	Organ pipe
Stenocereinae	*Stenocereus gummosus*	Agria
Stenocereinae	*Stenocereus alamocensis*	Cina
	Backebergia sp.	–
	Opuntia sp.	Prickly pear

according to the frequency of empty samples. If empty samples are present, the sampling size should be corrected to obtain a minimum of eight nonempty samples.

The use of traditional methods of identification of isolates of unknown species is limited by the fact that numerous yeast species have phenotypes so similar that they cannot be differentiated by these methods. Specialized groups of yeasts can usually be isolated by use of selective media containing particular carbon or nitrogen sources. These procedures also commonly used the appearance of cellular morphology and distinctive reactions based on a standardized set of fermentation and assimilation tests. The assays are laborious and sometimes ambiguous because of strain variability. However, the determination of physiological characteristics of yeasts will always be of importance in understanding their ecological niches; they are also useful in identifying intraspecific variation.

Molecular systematic methods are increasingly used for yeast identification. Among them, comparison of nuclear DNA (nDNA) relatedness has been very useful. Those studies require pairwise comparisons of isolates, and their resolution is limited to the level of closely related species ***(20)***.

Recently other molecular comparisons have been widely used for typing and assessing genomic diversity of microorganisms, including sequencing, restriction fragment length polymorphism (RFLP) ***(21,22)***, pulsed field gel electrophoresis (PFGE) ***(23)***, and random amplified polymorphic DNA-PCR (RAPD) ***(24)***. Sequencing appears to be the most important technique because strain comparisons are easily made, and with the selection of appropriate genes, both close and distant relationships can be resolved ***(20,25)***.

Phylogenetic analysis of ribosomal RNA (rRNA) and its template ribosomal DNA (rDNA) have been used extensively in recent years to assess both close and distant relationships among many kinds of organisms; however, the determination of complete sequences is laborious, and only a few hundred complete sequences are available for either the large (25S–28S) or the small (18S) rRNA subunits *(26)*. Partial sequences are relatively easy to determine by using sequencing methods. McCarroll et al. *(27)* and Lane et al. *(28)* have demonstrated that partial sequences of small rRNA subunits reveal essentially the same phylogenetic relationships as the complete sequence. In addition, much information is provided by short noncodifying sequences, such as internal ribosomal transcribed spacers (ITS) *(29,30)*.

The objective of this chapter is to describe procedures used for isolation of yeasts from their natural habitats and identify them. The work is focused on yeast ecology in Northwest Argentina.

2. Materials

2.1. Materials, Habitats

1. Materials for isolating yeasts associated with rotting citrus fruits: spoiled oranges, lemons, mandarins, limes, grapefruit, and so on, which proved to be infested with Mediterranean fruit fly.
2. Material for isolating yeasts from agricultural residues: corn silage, grapes, grape-must, wine, and so forth.
3. Material for isolating yeasts associated with algarrobo trees (*Prosopis* spp.) or with different species of cacti: exudates from trees or small trunks arising from one root, slime flux from white algarrobo tree *(Prosopis alba)*, rooting cacti (columnar cacti or *Opuntia* sp), and so on (*see* **Note 1**).

2.2. Culture Media

1. Isolation medium, YM (Difco Laboratories, Detroit, MI): 0.03 g/L Rose Bengal and 20 g/L agar, acidified to pH 3.5 with 1 *M* HCl.
2. Maintenance culture medium, YPED: 10 g/L yeast extract, 20 g/L peptone, and 20 g/L glucose, pH adjusted to 4.5 with 1 *M* HCl.
3. Medium for assimilation tests: yeast nitrogen base YNB without amino acids (w/aa) Difco (or yeasts carbon base, YCB) 6.7 g/L, 15 g/L agar, and 10 g/L of the compound to be tested.
4. Medium for fermentation tests: yeast extract 5 g/L plus 10 g/L of the sugar to be tested.

2.3. DNA Isolation

1. Lysis buffer TEC-SDS: 10 m*M* Tris-HCl, pH 8; 10 m*M* EDTA; 100 m*M* NaCl; and 2% (w/v) SDS.
2. 20 mg/mL Proteinase K in distilled water.

Table 2
Primers to Be Used

Primers	Sequence 5'-3'	References
ITS1	TCCGTAGGTGAACCTGCGG	Lott et al., 1993 ***(33)***
ITS3	GCATCGATGAAGAACGCAGCT	Lott et al., 1993 ***(33)***
ITS4	TCCTCCGCTTATTGATATGC	Lott et al., 1993 ***(33)***
NL-1	GCATATCAAAAGCGGAGGAAAAG	Kurtzman and Robnett 1998, ***(20)***
NL-4	GGTCCGTGTTTCAAGACGG	Kurtzman and Robnett 1998, ***(20)***

3. 3 *M* Sodium acetate pH 5.2.
4. TE-buffer-saturated phenol ***(32)***.
5. Chloroform:isoamyl alcohol, 24:1 (v/v).
6. Isopropanol and 70% ethanol.
7. RNAase A solution (stock 10 mg/mL) in distilled water ***(32)***.

2.4. Solutions and Devices for PCR Amplifications

1. Primers (**Table 2**).
2. *Taq* polymerase and 10X STR buffer (Promega).
3. Thermal cycler (e.g., Gene Amp PCR System 9700, Applied Biosystems, CA).
4. TAE buffer: 1X: 0.04 *M* Tris-acetate, 0.001 *M* EDTA, pH 8.0 ***(32)***.
5. 2% Agarose in 1X TAE buffer.
6. Ethidium bromide staining solution in 1X TAE buffer ***(32)***.
7. Molecular weight markers: 1 kb and 100 bp DNA Ladders (Promega).
8. Loading buffer (Promega).

2.5. Solutions and Devices for Sequence Analysis

1. Wizard® PCR Preps DNA Purification System (Promega).
2. BLAST tools at http://www.ncbi.nlm.nih.gov/ and DNAMAN 4.03 program.

2.6. Solutions and Devices for RFLP Analysis

1. *ITS1-NL-4* primers (**Fig. 1**).
2. Restriction enzymes *Bsp*DI, *Ssp*I, *Eco*RI, *Bam*HI, *Hae*III.
3. Molecular weight markers: 1 kb and 100 bp DNA Ladders (Promega).
4. 2–3% Agarose in 1X TAE buffer.

3. Methods

3.1. Habitat and Isolation

Samples obtained from different habitats (slime flux, exudates from trees, spoiled citrus fruits, and so on) should be 5–10 g. They are transported in sterile plastic bags. Samples are moistened if necessary, and then streaked on iso-

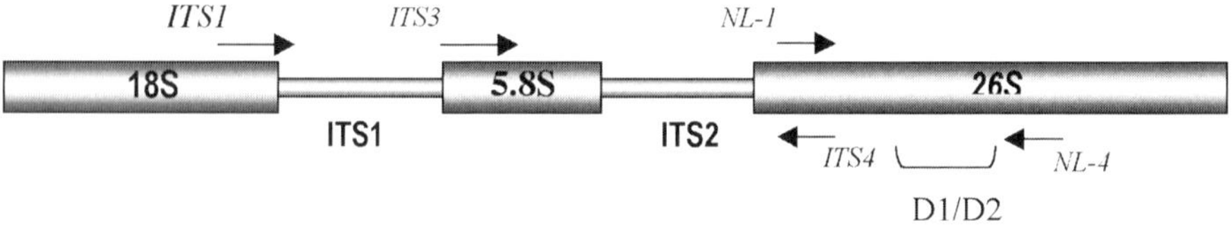

Fig. 1. Schematic representation of yeast rDNA operon.

lation medium (acidified YM). Inoculated plates are incubated at 24°C until growth is observed (4 or 6 d). Colonies are restreaked and picked to YEPD (*see* **Note 2**).

3.2. Assimilation Tests

Assimilation tests for carbohydrates and related compounds are done on Petri dishes of Wickerham's YNB medium containing 10 g/L of the compound to be tested. Ability to assimilate nitrate/nitrogen is determined using Wickerham's YCB containing nitrate as nitrogen source.

The determinations of assimilation patterns are done as follows: when yeast cultures have grown for 48–72 h on YEPD agar plates and colonies are well grown, cells are suspended in sterile distilled water in the wells of the sample plate of a multipoint inoculation apparatus.

Yeast strains are identified according to their saccharide and nitrogen assimilation patterns, using the keys and description of Yarrow ***(34)*** and the computerized yeast identification program devised by Barnett, Payne, and Yarrow ***(18)*** (*see* **Notes 3** and **4**).

3.3. Fermentation Tests

Fermentation tests are done according to the methods given by Yarrow ***(34)***, using culture tubes containing a Durham tube and 5 mL of the medium described in **Subheading 2.2.4.** All cultures must be tested for the ability to ferment glucose, and for the ability to ferment other sugars, if these are assimilated.

3.4. DNA Isolation

1. Transfer 1.5 mL of an overnight culture to a microcentrifuge tube and spin down 2 min. Discard the supernatant.
2. Resuspend the pellet in 750 µL of lysis buffer by repeated up and down pipetting. Add 15 µL of 20-mg/mL proteinase K, mix by inverting the tube carefully, and incubate 30 min to 2 h at 55°C (*see* **Notes 5** and **6**).
3. Add an equal volume of TE-buffer-saturated phenol and mix by inverting the tube several times.
4. Centrifuge (10,000*g*, 5 min). Transfer the aqueous phase (upper) to a new tube and repeat phenol extraction once.

5. Transfer the upper aqueous phase to a clean tube and add an equal volume of chloroform:isoamyl alcohol. Again mix well and centrifuge (10,000*g*, 5 min). Repeat this extraction three times.
6. Transfer the aqueous phase to a new tube. Add 1/10 vol of 3 *M* sodium acetate.
7. Add 0.6 to 1 vol of 2-propanol and mix gently until the DNA precipitates. If a low amount was recovered, precipitation could be improved by centrifugation (15,000*g*, 10 min).
8. Discard the supernatant and wash DNA with 500 µL of 70% ethanol to remove residual salts and isopropanol. Centrifuge (15,000*g*, 5 min), carefully discard the ethanol, and dry until ethanol has been removed (*see* **Note 7**).
9. Resuspend DNA in 100–200 µL of double-distilled sterile water and 1–2 µL RNAase A. Allow to dissolve at 37°C for several hours to overnight at 4°C (*see* **Note 8**).

3.5. PCR Amplifications

Reaction mixture: Prepare 20 µL for each sample plus an additional control tube that will not include template DNA. Each tube should contain 0.5 µL of isolated DNA, 2 µL of 10X STR buffer, 0.2 µL of *Taq* polymerase (1 U), and 0.2 µL of each primer (0.5 µ*M* final concentration). Double-distilled sterile water to 20 µL. Dispense 19.5 µL for all samples tested into PCR tubes and add samples (*see* **Notes 9** and **10**).

To achieve amplification of a fragment containing ITS1, 5.8S rDNA, and ITS2 sequences (**Fig. 1**), *ITS1* and *ITS4* primers have to be used. Primers *NL-1* and *NL-4* are used for 28S rDNA D1/D2 domain amplification. The following temperature profile is used with both primer combinations: initial denaturation at 94°C for 4 min, followed by 30 cycles, each of them consisting of: denaturation at 95°C for 1 min, primer annealing at 55°C for 1 min, primer extension at 70°C for 1 min, and a final extension step at 70°C for 7 min. In order to amplify a fragment including 5.8S rDNA and ITS2 sequences (**Fig. 1**), *ITS3* and *ITS4* have to be used as primers.

Evaluation of PCR fingerprint obtained: PCR-amplified products are separated by 2% agarose gel electrophoresis using 1X TAE electrophoresis buffer and appropriate markers in the range of 100 to 2000 bp. Electrophoresis is carried out at 20 V/cm for 3–4 h at room temperature; gels should be stained with 1 µg/mL ethidium bromide solution and observe under UV light.

3.6. Sequence Analysis

PCR-amplified products are recovered from agarose gel using a Wizard PCR Preps DNA Purification System (Promega). They can be sequenced to search for yeast rDNA sequence similarities by using BLAST tools. Phylogenetic trees can be constructed with the DNAMAN 4.03 program by using the neighbor-joining method and the Jukes-Cantor distance-correction method *(35)*. It is necessary to use a yeast that is well identified as out-group strain.

3.7. RFLP Analysis

PCR fragments containing sequences amplified with *ITS1-NL-4* primers (**Fig. 1**) from the problem strain are recovered from agarose gels as described above, and aliquots (15 μL) of each sample are separately digested with 10 U of *Bspd*I, *Ssp*I, *Eco*RI, *Bam*HI, and *Hae*III (Promega) according to the supplier's instructions. Resulting restriction fragments are separated in agarose 3%, and 100 and 1000 bp DNA Ladder (Promega) are used as molecular-weight markers.

4. Notes

1. Citrus fruits are produced in the Tucumán region of Northwest Argentina. The vectors of the yeast infection are very likely the larger fruit flies, *Anastrepha fraterculus* and *Ceratitis capilata,* rather than one or more of the *Drosophila* species. The two former species are endemic to the region *(15)*.

   Species of algarrobo in this part of Argentina are not forest trees, but grow as individuals or in rather small groups, often as several trunks arising from the same root. The region is arid or semi-arid and can expect approx 360 d of sunshine per year, which does not encourage dense tree growth. Slime fluxes on algarrobo differ in consistency from those on species such as poplar or willow, since they are much higher in sugar content and, in the hot, dry conditions of their environment, become very viscous, black, and sticky. Eventually they are completely solidified and brittle *(17)*.

   The European algarrobo (carob) is an important food crop *(31)*, and the South American algarrobo is a potential one. Beans and pods of the South American algarrobo are high in sugar and protein like those of the European carob. Sugars produced by algarrobo are used in cottage industries to produce a fermented beverage. Leguminous plants are known to produce a wide range of tannins, alkaloids, and other nitrogenous compounds, besides sugars, which undoubtedly influence the composition of the microflora of the slime fluxes as well as the food value of the seeds *(18)*.
2. It is important to take samples from citrus fruits showing evidence of decay caused by insect attack or other injury: brown, hardened rind, or soft spots which do not show evidence of fungal growth. For example, it is possible to plate out the decomposing flesh or the expressed juice obtained from directly beneath the rotting skin. When samples are taken of exudates from trees, the pesticide Cygon (dimethoate), 6.6 mg/L, should be added to the isolation medium for control of mites.
3. The authors have made an investigation of yeasts associated with spoilage of citrus fruits in the Tucumán region of Northwest Argentina. The range of yeasts found in decaying citrus fruits (oranges, mandarins, grapefruits, limes, and lem-

Table 3
Yeast Species Isolated From Algarrobo Tree Exudates in Northwest Argentina

Yeast species	Total isolates
Bullera variabilis	22
Candida famata	64
Candida insectorum	4
Candida magnoliae	4
Candida membranaefaciens	25
Candida rhagii	12
Candida sake	7
Cryptococcus albidus	59
Cryptococcus skinneri	9
Cryptococcus terreus	5
Debaryomyces hansenii	36
Kluyveromyces sp.	8
Pichia angusta (Hansenula polymorpha)	49
Pichia carsonii	3
Pichia ciferrii (Hansenula ciferrii)	24
Pichia farinosa	25
Pichia sydowiorum	3
Torulaspora delbrueckii	19
Unknown	1
Total	379

ons) was rather narrow, but differed from those isolated by Vasek et al. ***(13)*** to some extent. We noted that the latter workers isolated principally yeasts with very limited ability to assimilate carbon compounds, while the greater part of our isolates assimilated a moderately wide range of sugars and related compounds. In common with Vasek et al., we frequently isolated species of apiculate yeasts such as *Kloeckera apiculata.* It is also possible to find yeast strains like *Candida (Pichia) guilliermondii, Candida famata, Candida stellata, Pichia kluyveri, Pichia membranaefaciens, Geotrichum candidum,* other *Candida* spp., *Cryptococcus* sp., and so on.

4. It is possible to collect material from algarrobo tree exudates and rotting giant and *Opuntia* sp. cacti in Northwest Argentina, and isolate yeasts from them. Different species of Leguminosae (*Prosopis* or *Acacia*), as well as different species of Cactaceae (columnar cacti or *Opuntia*) are very common in Provincia de Tucumán (Tafí del Valle, Quilmes, Amaicha del Valle, Colalao del Valle), Provincia de Salta (Quebrada de Cafayate, Angastaco, Cachi), Provincia de Jujuy (Purmamarca), and Provincia del Chaco (Las Breñas, Quimilí, Añatuya). **Tables 3–5** show yeast species frequently found in those places.

Table 4
Yeast Species Isolated From Pods of *Prosopis* spp. and *Acacia* spp. in Northwest Argentina

		Prosopis		
Yeast species	*Acacia*	Cafayate	Amaicha	Purmamarca
Cryptococcus spp.	19	21	2	42
Candida famata	4	4	22	30
Candida ciferrii	2			2
Total	25	25	24	74

Table 5
Yeast Species Isolated From Columnar Cacti in Northwest Argentina

	Locations	
Yeast species	Quilmes	Cachi
Aureobasidium pullulans	2	2
Candida rugosa	8	8
Cryptococcus albidus	21	25
Pichia membranaefaciens	17	–
Sporidiobolus johnsonii	1	1
Sporobolomyces pararoseus	–	2
Torulaspora delbrueckii	–	22
Unknown	–	1
Total	49	51

5. When manipulating DNA, wear gloves to minimize the risks of DNAase contamination. In the case of PCR preparation, gloves should be powder-free because powder inhibits DNA polymerases. Precautions must be taken when handling dangerous solutions such as phenol, chloroform, and ethidium bromide. Material Safety Data Sheets can be found at www.sigma-aldrich.com.
6. For the isolation of DNA, it could be necessary to optimize lysis conditions when handling wild-type isolates. If there is poor clarification after lytic treatment, proteinase K treatment could be extended. Nevertheless, further phenol treatment ensures sufficient DNA suitable for PCR.
7. After two ethanol wash steps, liquid could be discard by inverting tubes and spinning down 1–2 min to allow residual liquid in the walls to be collected by carefully pipetting from the bottom of the tubes. Do not overdry DNA pellets; this can make resuspension difficult.

8. DNA samples can be stored for several weeks at 4°C. Aliquots of samples can be preserved at –20°C taking into account that repeated freezing and thawing of the samples can damage DNA.
9. DNA samples obtained by the procedure described above are usually suitable for PCR amplification. Concentration and quality should be tested by gel electrophoresis and absorbance measurements *(32)*. Nevertheless, OD measurements tend to overestimate DNA concentration. After electrophoretic evaluation, concentration of the samples should be adjusted to be similar by repeating **steps 5–8** of the DNA isolation procedure. Absence of amplification could be due to residual phenol or ethanol in the final preparations. To evaluate the reagents, a positive (amplifiable) control should be included in all assays.
10. Total PCR reaction volume should be loaded to allow detection of all bands, even the less intense ones. Inclusion of controls without template are important since possible PCR artifacts should be considered when analyzing bands of low size (100–300 bp). Electrophoresis runs of 2–3% agarose gels should be relatively slow (20–50 V), allowing a discrete band pattern to develop. Capture image systems are useful when several bands are obtained, to precisely determine the number and size of bands.

References

1. Torija, M. J., Rozes, N., Poblet, M., Guillamon, J. M., and Mas, A. (2001) Yeast population dynamics in spontaneous fermentations: comparison between two different wine-producing areas over a period of three years. *Antonie Van Leeuwenhoek* **79,** 345–352.
2. Spencer, J. F. T. and Spencer, D. M. (1997) Ecology: where yeasts live, in *Yeasts in Natural and Artificial Habitats* (Spencer, J. F. T. and Spencer, D. M., eds.), Springer-Verlag, Berlin, Heidelberg, pp. 33–58.
3. Barker, J. S. F. and Starmer, W. T. (1982) The cactus/yeast/*Drosophila* model system, in *Ecological Genetics and Evolution* (Barker, J. S. F. and Starmer, W. T., eds.), Academic Press, New York.
4. Maximova, I. A. (2000) Yeast biodiversity in forest ecosystems of European Russia, in *The Rising Power of Yeasts in Science and Industry, ISY 2000* (van Dijken, J. P. and Scheffers, W. A., eds.), Delft University Press, The Netherlands, p. 138.
5. Thomas-Hall, S., Watson, K., and Fell, J. (2000) Biodiversity of yeasts in Antarctica, in *The Rising Power of Yeasts in Science and Industry, ISY 2000* (van Dijken, J. P. and Scheffers, W. A., eds.), Delft University Press, The Netherlands, p. 139.
6. Soina, V. S., Vorobyova, E. A., Suzina, N. E., Dmitriev, V. V., Faizutdinova, R. N., Duda, V. I., et al. (2000) Viable ancient permafrost yeast, in *The Rising Power of Yeasts in Science and Industry, ISY 2000* (van Dijken, J. P. and Scheffers, W. A., eds.), Delft University Press, The Netherlands, p. 140.
7. Lachance, M. A. and Starmer, W. T. (1998) Ecology and yeasts, in *The Yeasts, a Taxonomic Study* (Kurtzman, C. P. and Fell, J. W., eds.), Fourth edition, Elsevier Science B.V., Amsterdam, New York, pp. 21–30.

8. Phaff, H. J. and Starmer, W. T. (1987) Yeasts associated with plants, insects and soil, in *The Yeasts, Volume I, Biology of Yeasts* (Rose and Harrison, eds.), Academic Press Inc., London, pp. 123–180.
9. Pechurkin, N. S. and Shirobokova, I. M. (2001) Closed artificial ecosystems as a means of ecosystem studies for earth and space needs. *Adv. Space Res.* **27,** 1497–1504.
10. Yarrow, D., de Vries, E., Theelen, B., Smith, M. T., Robert, V., Poot, G., et al. (2000) Yeast classification: new roads to old destinations?, in *The Rising Power of Yeasts in Science and Industry, ISY 2000* (van Dijken and Scheffers, eds.), Delft University Press, The Netherlands, pp. 82–83.
11. Kurtzman, C. P. and Sugiyama, J. (2001) Ascomycetous yeasts and yeast-like taxa, in *The Mycota VII, Part A. Systematics and Evolution* (Mc Laughlin, Mc Laughlin, and Lemke, eds.), Springer-Verlag, Berlin, Heidelberg, pp. 179–200.
12. De Camargo, R. and Phaff, H. J. (1957) Yeasts occurring in *Drosophila* flies and in fermenting tomato fruits in Northern California. *Food Res.* **22,** 367–372.
13. Vacek, D. C., Starmer, W. T., and Heed, W. B. (1979) Relevance of the ecology of citrus yeasts to the diet of *Drosophila. Microbial Ecol.* **5,** 43–49.
14. Nasca, A. J., Terán, A. L., Fernández, R. V., and Pasqualini, A. J., eds. (1981) *Animales perjudiciales y benéficos a los cítricos en el Noroeste Argentino.* Publinter S. A., Buenos Aires, Argentina, pp. 96–107.
15. Spencer, D. M., Spencer, J. F. T., de Figueroa, L. I. C., and Heluane, H. (1992) Yeasts associated with rotting citrus fruits in Tucumán, Argentina. *Mycol. Res.* **96,** 891–982.
16. Lachance, M. A. and Starmer, W. T. (1982) Evolutionary significance of physiological relationships among yeast communities associated with trees. *Can. J. Bot.* **60,** 285–293.
17. Spencer, D. M., Spencer, J. F. T., Fengler, E., and de Figueroa, L. I. C. (1995) Yeasts associated with algarrobo trees (*Prosopis* spp.) in Northwest Argentina: a preliminary report. *J. Ind. Microbiol.* **14,** 472–474.
18. Spencer, D. M., Spencer, J. F. T., de Figueroa, L. I. C., Garro, O., and Fengler, E. (1996) Yeasts associated with pods and exudates of algarrobo trees (*Prosopis* spp.) and species of columnar cacti in Northwest Argentina. *Appl. Microbiol. Biotechnol.* **44,** 736–739.
19. Altamirano, A., Vázquez, F., and de Figueroa, L. I. C. (2000) Isolation and identification of xylitol-producing yeasts from agricultural residues. *Folia Microbiol.* **45,** 255–258.
20. Kurtzman, C. P. and Robnett, C. J. (1998) Identification and phylogeny of ascomycetous yeast from analysis of nuclear large subunit (26S) ribosomal DNA partial sequences. *Antonie van Leeuwenhoek* **73,** 331–371.
21. Belloch, C., Barrio, E., Garcia, M. D., and Querol, A. (1998) Phylogenetic reconstruction of the yeast genus *Kluyveromyces:* restriction map analysis of the 5.8S rRNA gene and the two ribosomal internal transcribed spacers. *System. Appl. Microbiol.* **21,** 266–273.

22. Bruns, T. D., White, T. J., and Taylor, J. W. (1991) Fungal molecular systematics. *Ann. Rev. Ecol. Syst.* **22,** 525–564.
23. Bruns, T. D., White, T. J., and Taylor, J. W. (1991) Fungal molecular systematics. *Ann. Rev. Ecol. Syst.* **22,** 525–564.
24. Yuji, Oda, Yabuki, M., Tonomura, K., and Fukunaga, M. (1999) Sequence analysis of 18S–28S rRNA spacer region from *Saccharomyces kunashirensis, S. martiniae, S. rosinii,* and *S. transvaalensis. Curr. Microbiol.* **38,** 61–63.
25. Kurtzman, C. P. and Robnett, C. J. (1997) Identification of clinically important ascomycetous yeast based on nucleotide divergence in the 5' end of the large-subunit (26S) ribosomal DNA gene. *J. Clin. Microbiol.* **35,** 1216–1223.
26. Kurtzman, C. P. (1992) rRNA sequence comparisons for assessing phylogenetic relationships among yeasts. *Int. J. Syst. Bacteriol.* **42,** 1–6.
27. McCarroll, R., Olsen, G. J., Stahl, Y. D., Wose, C. R., and Sogin, M. L. (1983) Nucleotide sequence of the *Dictyostelium discoideum* small-subunit ribosomal ribonucleic acid infered from the gene sequence: evolutionary implications. *Biochemistry* **22,** 5858–5868.
28. Lane, D. J., Pace, B., Olsen, G. J., Sthal, D. A., Sogin, M. L., and Pace, N. R. (1985) Rapid determination of 16S ribosomal RNA sequences for phylogenetic analysis. *Proc. Natl. Acad. Sci. USA* **35,** 23–27.
29. James, S. A., Collins, M. D., and Roberts, I. N. (1996) Use of an rRNA internal transcribed spacer region to distinguish phylogenetically closely related species of the genera *Zygosaccharomyces* and *Torulaspora. Int. J. Syst. Bacteriol.* **46,** 189–194.
30. Turenne, C. Y., Sanche, S. E., Hoban, D. J., Karlowky, J. A., and Kabani, J. A. (1999) Rapid identification of fungi by using the ITS2 genetic region and an automated fluorescent capillary electrophoresis system. *J. Clin. Microbiol.* **37,** 1846–1851.
31. Marakis, S. G. and Karagouni, A. D. (1985) Screening of carob bean yeasts. Chemical composition of *Schizosaccharomyces versatilis* grown on aqueous carob extract. *Biotechnol. Lett.* **7,** 831–836.
32. Sambrook, J., Fritsch, E. F., and Maniatis, T. (1989) Gel electropheresis of DNA, in *Molecular Cloning: A Laboratory Manual*, 2nd ed. (Ford, N., Nolan, C., and Ferguson, M., eds.), Cold Spring Harbor Laboratory Press, Cold Spring Harbor, New York.
33. Lott, T. J., Kuykendall, R. J., and Reiss, E. (1993) Nucleotide sequence analysis of the 5.8S rDNA and adjacent ITS2 region of *Candida albicans* and related species. *Yeast* **9,** 1199–1206.
34. Yarrow, D. (1998) Methods for isolation, maintenance and identification of yeasts, in *The Yeasts, a Taxonomic Study,* Fourth edition (Kurtzman, C. P. and Fell, J. W., eds.), Elsevier Science B.V., Amsterdam, New York, Part IV, pp. 77–100.
35. Saitou, N. and Nei, M. (1987) The neighbor joining method, a new method for reconstructing phylogenetic trees. *Mol. Biol. Evol.* **4,** 406–425.

8

Aeromonads in Environmental Waters

Anavella Gaitan Herrera

1. Introduction

The normal habitat of *Aeromonas* species is the aquatic environment, and they represent a high percentage of the heterotrophic microorganisms in a variety of aquatic systems; for this reason, the potential public health significance cannot be ignored *(1)*. Their presence has been reported in sewage and sewage effluents, surface waters (fresh, estuarine, and marine), fish ponds, soils, natural mineral springs, stagnant water, chlorinated and unchlorinated drinking water, and especially fresh waters (*see* **Notes 1** and **2**). The motile species are often ubiquitous members of the aquatic ecosystem *(2)*. But all can be components of the microbial flora of aquatic animals, and may be pathogens of poikilotherms, homoiotherms, and even man *(3)*. They act as primary pathogens and make a significant contribution to the disease process in fish that they invade *(4)*. They may cause clinical dropsy, ulceration, and widespread tissue necrosis *(5)*. Losses may be substantial, and control is dependent on eliminating the underlying factors predisposing to infection. *A. salmonicida* and *A. hydrophila* occupy two well-differentiated ecological niches. *A. salmonicida* is a natural parasite of fish and other poikilotherms, and is normally found free in the water. *A. hydrophila* belongs to the fish microflora and is ubiquitous in all types of aqueous environments *(6)*.

The US Environmental Protection Agency (EPA) developed the drinking water Contaminant Candidate List (CCL), and *Aeromonas hydrophila* has been included in the EPA's proposed (April 1999) Unregulated Contaminant Monitoring Rule (UCMR), which requires assessment monitoring beginning in January 2001. The CCL serves as the primary source of priority contaminants for the EPA's drinking water program *(7)* (*see* **Note 2**).

From: *Methods in Biotechnology: Environmental Microbiology: Methods and Protocols*
Edited by: J. F. T. Spencer and A. L. Ragout de Spencer

Characteristics of genus *Aeromonas:* Gram-negative rods or cocco-bacilli (0.3–1.0 mm by 1.0–3.5 mm), singly, in pairs, or short chains; facultative anaerobes; oxidase positive; non-spore forming; motile with a single polar flagellum (*A. media* and *A. salmonicida* are psychrophilic and nonmotile); and resistant to the vibriostatic 0/129.

There are eleven named species of *Aeromonas,* and two of these have subspecies: *A. allosaccharophila, A. caviae, A. eucrenophila, A. hydrophila, A. jandaei, A. media, A. schubertii, A. sobria, A. trota, A. salmoncida* subsp. *achromogenes, A. salmoncida* subsp. *masoucida, A. salmoncida* subsp. *salmoncida, A. salmoncida* subsp. *smithia, A. veronii* subsp. *veronii, A. veronii* subsp. *sobria* ***(8)***.

Health effects: *Aeromonas* species have been associated with diarrheal illness and can cause infections and septicemia ***(9,10)***. The infections can occur in healthy people of all ages and are often acquired through foreign travel. Wound infections are relatively uncommon but can progress rapidly if not treated. The main species causing human infection are *A. hydrophila, A. caviae, A. veronii* subsp. *sobria,* and *A. media* ***(9)***. Some motile species associated with hemorrhagic septicemia in freshwater fish; *A. hydrophila, A. caviae,* and *A. sobria* are pathogenic for fish and amphibians ***(11)***.

2. Materials

1. Columbia blood agar, nutrient agar, and MacConkey agar (*see* **Note 3**).
2. Ampicillin dextrin agar has been recommended for the isolation of *Aeromonas* species from water, and ampicillin blood agar or Ryan's medium (based on modified XLD medium with ampicillin) have also been used. Media containing ampicillin cannot be used to isolate *A. trota,* which is sensitive to this antibiotic.

3. Methods

3.1. Biochemical Characteristics

Chemo-organotrophic facultative anaerobes have both respiratory and fermentative types of metabolism. They are oxidase, catalase, gelatinase, and DNase positive, reduce nitrates, and most strains ferment carbohydrates. Urease and phenylalanine tests are negative, and they are resistant to 2,4-diamino-6,7-di-isopropylpteridine (vibriostatic agent 0/129) (**Tables 1** and **2**). The species and subspecies can be differentiated biochemically, but such identifications are less reliable than methods based on genetic hybridization groups (*see* **Note 4**).

3.2. Cultural Characteristics

Aeromonas species can be isolated in waters with a wide range of physicochemical limits, i.e., waters with pH values between 5.2 and 9.8 and tempera-

Table 1
Differentiation Between the Motile *Aeromonas* Species by Biochemical Properties: *A. hydrophila, A. caviae,* and *A. sobria*

	A. hydrophila	*A. caviae*	*A. sobria*
Motility	+	+	+
Esculin hydrolysis	+	+	–
Growth in KCN broth	+	–	–
Gas from glucose	+	–	+
Voges-Proskauer			
Acetoin production	+	–	+/V
L-Histidine and L-arginine utilization	+	+	–
Acid from arabinose	+	+	–
L-Arabinose utilization	+	+	
Fermentation of salicin	+	+	–
H_2S from cysteine	+	–	+

tures lower than 10°C or as high as 45°C, 30°C being the thermal optimum. They can be isolated from water with very low organic matter content, as in oligosaprobiotic lakes and streams, and they can grow in polysaproiotic waters, such as sewage. *A. hydrophila* can be isolated from skin lesions, and occasionally from internal organs. *A. caviae* and *A. sobria* can be isolated from such affected fish (*see* **Notes 5** and **6**).

Most *Aeromonas* species grow well at 37°C, although their optimal growth temperature is 22–28°C, and their range is 4–42°C. *A. salmonicida* has a growth temperature range of 5–30°C, with an optimum of 20°C. The optimum pH range for *Aeromonas* species is 5.5 to 9, and they are not halophilic (4% NaCl). All species grow well in Columbia blood agar, nutrient agar, and MacConkey agar, but selective media are required for their isolation from contaminated food and water (*see* **Notes 7** and **8**).

3.3. Detection Methods

Aeromonas species grow on a wide range of culture media, and can be isolated from water on membrane filters of the type used for the enumeration of coliforms and *E. coli.* A number of cultural techniques have been used to selectively isolate aeromonads from water or sewage. Ampicillin dextrin agar has been recommended for the isolation of *Aeromonas* species from water, and ampicillin blood agar or Ryan's medium (based on modified XLD medium with ampicillin) have also been used. Media containing ampicillin cannot be used to isolate *A. trota,* which is sensitive to this antibiotic ***(12)*** (*see* **Notes 9** and **10**).

Table 2
Characteristics of Fermentative, Oxidase-Positive, 0/129 (150 μg) Resistant, Gram-Negative Rods

				Aeromonas salmonicida			
	Aeromonas hydrofila	*Aeromonas sobria*	*Aeromonas caviae*	Subsp. *salmonicida*	Subsp. *achromogenes*	Subsp. *masoucida*	Atypical
Motility	+	+	+	–	–	–	–
Growth at 37°C	+	+	+	–	–	–	–
Diffusible brown pigment	–	–	–	+	–	–	D
β-galactosidase	+	+	+	+	+	+	D
Arginine dihydrolase	+	+	+	+	+	+	D
Lysine decarboxylase	D	D	D	D	–	+	–
Ornithine decarboxylase	–	–	–	–	–	–	–
Simmon's citrate	D	D	D	–	–	–	–
H_2S production	+	+	–	–	–	+	D
Urease	–	–	–	–	–	–	–
Indole	+	+	+	–	D	+	D
Voges-Proskauer reaction	+	D	–	–	–	+	D
Gelatin hydrolysis	+	+	+	+	–	+	D
Aesculin hydrolysis	+	–	+	+	–	+	D
Gowth in KCN	+	–	+	–	–	–	•
Acid from:							
Glucose	+	+	+	+	+	+	+
Mannitol	+	+	+	+	–	+	D
Inositol	–	–	–	–	–	–	–
Sorbitol	D	D	D	–	–	–	–
Sucrose	+	+	+	–	+	+	D
Arabinose	+	+	+	+	–	+	D

D, variable reaction; •, not known.

4. Notes

1. Free available chlorine levels of 0.2–0.5 mg/L are considered sufficient to control aeromonad densities in water distribution systems. Lower levels may not prevent regrowth if other conditions are favorable for it to occur.
2. The presence of aeromonads on the coast is probably owing to outflows from wastewater discharges and rivers.
3. *A. hydrophila* can be isolated from feces of homeothermic animals; in human stools, 0.1–11.7%. It is also found in raw food—fish, meat, and vegetables. The presence of these organisms in food can be a source of human infection.
4. All species grow.
5. Aeromonads grow well on any complex medium such as nutrient agar or trypticase soy agar, as well as selective media for fecal coliforms such as MacConkey agar.
6. For most purposes in the water industry, identification to the level of *Aeromonas* species is sufficient, and this can be achieved using API 20E test strips or an equivalent biochemical system.
7. In farmed pond fishes, it is not uncommon to find, in spring or autumn especially, a condition characterized by brownish necrosis of fins or tail, with chronic hemorrhagic ulceration at the base of the fins or congestions of the subepithelias vessels. The motile aeromonad infection can at times cause devastating losses in both wild and farmed fish populations.
8. Exposure to waterborne aeromonads may be through ingestion of contaminated water or by contact with it from accidents during swimming or diving, through skin or wound infections due to impact injuries, lacerations, or punctures. There have also been descriptions of pneumonia after near-drowning incidents, and cases of bone infections after contact of water with open fractures and wounds.
9. Food may also be an important source of exposure; the possibility of this should be borne in mind if water is implicated as a source of infection. Aeromonads have been isolated from meat (particularly when packaged under modified atmospheres), poultry, sausages, fish and shellfish, raw milk, ice cream, and vegetables. The psychrophilic nature of some aeromonads allows them to grow well at refrigerator temperatures.
10. Water treatment should be aimed at reducing the possibility of regrowth, principally by removing organic compounds that could be used as carbon and energy sources. Distribution systems should be designed with short residence times if possible. Measures may be needed to prevent accumulation of sediments in sections of pipework with low water velocities.

References

1. Boulanger, Y., Lallier, R., and Cousineau, G. (1997) Isolation of enterotoxigenic *Aeromonas* from fish. *Can. J. Microbiol.* **23,** 1161–1164.
2. De Figuereido, J. and Plumb, J. A. (1971) Virulence of different isolates of *Aeromonas hydrophila* in channel catfish. *Aquaculture* **11,** 349–354.

3. EPA. (1998) Announcement of the Drinking Water Contaminant Candidate List. EPA-815-Z-98-001. *Fed. Reg.* **63,** 10273–10287. LeChevallier, M. W. (1998) Committee report: "Emerging Pathogens: Names to know and bugs to watch out for." Microbial Contaminant Research Committee, Amer. Water Works Assoc. (AWWA), Denver, CO.
4. Friedman-Huffman, D. and Rose, J. B. (1998) Waterborne pathogens: emerging issues, emerging treatments—A review. *Water Cond. and Purif.* **40,** 48–53. Centers for Disease Control. (1996) Surveillance for waterborne disease outbreaks, United States, 1993–1994. *Morbidity and Mortality Weekly Report* **45,** 22–1.
5. Hazen, T. C., Fliermaus, C. P., Hirish, R. P., and Esch, G. W. (1978) Prevalence and distribution of *Aeromonas hydripila* in the United States. *Appl. Environ. Microbiol.* **36,** 731–738.
6. Joseph, S. W., Daily, O. P., Hunter, W. S., Seidler, R. J., Allen, D. A., and Colwell, R. R. (1979) *Aeromonas* primary wound infection of a driver in polluted waters. *J. Clin. Microbiol.* **10,** 46–49.
7. Kaper, J. B., Lockman, H., and Colwell, R. R. (1981) *Aeromonas hydrofila:* ecology and toxigenicity of isolates from an estuary. *J. Appl. Bacteriol.* **50,** 359–377.
8. Ketover, B. P., Young, L. S., and Armstrong, D. (1973) Septicaemia due to *Aeromonas hydrofila:* clinical and immunological aspects. *J. Infect. Dis.* **127,** 284–290.
9. MacInnes, J. H., Trust, T. J., and Crosa, J. H. (1979) Deoxyribonucleic acid relationships among members of the genus *Aeromonas. Can. J. Microbiol.* **25,** 579–586.
10. Popoff. M. (1984) Genus III. *Aeromonas* Kluyver and van Niel (1936). In, *I. Bergey's Manual of Determinative Bacteriology,* Vol. 1 (Holt, J. G. and Krieg, N. R., eds.), Williams and Wilkins, Baltimore, MD, pp. 545–548.
11. Toranzo, A. E., Baya, A. M., Ronalde, J. L., and Herick, F. M (1989) Association of *Aeromonas sobria* with mortalities in adult gizzard shark *Dorosoma cepedianum* Leseur. *Journal of Fish Diseases* **12,** 439–448.
12. Inglis, V., Roberts, R. J., and Bromage, N. R. *Bacterial Diseases of Fish.* Institute of Aquaculture, University of Stirling, Stirling, Scotland.

9

Development of a Vital Fluorescent Staining Method for Monitoring Bacterial Transport in Subsurface Ground Water

Mark E. Fuller

1. Introduction

The ability to track or monitor individual bacterial cells in a given system has the potential to yield new data about the factors affecting the survival, movement, and growth of bacteria in a variety of environments and under a range of conditions. For field-scale studies, the methods have mainly focused on the use of nucleic-acid-binding fluorescent stains, most of which have detrimental effects on cell properties and functions, and also pose a health risk due to their inherent toxicity.

This chapter details a protocol for using a fluorescent protein stain, 5- (and 6-) carboxyfluorescein diacetate, succinimidyl ester (CFDA/SE, **Fig. 1A**) for labeling bacterial cells. The following information regarding CFDA/SE has been adapted from Molecular Probes literature *(1)*. CFDA/SE is initially cell permeant and nonfluorescent. After entering the cell, nonspecific esterases cleave the acetates and the central ester bond, generating a charged species that is much less cell permeant (**Fig. 1B**). The succinimidyl moeity then reacts with the amine groups of intracellular proteins to form stable fluorescent conjugates (**Fig. 1C**).

The cell-labeling protocol has been extensively tested in the laboratory with a wide range of bacterial strains (**Table 1**), and has been successfully scaled up and evaluated at the field scale during experiments examining bacterial transport in a shallow aquifer. The basic protocol involves cell growth and harvesting, cell staining, and poststaining washing and preparation of cells for use in experiments.

From: *Methods in Biotechnology: Environmental Microbiology: Methods and Protocols*
Edited by: J. F. T. Spencer and A. L. Ragout de Spencer

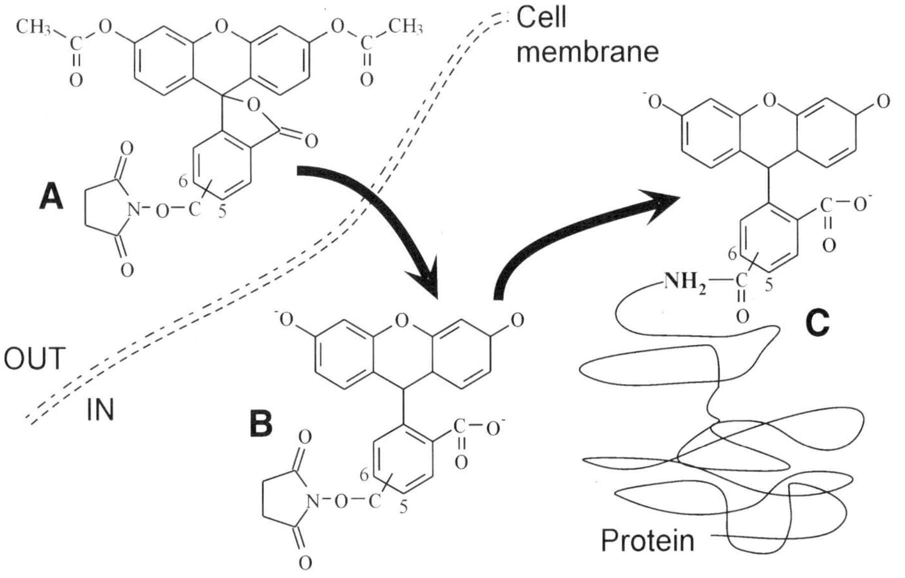

Fig. 1. Reaction schema of CFDA/SE during staining of cells.

2. Materials

2.1. Cell Growth and Harvesting

1. Basal salts medium (BSM) *(2)* augmented with a carbon source such as acetate or lactate, or a standard bacteriological medium, such as R2A, tryptic soy broth, or nutrient broth.
2. Phosphate buffered saline (PBS) containing (per L distilled H_2O): 0.24 g K_2HPO_4; 1.44 g NaH_2PO_4; 0.20 g KCl; 8.00 g NaCl. Adjust to pH 7.4 and sterilize by filtration or autoclaving.
3. Appropriate incubator or shaker to grow cells.
4. Centrifuge able to efficiently pellet bacterial cells.

2.2. Cell Staining

1. Anhydrous dimethyl sulfoxide (DMSO; Sigma, cat. no. D-8779).
2. 5- (and 6-) Carboxyfluorescein diacetate, succinimidyl ester (CFDA/SE; Fluka, cat no. 21888; Molecular Probes, cat. no. C-1157). Prepare a 50 m*M* solution in DMSO, and store frozen at –20°C.
3. Circulating water bath capable of relatively rapid cycling between 25°C and 40°C and culture agitation (*see* **Notes 1** and **2**).

2.3. Cell Washing and Preparation for Use

1. Artificial groundwater (AGW) containing (per L distilled H_2O): 70 mg $Ca(NO_3)_2 \cdot 4H_2O$; 60 mg $MgSO_4 \cdot 7H_2O$; 60 mg $NaHCO_3$; 29 mg $CaCl_2 \cdot 2H_2O$; 25 mg

Table 1
Bacterial Strains That Have Been Evaluated for Labeling With CFDA/SE

	Strain	Staining efficiency
Gram-negative	*Acinetobacter johnsonii*	High
	Alcaligenes eutrophus	Medium
	Comamonas testosteroni DA001	High
	Cytophaga pectinovora	High
	Enterobacter aerogenes	High
	Erwinia herbicola	High
	E. coli strain BL	High
	Flavobacterium odoratum	High
	Klebsiella sp. 1PC	High
	Ps. cepacia	High
	Ps. fluorescens	Low
	Ps. putida	Low
	Pseudomonas sp. LB300	Low
	Pseudomonas sp. Tol1A	High
	Rahnella aquitilis BFB	High
	Sphingomonas capsulata	High
Gram-positive	*Arthrobacter globiformis*	High
	Bacillus subtilis	High
	Micrococcus luteus	High
	Paenibacillus polymyxa FER-02	High
	Rhodococcus rhodocrous	High
	Streptomyces albus	Low
	Unidentified low G+C SO-B2	High
	Unidentified low G+C SO-9	High

Adapted from Fuller et al. *(2)*.

$CaSO_4{\cdot}2H_2O$; 10 mg KNO_3; 0.4 mg NaH_2PO_4. After adding chemicals, adjust to pH 6.0 with HCl, then sterilize by filtration (0.2 µm).

3. Methods

The entire procedure described here from growth through starvation is performed in the same 50 mL conical-bottom disposable centrifuge tube (Corning, cat. no. 25330-50). It assumes cells are grown to a density of 1×10^9 cells/mL prior to harvesting, and that staining is performed in a small circulating water bath placed on top of a magnetic stir plate.

3.1. Growth and Harvesting of Cells

1. Prepare 20-mL aliquots of sterile growth medium in 50-mL conical-bottom disposable centrifuge tubes.
2. Inoculate the culture from either a fresh colony from a streak plate or from a fresh liquid culture (0.5% v:v).
3. Grow the culture to late exponential/early stationary phase.
4. Harvest the cells by centrifugation (10,000*g*, 10 min, 4°C).
5. Discard the supernatant, and resuspend the cells in an equal volume of PBS.
6. Repeat **steps 4** and **5**, and harvest the cells as in **step 4**.
7. Discard the supernatant and resuspend the cells in 2 mL PBS to achieve a final cell concentration of 1×10^{10} cells/mL.

3.2. Staining of Cells (see Note 3)

1. Thaw and thoroughly mix the CFDA/SE stock solution.
2. Turn on the water bath and set the temperature to 25°C.
3. Add a small, sterile magnetic stir bar ("flea") to the concentrated culture.
4. Add 2 µL of the CFDA/SE stock solution per mL, yielding a final CFDA/SE concentration of 100 µ*M*.
5. Immediately cap the tube tightly and vortex the culture vigorously to disperse the stain.
6. Place the culture in the water bath and turn on the stir plate. Assure that the culture is mixing well.
7. Set the water bath temperature to 37–39°C.
8. When the temperature reaches the set point, adjust the temperature control back to 25°C.
9. Repeat the temperature cycling at least three to five times. The culture should become bright yellow as staining proceeds.
10. Remove a small aliquot of the culture (2 µL) and view using an epifluorescent microscope to confirm the cell staining has been effective. If cells are not fluorescing, or are only weakly fluorescent, the procedure should be repeated starting at **step 4**.

3.3. Washing and Starvation of Cells

1. When the staining procedure is complete, add 18 mL of AGW, mix, and retrieve the stir bar from the tube.
2. Harvest the cells by centrifugation (10,000*g*, 10 min, 4°C).
3. Decant the supernatant and resuspend cells in 20 mL of AGW.
4. Repeat **steps 2** and **3** twice more. The supernatant should become less and less yellow with each cycle.
5. Resuspend the final cell pellet in 20 mL AGW.
6. Incubate the cells with shaking for 48–72 h at the temperature at which the experiments using the cells will be performed (i.e., for experiments performed at typical *in situ* groundwater temperatures, stained cells should be incubated at 15°C).

3.4. Use of Cells

1. Harvest the cells by centrifugation (10,000*g*, 10 min, 4°C).
2. Decant the supernatant and resuspend the cells in equal volume of fresh AGW.
3. Repeat **steps 1** and **2** until the supernatant becomes absolutely colorless (*see* **Note 4**).
4. After the final centrifugation, resuspend the cells in a medium appropriate for the experiments to be performed (*see* **Note 5**).
5. Cells can be enumerated by direct epifluorescent microscopy, flow cytometry, and microplate spectrofluorometry, as described elsewhere ***(3,4)*** (*see* **Note 6**), as well as combined with other methods such as immunomagnetic separation (or ferrographic capture) ***(5)***.
6. If sample must be preserved prior to analysis, formaldehyde at a final concentration of 1–4% (w:v) is recommended. Glutaraldehyde is not recommended because of its tendency to increase cell autofluorescence.

4. Notes

1. The equipment used during the staining procedure should be able to keep cell cultures very well mixed, either by an integrated shaking mechanism, or by being able to be placed upon a magnetic stir plate. As an alternative, the use of bioreactor with integrated heating/cooling coils and mixing paddles has also proven very effective, especially for larger (10+ L) volumes of cells.
2. To bring the temperature down rapidly, the water bath should be equipped with a cooling coil attached to either a chiller or the cold-water tap.
3. The protocol described above is optimized for the specific bacterial strains used during the developmentof this staining method, although it has also proved generally applicable for the bacterial strains listed in **Table 1**. If this procedure does not prove effective for a given strain, two changes are suggested: (a) increase the CFDA/SE concentration during staining; (b) alter the staining temperature regime.

   With each change, it is also suggested that plate counts be performed before and after staining to assess the effects of the changes on cell culturability/viability.
4. An effective means to determine whether all the excess, noncell-associated stain has been removed is to measure the total fluorescence of the supernatant using a microplate or standard fluorometer.
5. The staining efficiency of the cells should be determined by counterstaining the cells with a general cell stain like DAPI (4',6-diamidino-2-phenylindole), and then enumerating the number of CFDA/SE+ and DAPI+ cells using epifluorescent microscopy or flow cytometry. The staining efficiency is defined as:

$$\text{Efficiency} = \frac{(\text{number of CFDA/SE-stained cells})}{(\text{number of DAPI-stained cells})} \times 100$$

6. The fluorescence intensity of this compound is pH-dependent, with the maximum fluorescence reaching a plateau at pH 8.5. Sample pH can be increased using phosphate buffer solutions.

References

1. Haugland, R. (1996) *Handbook of Fluorescent Probes and Research Chemicals,* sixth edition (Spence, M., ed). Molecular Probes, Inc., Eugene, OR.
2. Hareland, W. A., Crawford, R. L., Chapman, P. J., and Dagley, S. (1975) Metabolic function and properties of 4-hydroxyphenylacetic acid 1-hydroxylase from *Pseudomonas acidovorans. J. Bacteriol.* **121,** 272–285.
3. Fuller, M. E., Mailloux, B. J., Zhang, P., Streger, S. H., Hall, J. A., Vainberg, S. N., et al. (2001) Field-scale evaluation of CFDA/SE staining coupled with multiple detection methods for assessing the transport of bacteria *in situ. FEMS Microbiol. Ecol.* **37,** 55–66.
4. Fuller, M. E., Streger, S. H., Rothmel, R. K., Mailloux, B. J., Onstott, T. C., Fredrickson, J. K., et al. (2000) Development of a vital fluorescent staining method for monitoring bacterial transport in subsurface environments. *Appl. Environ. Microbiol.* **66,** 4486–4496.
5. Johnson, W. P., Zhang, P., Fuller, M. E., Scheibe, T., Mailloux, B. J., Onstott, T. C., et al. (2001) Ferrographic tracking of bacterial transport in the field at Oyster, VA. *Environ. Sci. Technol.* **35,** 182–191.

10

Bench Scale Flow Cell for Nondestructive Imaging of Biofilms

Eric S. Gilbert and Jay D. Keasling

1. Introduction

Microbial biofilms impact economically important processes ranging from water treatment to nosocomial infections. Understanding their ecology is a key step in learning how to manipulate them. A feature that changed scientists' understanding of microbial biofilms was the discovery of their three-dimensional heterogeneous structure, which occurred primarily with the advent of the confocal laser scanning microscope (CLSM) *(1)*. It was recognized that handling attached populations of cells prior to imaging would distort their structure, principally owing to shear forces at the air-liquid interface tearing the biofilm *(2)*. Consequently, flow cells that facilitated nondestructive imaging of biofilms were developed by several groups.

Several designs for flow cells have been reported *(2–7)*. Flow cells are currently commercially available from Stovall Life Science, Inc. (Greensboro, NC), based on a design reported in Christensen et al. *(7)*. This chapter focuses on the application of a flow cell design first reported in Cowan et al. *(8)* for imaging biofilms by CLSM.

The study of mixed microbial populations in biofilms has been facilitated by the availability of fluorescent markers. Biofilm-dwelling organisms can be labeled exclusively with fluorescent proteins *(8–10)*, a combination of fluorescent protein and fluorescent dyes *(11)* or antibodies *(12)*, or exclusively with fluorescently labeled nucleic acid probes or antibodies *(7,13,14)*. All of these labeling techniques have been used successfully in combination with flow cells to investigate the ecology, metabolism, and organization of microorganisms in biofilms.

From: *Methods in Biotechnology: Environmental Microbiology: Methods and Protocols*
Edited by: J. F. T. Spencer and A. L. Ragout de Spencer

2. Materials

2.1. Flow Cell Construction

1. Delrin® plastic sheeting, 2 mm thickness, or equivalent.
2. 1.0-mm Stainless steel tubing cut into sections (*see* **Subheading 3.1.**).
3. Silicone adhesive, e.g., General Electric RTV102 or equivalent.
4. 22 × 60-mm no. 1 coverslips.

2.2. Pumps and Tubing

2.2.1. Pumps

1. Cole Parmer (Vernon Hills, IL) Masterflex L/S® peristaltic pump drive, e.g., Cole Parmer, cat. no. 77521-40, or equivalent.
2. Ismatec Minicartridge pump head for Masterflex L/S® drive, e.g., Cole Parmer, cat. no. U-07623-10, or equivalent.

2.2.2. Tubing

1. 0.89-mm Internal diameter (id) silicone tubing (peroxide cured).
2. 3.2-mm id Pharmed tubing.
3. 0.89-mm id Pharmed three-stop tubing (e.g., Cole Parmer, cat. no. U95700-26).

2.2.3. Low-Pressure Chromatography Fittings (Bio-Rad, Inc., Hercules, CA)

1. 0.8-mm Barb to male and barb to female luers (Bio-Rad, cat. nos. 731-8224 and 731-8221).
2. 3.2-mm Barb to male and barb to female luers (Bio-Rad, cat. nos. 731-8223 and 731-8226).
3. 0.8-mm Male and female luer plugs (Bio-Rad, cat. nos. 731-8232 and 731-8233).
4. Three-way stopcock (Bio-Rad, cat. no. 732-8103).

2.3. Other Supplies

1. 0.22-µm Syringe filters (Millipore, Bedford, MA) or equivalent.
2. SYTO 59 soluble nucleic acid stain (Molecular Probes, Eugene, OR).
3. 10-L Water bath or similar size.
4. 1-L Autoclavable glass bottles with screw-top lids.

3. Methods

3.1. Constructing the Flow Cell

The bench scale parallel-plate flow cell was designed to approximate the dimensions of a standard microscope slide (**Fig. 1**). The frame of the flow cell is made from Delrin® plastic sheeting, an acetal-type polymer (Dupont, Wilmington, DE). It is 76.2 mm long, 25.4 mm wide, and 2 mm in height. Two 46 × 4-mm channels are cut into the piece of plastic. The channels have a parabolic curve at either end that starts to taper inward at 1.5 mm from the

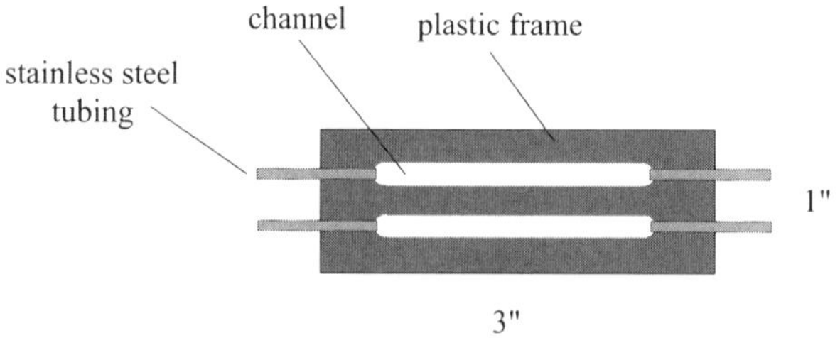

Fig. 1. Flow cell, top view. *See* **Subheading 3.1.** for details.

lengthwise edge. In order for fluid to enter and exit the channels, a 1.0-mm diameter centered hole is drilled into the lengthwise ends of each piece of plastic, leaving 0.5 mm above and below it. A 25.4-mm piece of stainless steel tubing with 1.0-mm outer diameter (od) is inserted into each hole, and subsequently a small amount of epoxy is touched to the outside of the hole to glue the tube in place, and also to seal it.

The channels are converted into chambers by adhering a 22 × 60-mm no. 1 cover glass to the plastic frame. This is accomplished by applying a bead of silicone adhesive along the surface of the frame. The cover glass is set on the adhesive, and then pressed into place using a lint-free wipe. Care must be taken not to slide the cover glass along the surface of the frame, as this will mark it with silicone. This can be done by securing the glass in place between the thumb and middle finger, and gently pressing on the glass with the index finger. It is recommended that latex gloves be worn to avoid leaving fingerprints on the glass. Allow the glass to sit for at least 20 min before repeating the process on the reverse side of the frame. The adhesive should be allowed to cure overnight. Subsequently attach 6-cm-long pieces of 0.8-mm id silicone tubing to the stainless steel tubes of the flow cell. At the ends of the silicone tubes, insert a barbed low-pressure chromatography fitting (Bio-Rad, Hercules, CA).

3.2. Tubing and Pump

To culture biofilms nondestructively, it may be necessary to pump fluid through the flow cell for several days. It is essential to use tubing and fittings that will not leak. The low-pressure chromatography fittings sold by Bio-Rad, Inc. (Hercules, CA) are well suited for this purpose. These fittings permit tubing with an id of 0.8, 1.6, or 3.2 mm to be coupled together through use of a standard-sized luer-type joint. The use of translucent silicone tubing is useful for monitoring the progress of liquid or air through the lines, but the selection of tubing material should be made based on the properties of the liquid medium to be used.

A peristaltic pump and associated pump head must be selected that can deliver medium at appropriate flow rates for the experiment. Pump manufacturers provide two- and three-stop tubing of different diameters for peristaltic pumps that deliver a wide range of flow rates. Flow rates for parallel-plate flow cells range from 2 mL/h *(15)* to higher rates (*see* **Note 1**). A pump with eight or more rollers will produce a more uniform flow through the channel.

3.3. Media Reservoir Bottles

Vessels to function as media reservoirs may be constructed from autoclavable glass bottles. To prepare the media bottles, drill four holes of 5-mm diameter into the bottle cap, and insert sections of glass tubing approx 80 mm in length (*see* **Note 2**). Secure the glass tubing in place and seal the holes using silicone adhesive. Repeat if necessary to completely cover the seam between the glass tube and the plastic cap. Once the adhesive has cured, attach 220-mm sections of 3.2-mm id Pharmed tubing (recommended because of its durability) to each of the four glass tubes protruding out of the cap. Insert barbed luer fittings into the Pharmed sections to allow the bottle to be connected to the rest of the system. To two of the glass tubes protruding through to the inside of the cap, attach lengths of 3.2-mm id Pharmed tubing of sufficient length to reach the bottom of the media bottle. One of these tubes will be the air inlet; the other will be the media outlet.

3.4. Assembling the Flow-Cell Circuit and Sterilizing the Components

Assemble all the necessary components for the flow-cell circuit prior to sterilization. Determine that the necessary pieces of tubing are of sufficient length to connect the components of the system. Insert low-flow chromatography fittings into the tubing to connect the separate pieces. It is recommended to use one male and one female fitting on each component of the system, thus allowing the pieces to be connected in a directional manner.

To sterilize tubing for the flow-cell circuit, disassemble the inspected flow-cell circuit, and autoclave the requisite pieces. Similarly, prepare liquid media for the experiment in the media bottles, cover the bottle openings, and autoclave. Autoclave the waste bottle or receptacle as well. To sterilize the channels of the flow cell, connect each one to a three-stop tubing/three-way valve segment (stationed in the peristaltic pump; **Fig. 2**, 9), and sterilize by pumping 0.2% bleach completely through the three-stop tubing and the flow-cell channels. Turn off the pump and allow the bleach to sit for a minimum of 20 min. Subsequently, rinse the bleach by pumping 125 mL of sterile water through each channel.

Assemble the sterilized components to build the flow-cell circuit. Attach a 0.22-μm syringe filter with luer fitting to the outflow of the media bottle (**Fig. 2**, 8). Also insert a 0.22-μm filter into the three-stop tubing for the aeration pump (**Fig. 2**, 7). Connect the segments of tubing to build the circuit by spray-

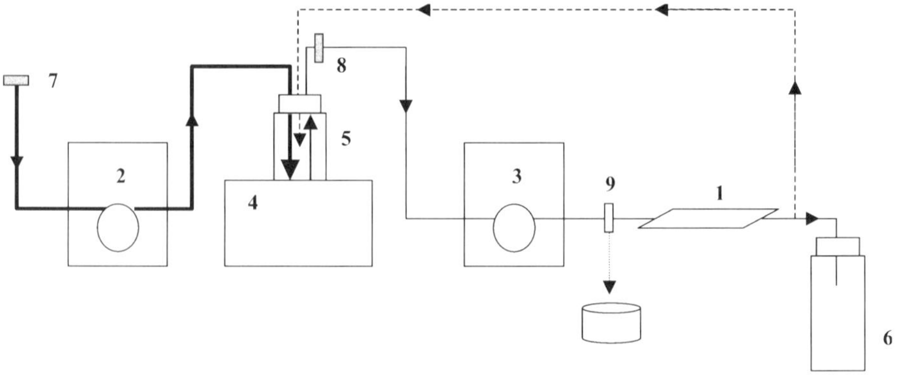

Fig. 2. Description of flow cell apparatus.

ing each chromatography fitting with 95% ethanol, followed by rapid attachment. Keep not-yet-connected components sterile by covering them with flame-sterilized aluminum foil.

3.5. Cultivating Biofilms

3.5.1. Recirculation Mode

To seed the sterilized flow cell, bacteria must be recirculated through each channel. Inoculate a media bottle to an initial optical density of 0.04–0.06 (600 nm) (*see* **Note 3**). Place the bottle in the preheated water bath **(Fig. 2**, 4). Connect the media bottle to the aeration pump **(Fig. 2**, 2), and connect the tubing leading to the flow cell to the 0.22-µm filter at the outlet of the media reservoir **(Fig. 2**, 8). Connect the tubing leading out of the flow cell to the second inlet in the media reservoir to make a complete circuit (**Fig. 2**, dashed line). Turn on the both the aeration pump and the media pump, and recirculate for a minimum of 2 h to allow bacteria to adhere to the cover glass. The aeration pump should be set to a rate that allows the bacteria to grow at a maximum rate, determined through prior experimentation. The media pump should circulate medium at a rate with enough shear stress to encourage colonization of the surface (*see* **Note 4**).

3.5.2. Switching to Continuous Mode

After a period of time in recirculation mode to allow cells to attach to the flow-cell surface, the system should be switched to a continuous-flow regime (**Fig. 2**, solid line). In continuous mode, liquid flows unidirectionally over the growing biofilm, providing nutrients and encouraging unattached cells to flow downstream to the waste bottle (**Fig. 2**, 6). Another media bottle should be prepared for each channel, to replace the one used during recirculation. The volume of medium in the bottle must be sufficient to keep the growing biofilm immersed in liquid overnight until it can be replaced (*see* **Note 5**). The replacement media bottle should have a 0.22-µm filter on the outflow to prevent contamination of the media reservoir. For continuous mode, the end of the tubing leading out of the flow cell (**Fig. 2**, dashed line) should be detached from the media reservoir and placed into the sterile waste bottle (**Fig. 2**, 6). The flow rate of the media pump should be reduced to a rate that will ensure that the level of liquid in the reservoir does not become low enough to permit air to enter the line leading to the flow cell (*see* **Note 6**).

3.5.3. Changing Bottles

For extended experiments, it is necessary to provide fresh medium to the growing biofilms. To change the bottles, first turn off the pumps. Place the fresh media bottle into the water bath, adjacent to the bottle being replaced. Spray the relevant fittings with ethanol, and quickly transfer the air tubing and the media tubing from the recirculation bottle to the new media reservoir bottle. Once the tubing has been securely attached to the new media bottle, turn on the aeration pump, and observe the formation of bubbles in the fresh media bottle. At this time, air in the lines must be pumped out of the media outlet tubing in the fresh media bottle. To accomplish this, open the three-way valve (**Fig. 2**, 9) to redirect the flow of liquid away from the flow cell. Place a receptacle under the opening to catch liquid that will be pumped out. Turn on the media pump, and observe the outflow of medium containing cells, followed by a volume of air, and finally the flow of fresh, clear medium. After it has been established that all of the air has been pumped out of the line, quickly turn the three-way valve to allow the fresh medium to flow again through the flow-cell channel.

3.6. Preparing the Biofilms for Imaging

3.6.1. Rinsing

If the biofilms are to be analyzed by CLSM, they must be rinsed and stained prior to imaging. To rinse, prepare a flask containing sterile 50 m*M* phosphate buffer, covered with aluminum foil. Disconnect the tubing leading to the flow cell from the media reservoir bottle. Observe the end of the tubing to see whether there is an air bubble in the line. If air is present, operate the pump in reverse mode to displace the air with medium (approx 3–5 s). Immerse the end of the tubing into the phosphate buffer, and run the pump in the forward direction for a minimum of 15 min. The rate of rinsing should not exceed the rate at which the pump was operated in continuous mode. A faster rate could result in damage to the biofilm.

3.6.2. Staining

To stain the biofilm, a suitable fluorescent dye must be pumped into each channel and subsequently rinsed out. Care must be taken during this procedure not to introduce air into the channels, as air bubbles can disrupt the biofilm structure. In theory, the minimum amount of dye that would be required to stain each channel would be equal to its volume, approx 0.37 mL, plus the volume of the tubing leading from the pump into the channel. In practice, the risk of introducing an air bubble into the channel exceeds the benefit of using the minimum amount of dye; thus, a volume of 0.6–1.0 mL is recommended.

Dilute the dye according to the manufacturer's guidelines, preparing it in a sterile microfuge tube, and set it in a suitable rack. Wear nitrile gloves when handling dyes, as they are often mutagenic. Stain each channel individually. Detach the end of the three-stop tubing from the tubing leading to the media pump. Briefly operate the pump in reverse mode to expel any air from the line. Insert the tubing into the microfuge tube containing the dye. Operate the pump in the forward direction, observing the dye flowing into the channel. Take care that the end of the tubing is fully immersed in liquid. When the channel has dye throughout, turn off the pump and allow the biofilm to stain (*see* **Note 7**).

3.6.3. Rinsing

To rinse the dye out of the channel, add 50 m*M* phosphate buffer to the microfuge tube containing the dye, taking care not to displace the tubing leading to the flow cell. Turn on the pump to start rinsing the dye. As the level of buffer in the microfuge tube drops, add more to replace it. For SYTO 59, 5 to 10 flow-cell volumes (approx 2–4 mL) of buffer is adequate to rinse out the dye. Once the channel is rinsed, turn off the pump and insert low-flow chromatography fitting plugs to the ends of the channels. Insert the first plug to the downstream side of the flow cell. At this point, the biofilm is ready for imaging.

3.6.4. In Situ *Imaging*

The flow-cell system is useful for imaging biofilms nondestructively *in situ*. Set up the flow-cell circuit such that the lengths of tubing leading into and out of the flow cell are long enough to permit the flow cell to be mounted on the stage of a microscope. Hold the flow cell in place with tape or clamps. Cells that produce a fluorescent protein such as *gfp* can be monitored using an epifluorescent microscope. Cell adhesion and biofilm development can then be observed in real time.

3.7. Reusing Flow Cells

After an experiment has been completed, reattach the flow cell to the media pump. Pump air through the channels to push out the bulk of the biofilm. Rinse the flow cell with 15% ethanol, to solubilize any remaining dye. After rinsing, the cover glasses attached to the plastic frame may be removed by sliding a razor blade underneath them. Use a lint-free wipe and rub off all remaining silicone adhesive. Once the plastic frame is dry, it may be reused with new cover glasses. Do not reuse the same pieces of cover glass for more than one experiment, as there are cells that are irreversibly bound to it.

4. Discussion

The flow-cell apparatus described in this chapter has several beneficial features. First, it produces biofilms that are cultivated nondestructively and that can be readily imaged by CLSM, as well as other microscopic techniques. Second, the flow cell is robust and can be reused many times. Third, the flow cell is low in cost, and it can be made using equipment found in many machine shops. Fourth, the system as described can be operated on the bench at a wide range of temperatures. Fifth, the modular nature of the system makes it flexible, allowing it to be adapted to a wide range of applications.

The system as described has been run with reliability using bleach as the sterilizing agent. It may be possible to sterilize the system by irradiation, although this has not been tested by the authors. The effectiveness of sterilizing agents can be determined by recirculating uninoculated rich medium through the flow cell apparatus for an extended period. The system would require modification to cultivate biofilms with volatile substrates.

5. Notes

1. The authors commonly use a flow rate of 0.84 mL/min during the recirculation phase of biofilm cultivation, and a flow of 0.35 mL/min after switching to a continuous (one-way) flow regime.
2. The authors prepare sections of glass tubing from 1-mL serological pipets.
3. The authors commonly use a 1-L media bottle containing 200 mL of growth medium.
4. The authors typically use a flow rate of 0.84 mL/min.
5. The authors commonly use a 1-L bottle containing 600 mL of growth medium.
6. The authors typically use a flow rate of 0.35 mL/min.
7. For SYTO 59 (Molecular Probes, Inc., Eugene, OR), a soluble red nucleic-acid stain frequently used by the authors to image biofilms, 10 min is usually sufficient.

References

1. Lawrence, J. R., Korber, D. R., Hoyle, B. D., Costerton, J. W., and Caldwell, D. E. (1991) Optical sectioning of microbial biofilms. *J. Bacteriol.* **173,** 6558–6567.
2. Busscher, H. J. and van der Mei, H. C. (1995) Use of flow chamber devices and image analysis methods to study microbial adhesion, in *Adhesion of Microbial Pathogens* (Doyle, R. J. and Ofek, I., eds.), Academic Press, New York, NY, pp. 455–477.
3. Davies, D. G. and Geesey, G. G. (1995) Regulation of the alginate biosynthesis gene *algC* in *Pseudomonas aeruginosa* during biofilm development in continuous culture. *Appl. Environ. Microbiol.* **61,** 860–867.
4. Kuehn, M., Hausner, M., Bungartz, H. J., Wagner, M., Wilderer, P. A., and Wuertz, S. (1998) Automated confocal laser scanning microscopy and semiautomated image processing for analysis of biofilms. *Appl. Environ. Microbiol.* **64,** 4115–4127.
5. Parsek, M. R. and Greenberg, E. P. (1999) Quorum sensing signals in development of *Pseudomonas aeruginosa* biofilms, in *Biofilms* (Doyle, R. J., ed.), Academic Press, New York, NY, pp. 43–55.
6. Kolenbrander, P. E., Andersen, R. N., Kazmerzak, K., Wu, R., and Palmer, R. J., Jr. (1999) Spatial organization of oral bacteria in biofilms, in *Biofilms* (Doyle, R. J., ed.), Academic Press, New York, NY, pp. 322–332.
7. Christensen, B. B., Sternberg, C., Andersen, J. B., Palmer, R. J., Jr., Nielsen, A. T., Givskov, M., et al. (1999) Molecular tools for study of biofilm physiology. *Methods Enzymol.* **310,** 20–42.
8. Cowan, S. E., Gilbert, E., Khlebnikov, A., and Keasling, J. D. (2000) Dual labeling with green fluorescent proteins for confocal microscopy. *Appl. Environ. Microbiol.* **66,** 413–418.
9. Bloemberg, G. V., Wijfjes, A. H., Lamers, G. E., Stuurman, N., and Lugtenberg, B. J. (2000) Simultaneous imaging of *Pseudomonas fluorescens* WCS365 populations expressing three different autofluorescent proteins in the rhizosphere: new perspectives for studying microbial communities. *Mol. Plant Microbe Interact.* **13,** 1170–1176.
10. Tolker-Nielsen, T. and Molin, S. (2000) Spatial organization of microbial biofilm communities. *Microb. Ecol.* **40,** 75–84.
11. Cowan, S. E., Gilbert, E., Liepmann, D., and Keasling, J. D. (2000) Commensal interactions in a dual-species biofilm exposed to mixed organic compounds. *Appl. Environ. Microbiol.* **66,** 4481–4485.
12. Palmer, R. J., Jr., Kazmerzak, K., Hansen, M. C., and Kolenbrander, P. E. (2001) Mutualism versus independence: strategies of mixed-species oral biofilms in vitro using saliva as the sole nutrient source. *Infect. Immun.* **69,** 5794–5804.
13. Guggenheim, M., Shapiro, S., Gmur, R., and Guggenheim, B. (2001) Spatial arrangements and associative behavior of species in an in vitro oral biofilm model. *Appl. Environ. Microbiol.* **67,** 1343–1350.

14. Daims, H., Nielsen, J. L., Nielsen, P. H., Schleifer, K. H., and Wagner, M. (2001) *In situ* characterization of *Nitrospira*-like nitrite-oxidizing bacteria active in wastewater treatment plants. *Appl. Environ. Microbiol.* **67,** 5273–5284.
15. Moller, S., Sternberg, C., Andersen, J. B., Christensen, B. B., Ramos, J. L., Givskov, M., et al. (1998) *In situ* gene expression in mixed-culture biofilms: evidence of metabolic interactions between community members. *Appl. Environ. Microbiol.* **64,** 721–732.

II

Fermented Milks

11

α-Galactosidase Assay in Fermented Soymilk Products

Marisa S. Garro, Graciela Font de Valdez, and Graciela Savoy de Giori

1. Introduction

The enzyme alpha-galactosidase (α-gal) (EC. 3.2.1.22) hydrolyzes the α-1,6-galactosidic bonds present in melibiose, raffinose, and stachyose. These carbohydrates, which are found in plants, cannot be metabolized by humans *(1)*, these sugars being responsible for intestinal disturbances, such as flatulence. Many attempts to reduce these antinutritional factors by soaking *(2)*, germination *(3)*, and water extraction at different times and ratios *(4)* have been made. Enzyme treatment of soymilk with α-gal from molds and yeasts is well documented *(5–8)*; however, little is known about this enzyme isolated from lactic acid bacteria. The presence of this enzyme has been reported in *Bifidobacterium* genera, whereas in lactic acid bacteria, it is produced only by some strains of *Lactobacillus, Leuconostoc,* and *Pediococcus* *(9–12)*. The method developed to study α-gal from several sources *(8,13)* is based on the absorbance measurement at 400 nm of the *p*-nitrophenol (pNP) released by the action of the enzyme (α-gal) upon its specific substrate, *p*-nitrophenyl-α-D-galactosidase (pNPG).

$$\text{p-nitrophenyl-}\alpha\text{-D-galactosidase} + H_2O \xrightarrow{\alpha\text{-galactosidase}} \text{D-galactose} + p\text{-nitrophenol}$$

Previous enzymatic determinations for α-gal were developed in cell-free extracts from microorganisms grown on broth. In this chapter we propose a new method for α-gal extraction from microorganisms grown on soymilk. The enzymatic technique modifications described here were carried out for α-gal determination in lactic acid bacteria and bifidobacteria.

From: *Methods in Biotechnology: Environmental Microbiology: Methods and Protocols*
Edited by: J. F. T. Spencer and A. L. Ragout de Spencer

2. Materials

Prepare all solutions with fresh, distilled water (dH_2O). To prevent the growth of microorganisms, sterilize the containers.

1. Sample: soymilk or fermented soymilk products (*see* **Note 1**).
2. McIlvaine buffer (citric acid–Na_2HPO_4): solution A (0.1 *M* citric acid): weigh 2.10 g citric acid monohydrate ($C_6H_8O_7{\cdot}H_2O$) and make to 100 mL with dH_2O. Solution B (0.2 *M* Na_2HPO_4): weigh 2.84 g Na_2HPO_4 and make to 100 mL with dH_2O. To prepare 100 mL McIlvaine buffer, pH 5.8, mix 39.55 mL solution A and 60.45 mL solution B and check the pH (glass electrode).
3. Glass beads (0.1 mm-diameter, Sigma).
4. 10 m*M* *p*-nitrophenyl-α-galactopyranoside (pNPG). Dilute 3 mg/mL pNPG in McIlvaine buffer, pH 5.8. Store at –20°C.
5. 0.25 *M* sodium carbonate (Na_2CO_3). Weigh 2.65 g Na_2CO_3 and make to 100 mL with dH_2O. Store at room temperature.
6. 10 m*M* *p*-nitrophenol (pNP). Weigh 34.7 mg pNP and make to 25 mL with McIlvaine buffer, pH 5.8.
7. Eppendorf tubes.
8. Micropipets and tips.
9. Water bath.
10. Spectrophotometer.

3. Methods

3.1. Enzyme Extraction

1. Transfer 1 mL of sample (soymilk or fermented soymilk products) to an Eppendorf tube.
2. Centrifuge at 12,000*g* for 5 min to separate soluble materials.
3. Discard the supernatant and add 1 mL McIlvaine buffer to wash the pellet; mix thoroughly by vortexing.
4. Repeat **step 2**.
5. Weigh the obtained pellet and add McIlvaine buffer to 1 mL.
6. Add glass beads (half of pellet weight) to resuspended pellet.
7. Vortex the mix 10 times for 20 min with 1-min intervals on ice.
8. Centrifuge at 12,000*g* for 10 min to remove glass beads, insoluble particles, and unbroken cells.
9. Transfer the supernatant carefully to a clean tube. It may be stored at 4°C up to 1 wk.

3.2. Enzyme Assay

1. The reaction mixture contains: 50 μL 10 m*M* pNPG, 10 μL McIlvaine buffer, pH 5.8, and 140 μL of the supernatant obtained in **Subheading 3.1.** Final reaction volume: 200 μL (*see* **Notes 2** and **3**).
2. Mix gently and incubate at 37°C for 15 min.

3. Stop the reaction by addition of 3 mL 0.25 *M* sodium carbonate. Final total volume: 3200 µL.
4. Measure the absorbance at 400 nm (A_{400}) against water.
5. Calculate the concentration of released pNP by means of a standard curve with pNP, which must be linear throughout a concentration range of 0.01 to 0.15 µmol/mL (*see* **Note 4**).
6. Results are expressed as units of enzyme (UE) (relative activity). For specific activity, the results should be expressed as UE per mg of protein (*see* **Note 5**).

4. Notes

1. Samples stored at –20°C are stable for long periods.
2. The supernatant volume (enzyme extract) adding to the reaction mix will depend on the enzyme activity. If the absorbance at 400 nm (A_{400}) is too high, falling outside of the linear portion of the standard curve of pNP, change the enzyme/McIlvaine buffer ratio (i.e., 80 µL enzyme + 70 µL buffer). The sum of both should be always 150 µL.
3. Reactive and sample blanks must also be prepared. Reactive blank: replace supernatant (enzyme extract) by buffer McIlvaine. Sample blank: replace p-PNPG by buffer McIlvaine. Last blank must be made for each sample.
4. The amount of pNP is calculated as follows:

$$\text{Activity (UE/mL)} = \frac{\Delta A_{400} \times \text{F } (\mu\text{mol/mL}) \times Vf \text{ (mL)} \times df}{15 \text{ min} \times V.enz \text{ (mL)}}$$

ΔA_{400}: A_{400} sample - A_{400} reactive blank (or higher sample blank obtained).

F: calculated factor from the standard curve of pNP.
Vf: Final total volume (3.2 mL).
V.enz: enzyme volume in the reaction.
df: dilution factor (if necessary).

5. One unit of enzyme (UE) is defined as the amount of enzyme released by 1.0 µmol of pNP from pNPG per mL per min under the assay conditions. The specific activity is expressed as UE per mg of protein. Protein can be estimated according to the method of Lowry ***(14)*** (or equivalent) with bovine serum albumin as standard.

References

1. Gitzelmann, R. and Auricchio, S. (1965) The handling of soy alpha-galactosides by a normal and a galactosemic child. *Pediatrics* **36,** 231–235.
2. Kawamura, S. (1966) Sugars and oligosaccharides in soybeans. Final report of research conducted under grant authorized by public law-480 for USDA. US Department of Agriculture (USDA), Oregon.

3. Kim, W. J., Smith, C. J. B., and Nakayama, T. O. M. (1973) The removal of oligosaccharides from soybeans. *Lebensm. Wiss. Technol.* **6,** 201.
4. Ku, S., Wei, L. S., Steinberg, M. P., Nelson, A. L., and Hymowitz T. (1976) Extraction of oligosaccharides during cooking of whole soybeans. *J. Food Sci.* **41,** 361–364.
5. McGhee, J. E., Silman, R., and Bagley E. M. (1978) Production of a-galactosidase from *Aspergillus awamori:* properties and action of p-NPG (-D) and galacto-oligosaccharides of soymilk. *J. Am. Oil Chem.* **55,** 244–247.
6. Sugimoto, H. and Van Buren, J. P. (1970) Removal of oligosaccharides from soymilk by an enzyme from *Aspergillus saitoi. J. Food Sci.* **35,** 655–660.
7. Bahl, O. P. and Agrawal, K. M. L. (1969) Glycosidases of *Aspergillus niger.* I. Purification and characterization of α- and β-galactosidases and β-*N*-acetylglucosaminidase. *J. Biol. Chem.* **244,** 2970–2978.
8. Church, F. C., Meyers, S. P., and Srinivasan, V. R. (1980) Isolation and characterization of alpha-galactosidase from *Pichia guilliermondii*, in *Developments in Industrial Microbiology* (Underkofler, L. A. and Wulf, M. L., eds.), Society for Industrial Microbiology, Arlington, VA, vol. 21, pp. 339–348.
9. Mital, B. K. and Steinkraus, K. H. (1975) Utilization of oligosaccharides by lactic acid bacteria during fermentation of soymilk. *J. Food Sci.* **40,** 114–118.
10. Garro, M. S., de Giori, G. S., de Valdez, G. F., and Oliver, G. (1994) Utilization of soymilk galactosaccharides by lactobacilli. *Microbiologie-Aliments-Nutrition* **12,** 61–66.
11. Garro, M. S. (1995) Behavior of lactic acid bacteria and related micro-organisms on soymilk. Ph.D. Thesis. National University of Tucumán (UNT), Argentina.
12. Gonzalez, C. F. and Kunka, B. S. (1986) Evidence for plasmid linkage of raffinose utilization and associated α-galactosidase and sucrose hydrolase activity in *Pediococcus pentosaceous. Appl. Env. Microbiol.* **51,** 105–109.
13. Bergmeyer, H. U., Gawehn, K., and Grassl, M. (1974) Enzymes as biochemical reagents. α-galactosidase, in *Methods of Enzymatic Analysis* (Bergmeyer, H. U. and Gawehn, K., eds.), vol. 1, Academic Press, Inc., NY, pp. 455–456.
14. Lowry, O. H., Rosebrough, N. J., Farr, A. L., and Randall, R. J. (1951) Protein measurement with the folin phenol reagent. *J. Biol. Chem.* **193,** 265–275.

12

Considerations to Avoid the Overestimation of Exopolysaccharides Produced by Lactic Acid Bacteria

María Inés Torino and Graciela Font de Valdez

1. Introduction

Lactic acid bacteria (LAB) producing exopolysaccharide (EPS) are of commercial interest especially because of their "food-grade" nature and their application as starter cultures in the food industry *(1,2)*. Milk has been commonly used as a culture medium for EPS production. However, the isolation of polysaccharides from coagulated milks is tedious, time-consuming, and includes the risk of degrading the polymer during manipulation *(3)*. When a complex medium is used, the isolated EPS is always heavily contaminated with other medium components (the so-called EPS reacting material), such as beef extract, peptones, and yeast extract, which remain after dialysis. Thus, the amount of EPS measured in complex media is overestimated. To avoid this problem, some practical considerations are given, as well as evidence of the interference between the EPS reacting material (ERM) and the EPS produced by starter cultures.

2. Materials

2.1. Growth Media

Complex medium: MRS broth *(4)* (*see* **Note 1**): 10 g/L polypeptone, 10 g/L meat extract, 5 g/L yeast extract, 20 g/L glucose, 5 g/L sodium acetate, 1.08 g/L Tween-80, 2 g/L ammonium citrate, 2 g/L K_2HPO_4, 0.2 g/L $MgSO_4 \bullet 7H_2O$, and 0.05 g/L $MnSO_4 \bullet 4H_2O$; pH 6.4 ± 0.2. Sterilize by autoclaving at 121°C for 20 min.

From: *Methods in Biotechnology: Environmental Microbiology: Methods and Protocols*
Edited by: J. F. T. Spencer and A. L. Ragout de Spencer

2.2. Media for Detection of EPS Reacting Material (ERM)

1. Complete complex medium: MRS (*see* **Note 2**).
2. Simplified complex media:
   a. Basal medium (MB): 20 g/L glucose, 5 g/L sodium acetate, 1.08 g/L Tween-80, 2 g/L ammonium citrate, 2 g/L K_2HPO_4, 0.2 g/L $MgSO_4 \cdot 7H_2O$, and 0.05 g/L $MnSO_4 \cdot 4H_2O$; pH 6.4 ± 0.2 (*see* **Note 3**). Sterilize at 121°C for 20 min.
   b. MB containing peptone (MBp), yeast extract (MBye), or beef extract (MBbe) at the concentrations present in MRS (*see* **Note 4**). Sterilize at 121°C for 20 min.

2.3. Media for Total EPS Determination

1. Complete complex medium: MRS (*see* **Note 5**).
2. Media MB with the addition of either peptone, yeast extract, or beef extract (*see* **Note 6**). Sterilize at 121°C for 20 min.

2.4. Materials for Cell Growth

1. An active culture grown in complex MRS broth for 16–18 h (overnight) at the usual growth temperature.
2. Sterile saline solution (NaCl 0.85% w/v). Sterilize at 121°C for 20 min.
3. MB glucose with the individual contaminant compounds of the MRS broth to be evaluated, as described in **Subheading 2.3.**
4. Sterile pipets.

2.5. Materials for the Isolation of EPS and ERM

1. LAB grown in the different complex media (**Subheading 2.4., step 3**) for about 24 h (*see* **Note 7**) at the usual growth temperature.
2. Uninoculated complex media as control (**Subheading 2.2.**).
3. Solution of 50% (w/v) trichloroacetic acid (TCA) (*see* **Note 8**).
4. Ethanol 95%. Store at 4°C before using.
5. Distilled water (dH_2O).
6. Dialysis sacks (Sigma) that retain compounds with molecular weights greater than 12,000 (*see* **Note 9**).
7. Dialysis sack closures (Sigma) (*see* **Note 10**).
8. Waterproof marker.

2.6. Materials for the EPS Quantification by Using the Phenol-Sulfuric Method

1. Very clean glass tubes (*see* **Note 11**).
2. Distilled water.
3. Phenol reagent (80% w/v). To prepare 10 mL, weigh 8 g phenol using a 10-mL test-tube, dissolve at 37°C, and then add dH_2O to the final volume. Phenol is toxic and must be carefully prepared.
4. Concentrated sulfuric acid (*see* **Note 12**).

2.7. Materials for the Separation of the ERM and the EPS Produced by LAB: Filtration Chromatography

1. *Moving phase:* 50 m*M* Tris-HCl buffer, pH 7.0: to prepare 1 L buffer, dissolve 6.05 g of Tris base in 250 mL dH_2O, adjust to pH 7.0 with pure HCl, and make up to 1000 mL with dH_2O.
2. *Stationary phase:* Sepharose 4B matrix (*see* **Note 13**) dissolved in Tris-HCl buffer (*see* **Note 14**): add glass wool to cover (approx 0.5 cm) the bottom of the glass column (1 m × 1.2 cm) to support the matrix, and pour the swollen sepharose smoothly to fill up the column, which must be in a vertical position (*see* **Note 15**). The gel suspension must be deaerated at low pressure before being packed into the column. It is of particular importance to avoid the inclusion of air bubbles in the packed matrix. Once packed and washed, the column must be calibrated with a mix of known molecular weight (MW) dextrans (0.25 mg/mL each), using Dextran Blue (DB) to determine the death volume (*see* **Note 16**).
3. Clean pipets.
4. Total lyophilized EPS, which has been isolated from the different cultures and suspended in 1 mL dH_2O (*see* **Note 17**).
5. Distilled water.
6. Fractions collector.
7. Clean glass tubes to collect the eluted liquid.
8. Materials to check the eluted fractions by using the phenol-sulfuric method (**Subheading 2.6.**).

3. Methods

3.1. Isolation of the EPS and the ERM From Complex Media (5)

1. Grow the LAB under study in 100 mL (*see* **Note 18**) of MRS broth at the usual growth temperature for 16 h.
2. Harvest the cells aseptically by centrifugation (10,000*g*, 5 min at 16°C) (*see* **Note 19**) and wash the pellet twice with 20 mL sterile saline solution. Suspend the cells in 10 mL saline solution (*see* **Note 20**).
3. Inoculate 100 mL (*see* **Note 15**) of the different simplified media (**Subheading 2.4.**) with the cell suspension to reach an initial A_{560} of 0.1; mix, and incubate at the usual temperature for 24 h (*see* **Note 7**). Incubate an equal volume of the uninoculated media (control of ERM).
4. Remove the cells by centrifugation (*see* **Note 21**), and keep the supernatants containing the EPS. Record the volume (mL) of supernatant recovered for calculation.
5. Add 25 mL of 50% TCA to reach a final concentration of 10% in both cultures and uninoculated media (*see* **Note 22**); mix, and store at 4°C for a 30 min minimum to assure the elimination of proteins.
6. Centrifuge (10,000*g*, 10 min at 4°C), and recover the EPS from the supernatants by precipitation with 3 vol of cold 95% ethanol at 4°C for 24 h (*see* **Note 23**).

7. Centrifuge, discard the supernatant, and dissolve the precipitated EPS with 20 mL of distilled water. Repeat **steps 5** (adding 5 mL of 50% TCA) and **6**.
8. Centrifuge, dissolve the precipitated EPS with 3–5 mL of distilled water, and pour it into dialysis sacks (*see* **Note 24**).
9. Gently apply clamping closures to the dialysis sacks, providing a leak-proof seal without knots, and label the closures with a waterproof marker (*see* **Note 25**).
10. Dialyze while stirring against distilled water at 4°C for 48–72 h (*see* **Note 26**). Record the volume retained inside the sacks (containing the EPS) for calculations.
11. Store the samples at –20°C until determination of both the total EPS and the ERM (*see* **Note 27**).

3.2. Quantification of EPS (6)

1. Add to very clean glass tubes: 800 µL sample (dialyzed EPS solution) (*see* **Note 28**); 40 µL phenol reagent. Mix by vortexing. Add 2 mL sulfuric acid (*see* **Note 29**). Mix by vortexing.
2. Prepare a reagent blank using 800 µL of distilled water instead of sample.
3. Read the absorbance at 490 nm (A_{490}) against reagent blank (*see* **Note 30**).
4. Perform the determinations in triplicate (*see* **Note 31**).
5. Express the results as mg/L of EPS (*see* **Note 32**).
6. Determine the real amount of EPS and the contamination level (expressed as % of the total amount of isolated EPS) of each evaluated compound (ERM) (*see* **Note 33**).

3.3. Separation of the ERM and the EPS Produced by LAB: Filtration Chromatography (7)

1. Wash the column with Tris-HCl buffer for about 24 h at high flow rate until the elution liquid reaches pH 7.0.
2. Set the fractions collector according to the supplier's instructions to the following parameters: 2 mL (40 drops) of eluent collected in each tube, a flow rate of the elution buffer of 0.1 mL/min (*see* **Note 34**).
3. Apply carefully 1 mL of sample (minimum 0.3 mg/mL total EPS) along the column wall in a layer over the gel surface, below an eluent layer a few centimeters thick.
4. Start the flow slowly at the fixed rate to allow samples to penetrate into the gel bed. Continue until the gel surface is just wet.
5. Pass the sample stepwise into the column while applying another eluent layer, and repeat the operation.
6. Test the eluted fractions by the phenol sulfuric method (*see* **Note 35**).
7. Plot the absorbance values (A_{490}) against the elution fractions (tested tubes) (*see* **Note 36**).

4. Notes

1. MRS broth was arbitrarily chosen as the complex medium to focus this work. Nevertheless, the culture medium will depend on the laboratory under study.
2. The uninoculated MRS broth is employed as a control; the value obtained corresponds to 100% of ERM.
3. The use of MB medium has three objectives: (1) to confirm the complete flow out of the sugar present in the culture medium during the dialysis step; (2) to determine whether the ERM actually remains inside the dialysis sack; and (3) to determine whether the positive reaction obtained with the phenol-sulfuric reagent is due to the ERM itself or to free sugars.
4. Other nitrogen and vitamin sources such as triptone, caseine hydrolysate, casitone, casaminoacids, Yeast Nitrogen Base, and so on, may also be added to the culture medium instead of peptone, yeast extract, and beef extract, depending on the nutritional requirements of the EPS-producer microorganism. In this case, the level of contamination should also be evaluated.
5. The complete complex MRS medium is arbitrary employed as the control for total EPS (EPS from bacterial origin plus ERM), and it is given a value of 100%. It is important to consider the consumption of peptone, yeast extract, and beef extract by the microorganism during growth; thus, the contamination level derived from these ERM is lower at the end of fermentation. The most appropriate procedure would be to subtract the amount of ERM from the total amount of EPS determined after cell growth.
6. The simple omission technique allows determination of the role that each component (e.g., peptone, yeast extract, or beef extract) plays, and also which one should be omitted from the culture medium or replaced by another compound (*see* **Note 4**). This technique is mainly useful in cases of high EPS production.
7. The maximal EPS production is generally obtained during the stationary growth phase. In this study, the isolation of EPS was performed after 24 h according to previous studies *(8,9)*. The kinetics of EPS production should be determined for the strain of interest.
8. The high concentration (50%) of the stock TCA solution is of practical utility; it avoids increasing the final volume. It is kept at 4°C.
9. The cutoff of the dialysis sacks depends on the molecular weight of the EPS isolated. LAB usually produces EPS of approx 10^4–10^6 MW.
10. Best results are obtained when selecting a closure that is 4–10 mm wider than the flat width of the dialysis tubing.
11. The phenol sulfuric method is very sensitive, but precautions should be taken with the materials (tubes, pipets), which must be very clean. When using new glass tubes, they have to be carefully washed with acidulated water and rinsed at least three times with distilled water before using. Once used, keep them with the reaction mixture. When needed, discard the reaction mixture and rinse several times with water and three times with distilled water. Remember that phenol is toxic and must be carefully prepared.

12. The sulfuric acid used must be of high quality. For safety, use a pipet pump to avoid irritation by acid.
13. The selection of the gel matrix will mainly depend on the molecular weight of the substances to be withdrawn. In this work, the EPS produced is of a higher MW than the contaminants, and a good separation between them is obtained. If the bacterial EPS is of low molecular weight, it could not be separated from the ERM by using this matrix.
14. The column should be packed only with completely swollen gel. Columns must be packed at the same temperature at which the chromatography is performed. It is of utmost importance to avoid the inclusion of air bubbles in the packed matrix.
15. An appropriately prepared gel bed packed into the column is of primary importance in gel chromatography.
16. In the gel filtration chromatography, substances with high MW such as DB (MW $= 2 \times 10^6$) pass quickly through the matrix and are collected in the first tubes (low elution fraction).
17. To be detected, the total EPS concentration in the elution volume (including both bacterial EPS and ERM) must be at least 0.3 mg/mL. It is important to consider that the sample is diluted when passing through the column.
18. The volume of culture medium to be used will depend o the infrastructure of the laboratory (refrigerator, centrifuge, availability of great amounts of ethanol and TCA, dialysis flasks, and so on).
19. The presence of EPS (especially as capsule) usually decreases the adhesion of the cell pellet, and the centrifugation time should be extended.
20. The inoculum is prepared in a low volume in order to minimize the dilution of the culture medium by inoculation.
21. The amount of EPS produced by laboratory is tiny (usually on the order of mg/L), and the isolation of EPS from small volumes of culture implies high error levels.
22. By adding 25 mL of 50% TCA to 100 mL broth (either supernatant or uninoculated medium), a final volume of 125 mL is reached, given the desired final concentration of 10% TCA (employ the formula: $N1 \times V1 = N2 \times V2$). It is important that the supernatant volume be recorded.
23. The precipitation period mainly depends on the MW of the EPS. In general, good results are obtained with 24-h precipitation periods, although longer periods may be employed. Nevertheless, this parameter should be standardized for each EPS-producing strain.
24. Cut each dialysis sack in half, and put them in dH_2O (approx 30 min) before use.
25. To ensure the adequate clamping of closures and integrity of dialysis sacks, test them with dH_2O.
26. The dialysis is one of the critical steps in the EPS isolation, during which residual sugars of the culture medium are removed. Dialyze only four dialysis sacks in a 2-L flask or jar, changing the dH_2O twice a day. The period of dialysis has to be standardized before the assay.

27. Store the isolated EPS freeze-dried at 4°C to avoid contamination.
28. The amount of sample depends of the EPS concentration. Usually, 50 µL (plus 750 µL of distilled water to reach the final volume of 800 µL) gives appropriate A_{490} values.
29. The sulfuric acid must be quickly added in the middle of the reaction mixture.
30. The A_{490} of the blank reagent must be less than 0.1. The difference in A_{490} between the sample and the reaction blank should be at least 0.2 Absorbance Units (AU) to achieve accurate results. If this difference is lower than 0.2 AU, increase the amount of sample.
31. Because the phenol sulfuric method is highly sensitive, a single determination is not acceptable. The A_{490} values obtained for each sample differing by 0.1 AU or less are averaged.
32. For calculation, *see* **ref. *10***.
33. Example:

Real bacterial EPS (mg/L) = total EPS (from fermented culture) – ERM (from each uninoculated medium)

Contamination level (%) = [ERM (uninoculated medium) × 100]/total EPS

34. The number of tubes in the collector will vary according to the MW of the bacterial EPS, desired resolution separation between contaminants and bacterial EPS, flow rate, and so on. Sixty tubes are appropriate for the first chromatography works, because it allows collection of all the compounds that usually are present in the samples poured into the column.
35. The MW of the EPS produced by LAB is usually lower than that of the DB. Once the column is calibrated with the dextrans mix, record the elution fraction of DB (approx tubes 10–12). The peaks corresponding to the bacterial EPS and ERM are usually obtained in the higher elution fractions than that corresponding to DB.
36. **Figure 1** shows the purification of the EPS produced by *Lactobacillus helveticus* ATCC 15807 (peak A) and the ERM (peak B), by using sepharose 4B chromatography. The EPS and the ERM were isolated from a complex medium containing peptone, triptone, and yeast extract. The high contamination level of the medium with ERM can be seen. The MW of both bacterial EPS and ERM may be determined by using a semilog plot of known MW dextrans against the eluted fractions.

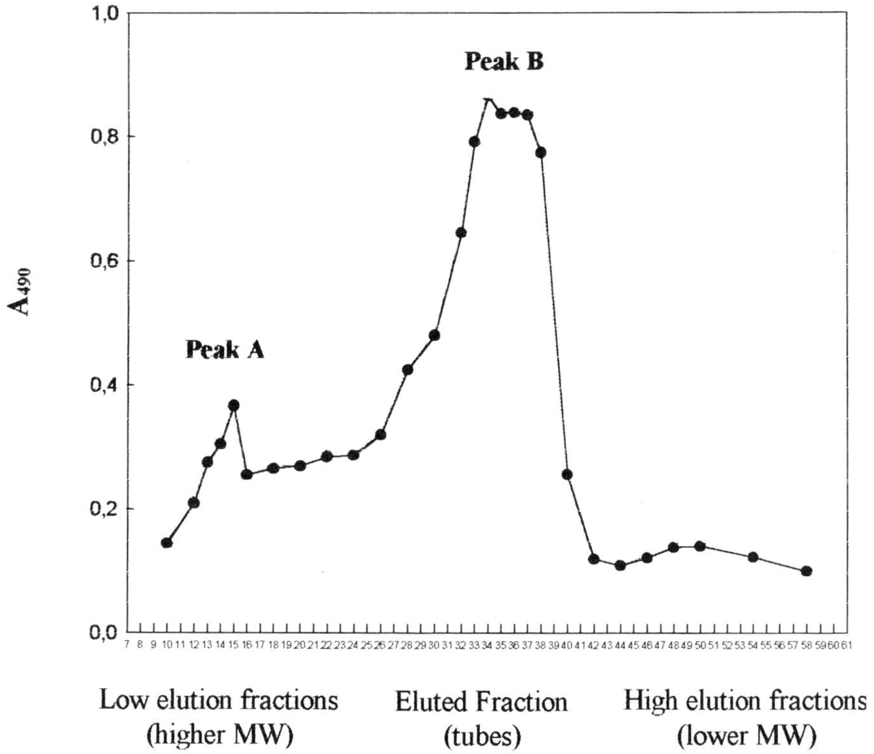

Fig. 1. Separation of the EPS produced by *L. helveticus* ATCC 15807 (peak A) from the ERM (peak B).

References

1. Sutherland, I. W. (1990) Food usage of polysaccharides, in *Biotechnology of Microbial Polysaccharides* (Sutherland, I. W., ed.), Cambridge University Press, Cambridge, UK, pp. 117–125.
2. Broadbent, J. R., Mcahon, J., Oberg, C. J., and Welker, L. (2001) Use of exopolysaccharde-producing cultures to improve the functionality of low fat cheese. *Int. Dairy J.* **11,** 433–439.
3. Cerning, J., Bouillanne, C., Landon, M., and Desmazeaud, M. (1992) Isolation and characterization of exopolysaccharides from slime-forming mesophilic lactic acid bacteria. *J. Dairy Sci.* **75,** 692–699.
4. De Man, J., Rogosa, M., and Sharpe, M. (1960) A medium for the cultivation of lactobacilli. *J. Appl. Bacteriol.* **23,** 130–135.
5. De Vuyst, L., Vanderveken, F., Van de Ven, S., and Degeest, B. (1998) Production by and isolation of exopolysaccharides from *Streptococccus thermophilus* grown in a milk medium and evidence for their growth-associated biosynthesis. *J. Appl. Microbiol.* **84,** 1059–1068.

6. Dubois, M., Gilles, K. A., Hamilton, J. K., Rebers, P. A., and Smith, F. (1956) Colorimetric method for determination of sugars and related substances. *Analytical Chemistry* **28,** 350–356.
7. Kremmer, T. and Boross, L. (1979) *Gel Chromatography: Theory, Methodology, Applications*. John Wiley & Sons, Chichester, pp. 119–154.
8. Torino, M. I., Mozzi, F., Sesma, F., and Font de Valdez, G. (2000) Effect of stirring on growth and phosphopolysaccharide production by *Lactobacillus helveticus* ATCC 15807 in milk. *Milchwiss.* **55,** 204–206.
9. Torino, M. I., Taranto, M. P., Sesma, F., and Font de Valdez, G. (2001) Heterofermentative pattern and exopolysaccharide production by *Lactobacillus helveticus* ATCC 15807 in response to environmental pH. *J. Appl. Microbiol.* **91,** 846–852.
10. Mozzi, F., Torino, M. I., and Font de Valdez, G. (2000) Identification of exopolysaccharide-producing lactic acid bacteria: a method for the isolation of polysaccharides in milk cultures. *Methods in Biotechnology* (Spencer, J. F. T. and Ragout de Spencer, A., eds.), vol. 14, chapter 22: *Food Microbiology Protocols*, Humana Press, Totowa, NJ, pp. 183–190.

13

Determination of Oligosaccharides in Fermented Soymilk Products by High-Performance Liquid Chromatography

Marisa S. Garro, Graciela Font de Valdez, and Graciela Savoy de Giori

1. Introduction

Soybean is one of the most economical and valuable agricultural commodities because of its unique chemical composition. However, soy products have low consumer acceptability owing to the presence of high levels of undigested sugars (e.g., stachyose, raffinose). These oligosaccharides are degraded by the human intestinal bacteria to carbon dioxide, hydrogen, and methane, which produce flatulence and abdominal pain *(1)*. Therefore, processes including fermentation of soymilk with fungi *(2,3)* or lactic acid bacteria *(4–6)* have been examined as means to avoid such problems and to improve the acceptability of soybean. Analysis of the carbohydrate contents of soymilk and soymilk products are very important for nutritional evaluations. High-performance liquid chromatography (HPLC) provides a rapid and simple method for assessing the type and quantity of sugar present in these products. The column used in the method applied here, separates compounds using a combination of size exclusion and ligand exchange mechanisms. For the determination of sugars by HPLC, it is necessary to previously remove the potential interferences present in the substrate, such as proteins.

2. Materials

Sample: soymilk or fermented soymilk products (*see* **Note 1**).

From: *Methods in Biotechnology: Environmental Microbiology: Methods and Protocols*
Edited by: J. F. T. Spencer and A. L. Ragout de Spencer

2.1. Sample Preparation

1. Solution A: 1.8% $Ba(OH)_2$. Dissolve 1.8 g $Ba(OH)_2$ in 100 mL warm distilled water (dH_2O) with gentle stirring. Store at room temperature. This solution is stable at least 6 mo (*see* **Note 2**).
2. Solution B: 2% $ZnSO_4$. Dissolve 2 g $ZnSO_4$ in 100 mL dH_2O with gentle stirring. Store at room temperature. This solution is stable at least 6 mo.
3. Tubes and pipets.
4. Vortex.
5. Centrifuge.
6. Filter paper.

2.2. Carbohydrates Quantification

1. Liquid chromatograph: Gilson system™ equipped with a pump model 305, an LKB differential refractometer model 2142, and a Shimadzu recorder/integrator Chromatopac model CR601, or equivalent systems.
2. Eppendorf column heater and temperature control, model TC-50 or equivalent.
3. Phenomenex REZEX RSO-oligosaccharide column (200 × 10 mm inner diameter) with guard column, or equivalent.
4. Sample loop: 20 µL.
5. Double-distilled water, HPLC grade.
6. Hamilton syringe.
7. Analytical balance.
8. External standards: stachyose, raffinose, saccharose, melibiose, galactose, glucose, and fructose. Solutions at 1.5 and 3.0 mg/mL of each carbohydrate were individually prepared. Accurately weigh 1.5 mg or 3.0 mg of each sugar, transfer to 1-mL volumetric flask, and add HPLC-grade double-distilled water to volume. Fresh solutions are necessary for equipment calibration.
9. 0.45-µm Filter and filtration apparatus.

3. Methods

In order to reduce potential interferences in the sample, an extracting protein protocol must be applied (*see* **Note 3**).

3.1. Sample Preparation

1. Transfer 1 mL of sample (fermented or nonfermented soymilk) to a clean tube.
2. Add 2 mL solution A slowly with agitation by vortexing.
3. Add 2 mL solution B slowly with agitation by vortexing.
4. Vortex vigorously and allow to stand for 10 min.
5. Centrifuge the mix at 3500*g* for 5 min.
6. Transfer the supernatant to a clean tube carefully with a pipet (*see* **Note 4**).
7. Filtrate the supernatants through a 0.45-µm filter before analyzing by HPLC.

Table 1
Retention Times for Soymilk's Sugars on REZEX RSO-Oligosaccharide (Phenomenex) Analysis Column

Carbohydrates	Retention times
Stachyose	27.501
Raffinose	31.325
Melibiose	35.528
Saccharose	37.250
Glucose	43.180
Galactose	46.245
Fructose	46.495

See **Note 5**.

3.2. Carbohydrates Quantification

1. Analyze 20 µL of supernatant on an REZEX RSO-oligosaccharide column with guard column.
2. Set the column temperature control at 70°C.
3. Use a mobile phase of double-distilled water, HPLC-grade, at a flow rate of 0.3 mL/min (isocratic elution).
4. Determine the concentration of each carbohydrate by using a refractive index (RI) detector.

3.2.1. Calibration

1. Inject 20 µL standard solution into chromatograph.
2. Determine retention times.
3. Measure peak areas (or heights).
4. Check reproducibility.

3.2.2. Quantification

The external standard method was used *(7)*. A computing integrator may be calibrated to directly calculate each sugar concentration. If the calibration table is not used, calculate the percent of each sugar as follows:

1. Concentration gradient of standard solution of each sugar must be chromatographed (*see* **Note 6**).
2. Plot of peak area (or height) vs concentration is constructed. If the system is operating properly, the calibration plot should result in a straight line (within a given concentration range) intercepting the origin. The slope of the linear part of the curve is the response factor (*R*).

$$R = \frac{\text{Peak area (or height)}}{\text{Amount (concentration)}}$$

3. The response factor is used to calculate the amount of the component X in the unknown sample:

$$X = \frac{\text{Area (or height)} \times 1/R \times \text{total sample volume}}{\text{Volume injected}}$$

4. Notes

1. Samples stored at –20°C are stable for long periods.
2. To ensure a clarified solution it may be necessary to heat and then to filter.
3. Alternative methods to remove interfering substances in the sample (proteins and so on) can be used. Final pH of the extracts must be neutral to be compatible with the column pH range.
4. The supernatants can be frozen at –20°C until they are used, no more than 1 mo.
5. Galactose and fructose are difficult to separate under the conditions stated in **Subheading 3.2.** To avoid overlapping peaks, galactose plus fructose should be informed together (retention time for galactose plus fructose is 46.502).
6. Points on the calibration curve should be checked daily by chromatographing the standard solutions. If deviations are encountered, a new calibration plot should be constructed.

References

1. Cristofaro, E., Mottu, F., and Wuhrmann, J. J. (1974) Involvement of the raffinose family of oligosaccharides in flatulence, in *Sugars in Nutrition* (Sipple, H. L. and McNutt, K., eds.), Academic Press, New York, p. 313.
2. Lee, Y. C. and Wacek, V. (1970) Galactosidases from *Aspergillus niger. Arch. Biochem. Biophys.* **138,** 264–271.
3. Sugimoto, H. and Van Buren, J. P. (1970) Removal of oligosaccharides from soymilk by an enzyme from *Aspergillus saitoi. J. Food Science* **35,** 655–660.
4. Mital, B. K. and Steinkraus, K. H. (1975) Utilization of oligosaccharides by lactic acid bacteria during fermentation of soymilk. *J. Food Science* **40,** 114–118.
5. Garro, M. S., de Giori, G. S., de Valdez, G. F., and Oliver, G. (1994) Utilization of soymilk galactosaccharides by lactobacilli. *Microbiologie-Aliments-Nutrition* **12,** 61–66.
6. Garro, M. S. (1995) Behavior of lactic acid bacteria and related micro-organisms on soymilk. Ph.D. Thesis. National University of Tucumán (UNT), Argentina.
7. Krstulovic, A. M. and Brown, P. R. (1982) Quantitative analyses using RPLC, in *Reversed-Phase High-Performance Liquid Chromatography*, John Wiley & Sons, New York, pp. 213–214.

14

Evaluation of Minimal Nutritional Requirements of Lactic Acid Bacteria Used in Functional Foods

Elvira M. Hébert, Raúl R. Raya, and Graciela Savoy de Giori

1. Introduction

The lactic acid bacteria (LAB) are important industrial microorganisms because of their role in food fermentation, especially dairy products. In addition, it has been showed that LAB exhibit a range of physiological and therapeutic effects in consumers of these foods, including immune stimulation, pathogen exclusion, production of bioactive peptides, and others *(1)*. Foods containing such bacteria are considered "functional foods," foods that promote health beyond providing basic nutrition.

LAB are strictly fermentative and have complex nutritional requirements, needing to be supplied with carbohydrates, amino acids, peptides, fatty acid esters, salts, nucleic acid derivates, and vitamins. The LAB are unable to grow on simple mineral media supplemented with a carbon source. These organisms are auxotrophic for several amino acids *(2,3)*. However, some residual genetic material for biosynthesis is still present, since mutagenesis can render some lactobacilli prototrophic *(3)*; thus loss of the ability to synthesize an amino acid may be conditional rather than absolute. For example, the alanine requirement of *Lactobacillus delbrueckii* subsp. *delbrueckii* ATCC 9649 can be bypassed by the addition of vitamin B_6 to the medium *(4)*. The search for nutrients, and especially amino acids, constitutes a real challenge for LAB. Chemically defined media have been developed for these bacteria *(5,6)* for different purposes, including investigations of the nutritional requirements of bacteria cells *(2,3)*, identification of specific components, and isolation of auxotrophic mutants *(3)*. A defined medium that supports sustained growth of a LAB at a

From: *Methods in Biotechnology: Environmental Microbiology: Methods and Protocols*
Edited by: J. F. T. Spencer and A. L. Ragout de Spencer

constant specific growth rate is required for metabolic investigations, including the isolation of bioactive molecules.

In this chapter, we describe a chemically defined medium (CDM) that allows an optimal growth of different LAB; mainly those used for the production of functional foods, such as *Lactobacillus casei, Lactobacillus helveticus, Lactobacillus plantarum, Lactobacillus reuteri, Lactobacillus fermentum, Lactobacillus delbrueckii* subsp. *Lactis,* and *Lactobacillus acidophilus.* This CDM allows the study of minimal nutritional requirements of these LAB.

2. Materials

1. MRS broth *(7)*: 10 g/L polypeptone, 10 g/L meat extract, 5 g/L yeast extract, 20 g/L glucose, 5 g/L sodium acetate, 1.08 g/L Tween-80, 2 g/L ammonium citrate, 2 g/L K_2HPO_4, 0.2 g/L $MgSO_4$•$7H_2O$, 0.05 g/L $MnSO_4$•$4H_20$, pH 6.4 ± 0.2. Sterilize in an autoclave at 121°C for 15 min.
2. Saline solution (0.8% NaCl): dissolve 4 g NaCl in 500 mL distilled water (dH_2O). Sterilize in an autoclave at 121°C for 15 min.
3. Phosphate buffer: weigh 3 g K_2HPO_4 and 3 g KH_2PO_4 and make to 100 mL with dH_2O. Sterilize in an autoclave at 121°C for 15 min.
4. 200 g/L glucose: dissolve 20 g glucose in 100 mL dH_2O. Sterilize in an autoclave at 121°C for 15 min.
5. 2 mL/L Tween-80: mix 2 mL Tween-80 and 98 mL dH_2O with gentle stirring. Sterilize in an autoclave at 121°C for 15 min.
6. 500 g/L Sodium acetate: dissolve 5 g sodium acetate in 10 mL dH_2O. Sterilize in an autoclave at 121°C for 15 min (*see* **Note 1**).
7. 100 g/L Triammonium citrate: dissolve 1 g triammonium citrate in 10 mL dH_2O. Sterilize in an autoclave at 121°C for 15 min (*see* **Note 1**).
8. 20 g/L $MgSO_4$•$7H_2O$: dissolve 0.2 g $MgSO_4$•$7H_2O$ in 10 mL dH_2O. Sterilize in an autoclave at 121°C for 15 min (*see* **Notes 1** and **2**).
9. 2.5 g/L $MnSO_4$•$4H_2O$: dissolve 25 mg $MnSO_4$•$4H_2O$ in 10 mL dH_2O. Sterilize in an autoclave at 121°C for 15 min (*see* **Notes 1** and **2**).
10. 2 g/L $FeSO_4$•$7H_2O$: dissolve 20 mg $FeSO_4$•$7H_2O$ in 10 mL dH_2O. Sterilize by filtration through a cellulose nitrate membrane (0.22-μm pore size) (*see* **Notes 1** and **2**).
11. Separate 100-fold-concentrated solutions of each amino acid, base, and vitamin (*see* **Table 1**), except the trytophan solution, which is 25-fold concentrated (*see* **Note 3**).

All stock solutions are prepared in distilled water unless otherwise indicated. The amino acid solutions can be stored at –20°C for 1 mo, except for the cysteine solution, which must be freshly prepared (*see* **Note 4**), and the tyrosine solution, which is stored at room temperature for at least 1 wk. All the solutions must be sterilized by filtration through a cellulose nitrate membrane (0.22-μm pore size) into the sterilized (20 min, 121°C) culture vessel unless otherwise

Table 1
Composition of the Chemically Defined Medium (CDM)

Constituent	Stock solution concentration (g/L)	Final concentration (g/L)	Constituent	Stock solution concentration (g/L)	Final concentration (g/L)
Glucose	200	10	*Bases*		
			Guanine	1	0.01
Sodium acetate	500	5	Adenine	1	0.01
Ammonium citrate	100	1	Xanthine	1	0.01
KH_2PO_4/ K_2HPO_4	60	6	Thymine	1	0.01
$MgSO_4$•$7H_2O$	20	0.2	Uracil	1	0.01
$MnSO_4$•$4H_2O$	2.5	0.025			
$FeSO_4$•$7H_2O$	2.0	0.020	*Vitamins*		
			Nicotinic acid	0.1	0.001
Tween-80	20	1	Thiamine	0.1	0.001
			Biotin	1	0.01
Amino acids			Vitamin B_{12}	0.1	0.001
L-Alanine	10	0.10	Ca-pantothenate	0.1	0.001
L-Arginine	10	0.10	Pyridoxal	0.2	0.002
L-Asparagine	20	0.20	Riboflavin	0.1	0.001
L-Aspartic acid	20	0.20	Folic acid	0.1	0.001
L-Cysteine	20	0.20	*p*-Aminobenzoate	1	0.01
L-Glutamine	20	0.20	Inosine	0.5	0.005
L-Glutamic acid	20	0.20	Orotic acid	0.5	0.005
Glycine	10	0.10			
L-Histidine	10	0.10			
L-Isoleucine	10	0.10			
L-Leucine	10	0.10			
L-Lysine	10	0.10			
L-Methionine	10	0.10			
L-Phenylalanine	10	0.10			
L-Proline	10	0.10			
L-Serine	10	0.10			
L-Threonine	10	0.10			
L-Tryptophan	4	0.10			
L-Tyrosine	10	0.10			
L-Valine	10	0.10			

indicated (*see* **Note 5**). For the following compounds special care must be taken:

1. Tyrosine must be dissolved in 0.2 *N* NaOH.

2. To dissolve the glutamic acid, bring the pH to 7.0 with 1 *N* NaOH.
3. Phenylalanine is soluble in 0.05 *N* NaOH.
4. Prepare the solution of cysteine in 0.1 *N* HCl (*see* **Note 4**).
5. Guanine and xanthine are soluble in 0.2 *N* NaOH.
6. Uracil is soluble in 0.1 *N* NaOH.
7. Dissolve the riboflavin in concentrated HCl and add distilled water to prepare a solution containing 0.1 mg/mL of riboflavin. It is destroyed by visible light, especially at neutral pH or above; aqueous solution should be stored in the dark. Riboflavin solution is stable to autoclaving.
8. *p*-Aminobenzoic acid (PABA) is soluble in hot water.
9. Folic acid is soluble in diluted mineral acids and alkaline solutions; alkaline solutions are relatively stable. Store in a dark bottle. The vitamin is stable to autoclaving.
10. Biotin can be heat-sterilized.
11. Vitamin B_{12} coenzyme (cyanocobalamin coupled to adenine nucleoside) is light sensitive and should be filter sterilized. In contrast, vitamin B_{12} (cyanocobalamin) itself is relatively stable to both light and autoclaving. Aqueous solutions are light sensitive and form hydroxocobalamin; prolonged exposure causes irreversible destruction. Solutions are most stable at pH 4.0–6.0; may be autoclaved at 120°C with only slow decomposition. Vitamin B_{12} activity is completely destroyed by heating in alkaline solution, 0.1 *N*, 100°C, 10 min.
12. The coenzyme forms of the vitamins NAD and NADP are destroyed by autoclaving. Nicotinic acid and its amide (nicotinamide) are stable to autoclaving to neutral pH.
13. Calcium pantothenate is stable to autoclaving.
14. Thiamine is cleaved into its component moieties when aqueous solutions at pH 5.0 or above are autoclaved; therefore the solutions should be filter sterilized or autoclaved separately at pH below 5.0 to maintain the intact vitamin.
15. Pyridoxal solution is light sensitive, especially at alkaline pH.

3. Methods

3.1. Preparation of Chemically Defined Medium (CDM)

The glassware must be carefully washed in a free-rinsing detergent, repeatedly rinsed in distilled water, and air dried.

The composition of the complete CDM is detailed in **Table 1**, and it was adapted from that described by Hébert et al. *(2)*. It essentially contains 44 components, including one carbohydrate source, 20 amino acids, 11 vitamins, 3 metallic ions, and 5 nucleic acid bases (*see* **Note 6**).

To prepare 1 L of CDM, add the following solutions in sequence:

1. 375 mL Distilled water.
2. 100 mL Phosphate buffer.
3. 10 mL Sodium acetate solution.

4. 10 mL Ammonium citrate solution.
5. 10 mL of each vitamin solution (*see* **Table 1**).
6. 10 mL of each amino acid solution, except tryptophan (25 mL).
7. 10 mL of each base solution (*see* **Table 1**).
8. 50 mL Tween-80
9. 10 mL $MgSO_4•7H_2O$ solution.
10. 10 mL $FeSO_4•7H_2O$ solution.
11. 10 mL $MnSO_4•4H_2O$ solution.
12. 50 mL 20% (w/v) glucose.

Adjust to pH 6.4 with 1 *N* HCl or 1 *N* NaOH. Sterilize the medium by filtration through a cellulose nitrate membrane (0.22-μm pore size). The medium must be used within 24 h.

3.2. Inoculum Preparation

1. Activate the strain to be tested by daily transfers in MRS broth for 3–4 successive days. Inoculate the microorganisms (1%) in 100 mL MRS broth (*see* **Note 7**) and incubate overnight at 30°C (mesophilic bacteria) or 37°C (thermophilic bacteria).
2. Collect cells by centrifugation at 10,000*g*.
3. Wash the pellet three times with 50 mL of sterile 0.8% NaCl (*see* **Note 8**).
4. Homogenize pellet in 10 mL of sterile 0.8% NaCl before proceeding.
5. Inoculate the cell suspension in sterile CDM at initial optical density at 560 nm (OD_{560}) of 0.1.

3.3. Analytical Methods to Estimate the Cell Growth

Bacterial growth can be monitored by measuring OD_{560}, and these measurements can be used for correlation with cell dry weight determinations. Harvest the cells by filtration (0.2-μm pore size), wash twice with distilled water, and dry to a constant weight at 60°C under partial vacuum (200 mmHg). Correlate a change of 1 U of optical density with the equivalent of dry matter. The cell growth can also be estimated by serial dilution in saline solution and plating in MRS agar.

3.4. Evaluation of Nutritional Requirements

To identify the absolute nutritional requirements of a LAB, each of the components of the complete CDM must be omitted in turn. Calculate the maximal growth rate (μ) in complete CDM (μ_{CDM}) and in CDM without the nutrient under study (μ_X). Then, determine the relation (μ_X/μ_{CDM}) × 100. A constituent is considered as essential when the growth rate of the analyzed strain in CDM deprived of the constituent under study is lower than 15% of that obtained in the complete CDM. On the contrary, when the growth rate represents 80% (or more) of that obtained in the complete CDM, the component under study is considered as nonessential. When the growth rate is between 15% and 80% of that obtained in the complete CDM, the component under study is considered as stimulatory. The following three tests must be made: (1) cell growth in CDM without addition of amino acids (negative control), (2) cell growth in the complete CDM, and (3) cell growth in CDM deprived of the constituent under study.

3.4.1. Mineral Salts

The nutritionally essential metal ions serve bacteria in a number of functions: (1) as activators or cofactors of a variety of enzymes (e.g., potassium, magnesium, and manganese), (2) in membrane transport (e.g., potassium and sodium), and (3) as components of molecules or structural complexes. With complex media, all these ions are present as contaminants in medium components such as peptones and yeast extracts, and no supplementation is required for routine growth.

It is recommended to evaluate the single omission of each mineral salt separately and then analyze the multiple omissions of them.

3.4.2. Choice of Buffer

A buffer to be used in a defined medium must (1) have a high buffering capacity in the relevant pH area, (2) not be metabolized, and (3) not give precipitation of medium components.

To analyze the omission of phosphate buffer, replace it by another buffer such as 2-(*N*-morpholino)ethanesulfonic acid (MES), which is useful for pH 5.5 to 6.7 and is used at 0.01 (*see* **Note 9**) to 0.05 *M*, or succinate at a concentration of 0.1 *M* (11.9 g/L). Succinic acid fulfills the buffer criteria mentioned previously, being a reduced dicarboxylic acid with $pKa_1 = 4.21$ and $pKa_2 = 5.64$.

3.4.3. Amino Acid, Base, and Vitamin Requirements

To evaluate the amino acid requirements, follow the growth of the strains to be tested in CDM in which individual amino acids or an entire metabolic family of amino acids has been removed. The different families of amino acids are listed in **Fig. 1.**

To analyze the base and vitamin requirements, evaluate the growth of the bacteria in CDM in which individual or multiple bases (purines and pyrimidines) or vitamins have been omitted.

Performing an assay (e.g., requirement of nicotinic acid):

1. Prepare the CDM medium without nicotinic acid as described in **Subheading 3.1.**
2. Prepare the inoculum as described in **Subheading 3.2.**

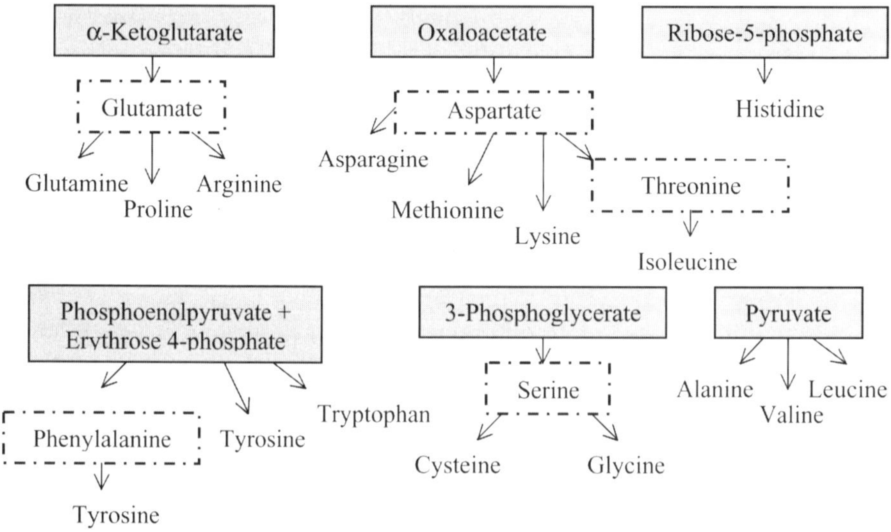

Fig. 1. Biosynthetic families of amino acids. The metabolic precursors are shaded and the intermediate amino acids are surrounded with dotted lines.

3. Inoculate CDM medium in which nicotinic acid was omitted, and the negative and positive controls. Positive control consists of culturing the strain under study in the complete CDM (contains all vitamins), whereas negative control consists of incubating the strain in the complete CDM deprived of amino acids.
4. Incubate at optimal temperature for 48 h.
5. Analyze the cell growth as described in **Subheading 3.3.** For the turbidimetric method, zero a spectrophotometer (at 560 nm) with an uninoculated assay medium tube blank. Then read the OD of the inoculated tube, and plot increasing optical density vs. time.
6. Calculate the maximal growth rate (μ) in complete CDM (μ_{CDM}) and in CDM without nicotinic acid (μ_X).
7. Calculate the relationship (μ_X / μ_{CDM}) × 100. If the result is lower than 15%, nicotinic acid is essential. If the growth rate represents >80% of that obtained in the complete CDM, nicotinic acid is nonessential. If the result is between 15% and 80%, nicotinic acid is stimulating.

3.5. Design of a Minimal Medium

To develop a minimal medium, remove simultaneously for the complete CDM the components neither essential nor stimulating for growth as determined by the single or multiple omission technique. If the strain is unable to grow, indicating that at least one essential nutrient was lacking despite the fact it has not been detected as such using the single-omission technique, the complete CDM must be simplified gradually, by removing unessential compounds of each building block step by step.

The last step of the formulation of the minimal CDM consists of the optimization of the concentration of the nutrients. The concentration of each component must be decreased progressively, and the optimal concentration is defined as the lowest concentration that did not affect the final population. The optimal concentration reflects the actual need of the LAB for each component.

3.6. Interpretation of Nutritional Requirements

LAB require several of the naturally occurring L-amino acids. Even in nonconditional auxotrophs, the addition of an excess of one amino acid may result in an increased requirement for another amino acid. This effect can be derived either from the competition of the two amino acids for a single transport system or the regulation of certain amino acid biosynthetic genes. The antagonistic interrelationships in the competition for a transport system can be avoided by the addition of an appropriate peptide of the limiting amino acid, because the peptide is absorbed via an independent transport system. The genes coding for the enzymes necessary for the synthesis of the branched-chain amino acids (Ile, Val, and Leu) are clustered *(8)* and under common regulatory control.

Vitamins play catalytic roles within the cell, usually as components of coenzymes or as prosthetic groups of enzymes. The functions of several vitamins are as follows:

PABA serves as a precursor for the biosynthesis of folic acid and thereby also influences the metabolism of thymine, methionine, serine, the purine bases, and vitamin B_1.

Folic acid functions as a cofactor in the synthesis of thymine, purine base, serine, methionine, and pantothenic acid.

Biotin participates in many biosynthetic reactions that require CO_2 fixation, including synthesis of oxalacetate and of fatty acids.

Nicotinic acid is involved in coenzyme biosynthesis and, as part of NAD and NADP, functions in redox reactions in cellular metabolism.

Panthotenic acid is a component of CoA and the acyl carrier protein (ACP).

Riboflavin is a component of flavin coenzymes, which function in redox reactions.

The coenzyme form of thiamine, thiamine pyrophosphate (TPP), functions in the decarboxylation of α-keto acids and in the transketolase reaction.

Pyridoxal is the prosthetic group of different enzymes and is involved in catalysis, especially in the synthesis and degradation of the α-amino acids. This compound acts as a catalytic agent of transaminases, and therefore might be involved in the biosynthesis of inosinic acid (an intermediate in the biosynthetic pathway of purines) from 5'-phosphoribosyl pyrophosphate.

The coenzyme form of Vitamin B_{12} (cyanocobalamin) functions in several isomerization reactions and also in the biosynthesis of deoxyribonucleosides and methionine.

4. Notes

1. The solution is 100-fold concentrated.
2. Most metals, in particular the divalent cations Ca^{++}, Fe^{++}, Mn^{++}, and Mg^{++}, tend to form insoluble hydroxides or phosphates at neutral to alkaline pH, making these elements unavailable for use by bacteria. The formation of precipitates of this type in nonchelated or poorly designed media is spontaneous and hastened by heat (autoclaving).
3. To identify the nutritional requirements of bacteria, each solution must be tested separately. To use a minimal medium for metabolic investigations, it is possible to prepare a metal mixture, a vitamin mix, and a pool of amino acids to supplement a defined medium. All mixtures should be stored in small portions so that they can be easily thawed. The medium needs to be made fresh and should not be stored longer than 1 d.
4. Cysteine should be made fresh because it is quickly oxidized to cystine crystals on contact with air. Alternatively, the stock solution can be overlaid with sterile mineral oil.
5. Except for glutamine, cysteine, and tryptophan, which must be filter-sterilized, amino acids are stable to autoclaving. Glutamine is rapidly converted in boiling neutral aqueous solution to the ammonium salt of pyrrolidonecarboxylic acid.
6. All chemicals should be reagent grade or better.
7. The bacterial cells must be previously cultured in a complex medium to assure good growth.
8. Cells must be washed to minimize carryover of residual nutrients during inoculation. This step is critical, and failure to perform it properly causes most assay failures.
9. MES buffer has an inadequate buffering capacity when the pH of the medium at the end of fermentation is approx 3.8. In this case, succinate is the appropriate buffer. The use of phosphate buffer can result in precipitation of the medium, particularly with bivalent cations like Ca^{++}; hence, this buffer is not suitable.

References

1. Klaenhammer, T. R. (1998) Functional activities of *Lactobacillus* probiotics: genetic mandate. *Int. Dairy J.* **8,** 497–505.
2. Hébert, E. M., Raya, R. R., and De Giori, G. S. (2000) Nutritional requirements and nitrogen-dependent regulation of the proteinase activity of *Lactobacillus helveticus* CRL 1062. *Appl. Environ. Microbiol.* **66,** 5316–5321.

3. Morishita, T., Deguchi, Y., Yajima, M., Sakurai, T., and Yura, T. (1981) Multiple nutritional requirements of lactobacilli: genetic lesions affecting amino acid biosynthetic pathways. *J. Bacteriol.* **148,** 64–71.
4. McNutt, W. S. and Snell, E. E. (1950) Pyridoxal phosphate and pyridoxamine phosphate as growth factors for lactic acid bacteria. *J. Biol. Chem.* **182,** 557–562.
5. Jensen, P. R. and Hammer, K. (1993) Minimal requirements for exponential growth of *Lactococcus lactis. Appl. Environ. Microbiol.* **59,** 4363–4366.
6. Poolman, B. and Konings, W. N. (1988) Relation of growth of *Streptococcus cremoris* to amino acid transport. *J. Bacteriol.* **170,** 700–707.
7. De Man, J. C., Rogosa, M., and Sharpe, M. E. (1960) A medium for the cultivation of lactobacilli. *J. Appl. Bacteriol.* **23,** 130–135.
8. Godon, J. J., Chopin, M. C., and Ehrlich, S. D. (1992) Branched chain amino acid byosynthesis genes in *Lactococcus lactis* subsp. *lactis. J. Bacteriol.* **174,** 6580–6589.

III

Nucleic Acids, Recovery, and Determination

15

A Simple Method for Obtaining DNA Suitable for RAPD Analysis From *Azospirillum*

Raúl O. Pedraza and Juan C. Díaz Ricci

1. Introduction

The random amplified polymorphic DNA (RAPD) technique has increasingly been used in the last decade as a simple, low-cost, and time-effective technique for the analysis of genomic polymorphism among related organisms. This method proved to be particularly useful in genomic fingerprinting and phylogenetic studies of microorganisms that present taxonomic problems or a large diversity within the taxon, such as clinical and soil microorganisms *(1–5)*. However, this technique presents persistent problems of reproducibility that have been attributed to a high sensitivity to subtle procedure changes and to the quality of the DNA used as template *(6–8)*. For this reason, other techniques that are more time and budget demanding, such as restriction fragment-length polymorphism (RFLP) and amplified fragment-length polymorphism (AFLP) have been used as valid and more reliable alternatives to assess genomic diversity *(9,10)*. Although DNA fingerprints obtained by RFLP and AFLP yield excellent results, often cost considerations make RAPD the technique of choice. Additionally, because the RAPD method is much faster and simpler than RFLP and AFLP, it is also well-suited when large numbers of specimens, or microbial isolates that require previous laborious manipulations, have to be analyzed. Thus, any procedural simplification of tedious and time-consuming steps is welcome, provided that the information quality does not decrease due to technical artifacts or loss of reproducibility; the latter is precisely the most sensitive issue of the RAPD technique, attributed mainly to DNA quality. Hence, obtaining DNA of such a quality that it can be directly used in RAPD

From: *Methods in Biotechnology: Environmental Microbiology: Methods and Protocols*
Edited by: J. F. T. Spencer and A. L. Ragout de Spencer

experiments is the most critical and time-consuming step required in this technique, which has to be carefully adjusted and optimized for every case under study.

The genus *Azospirillum* belongs to a group of rhizosphere bacteria often referred to as plant-growth-promoting rhizobacteria ***(11)***. They are free-living nitrogen-fixing bacteria closely associated with grasses ***(12)***. Owing to the increasing interest in making world agriculture sustainable, they are used as inocula for enhancing crop productivity and to replace chemical nitrogen fertilizers ***(13)***.

With the aim of assessing azospirilla diversity in soils, a short protocol based on direct cell disruption ***(14)*** was tried in our laboratory; although it proved to work quite well with specific gene primers (15- and 20-mer oligonucleotides), the method failed to give reliable profiles when using the random decanucleotide primers utilized in the RAPD technique. In this chapter, we present a modification of the aforementioned protocol ***(14)***. The method involves DNA extraction based on cell lysis by thermal disruption of bacterial suspensions, which does not require any extra purification step and yields a DNA suitable for RAPD experiments. This technique has been successfully used in our laboratory to assess the diversity of *Azospirillum* sp. and other soil microorganisms present in the area and associated with sugarcane plantations.

2. Materials

1. Strain: *Azospirillum brasilense* Sp7 (ATCC 29145) (EMBRAPA-CNPq, Seropédica Km 47, R.J., Brazil).
2. LB medium ***(15)***: 10.0 g tryptone, 5.0 g yeast extract, 3.0 g NaCl, and distilled water to 1000 mL, pH 7.0, adjusted with NaOH. For solid medium, add 15.0 g agar.
3. N-free malate medium (NFb) ***(16)***: 5.0 g malic acid, 0.5 g K_2HPO_4, 0.2 g $MgSO_4{\bullet}7H_2O$, 0.1 g NaCl, 0.02 g $CaCl_2{\bullet}2H_2O$, 2.0 mL micronutrients solution, 2.0 mL bromothymol blue (0.5% w/v in 0.2 *N* KOH), 4.0 mL FeEDTA (1.64% w/v, aqueous), 1.0 mL vitamins solution, 4.5 g KOH, and distilled water to 1000 mL, pH 6.5–6.8, adjusted with NaOH. For semisolid NFb medium add 1.75 g agar.
4. Micronutrients solution ***(16)***: 0.04 g $CuSO_4{\bullet}5H_2O$, 1.20 g $ZnSO_4{\bullet}7H_2O$, 1.40 g H_3BO_3, 1.00 g $NaMoO_4{\bullet}2H_2O$, 1.175 g $MnSO_4{\bullet}H_2O$, and distilled water to 1000 mL.
5. Vitamins solution ***(16)***: 10 mg biotin, 20 mg piridoxal-HCl, and distilled water to 100 mL. Store at 4°C.
6. Chelex 100 resin (Bio-Rad).
7. Wizard® Genomic DNA Purification Kit (Promega).
8. Spectrophotometer.

9. RAPD mixture: $MgCl_2$, 0.1 m*M* of each dNTP (deoxyribonuleotides), 0.2 µ*M* of random-sequence 10-mer primers (OPJ series from Operon Technologies), 0.75 U *Taq* polymerase (Promega), and 2 µL *Taq* buffer (10X).
10. Thermocycler.
11. 0.5X Tris-borate-EDTA buffer (TBE).
12. Agarose and electrophoresis equipment.
13. UV transiluminator analyzer.

3. Methods

The methods described below outline (1) *Azospirillum* DNA extraction by cellular thermal disruption based on the direct lysis of bacterial suspensions, (2) other DNA extraction techniques to compare the quality and reproducibility of the DNA obtained by the thermal cellular technique, and (3) the RAPD method used in this assay.

3.1. Cell Culture of Azospirillum and DNA Extraction by Cellular Thermal Disruption

1. The starting material is an overnight culture of *A. brasilense* Sp7 in NFb liquid medium (DO_{600} = 1) (*see* **Note 1**).
2. Inoculate 10 µL of the starting culture into 6 mL of LB broth.
3. Inoculate 10 µL of the starting culture into 6 mL NFb liquid medium supplemented with 1 g/L NH_4Cl and into 5 mL NFb semisolid medium.
4. Likewise, plate a loopful of the inoculum on LB solid medium.
5. Incubate liquid cultures overnight at 30°C in water-bath shaker at 200 rpm.
6. Incubate solid and semisolid cultures at 37°C for 48 h in oven.
7. Pick up with a sterile toothpick one or two colonies from solid culture and suspend it thoroughly in 30 µL of double-distilled sterile water contained in a 1.5-mL microtube.
8. Put 6 µL of cell suspensions coming from liquid or semisolid media in 30 µL of double-distilled sterile water in a 1.5-mL microtube.
9. Place the microtubes in upright position into a container of tap water and boil at 95°C for 10 min (*see* **Note 2**).
10. After thermal treatment, remove the microtubes from the water and cool down at room temperature (25°C). DNA template samples obtained by this method can be stored at –20°C for months to be used in further experiments.

3.2. Other DNA Extraction Techniques to Compare the Quality and Reproducibility of the DNA Obtained by Thermal Cellular Method

In order to test the influence of the DNA quality on the reproducibility of RAPD profiles, experiments and controls were carried out with DNA coming from the same cultures of *A. brasilense* Sp7 previously described, but extracted and purified according to the following protocols: Dellaporta et al. *(17)*; Walsh et al. ***(18)***, which includes the use of Chelex 100 resin (BioRad); and the Wizard™ Genomic DNA Purification Kit (Promega). The DNA was quantified with a Beckman 7000 spectrophotometer.

3.3. RAPD Assay

3.3.1. PCR Reaction

Reaction mix contains: 0.5 m*M* $MgCl_2$, 0.1 m*M* of each dNTP (deoxyribonuleotides), 0.2 µ*M* of random-sequence 10-mer primers (*see* **Note 3**), 0.75 U of *Taq* polymerase (Promega), 2 µL *Taq* buffer (10X), 1 µL of the suspension obtained by thermal cell disruption or 20 ng DNA obtained with other methods (*see* **Note 4**). Final volume: 20 µL.

3.3.2. Amplification Program

Denaturation, 30 s at 92°C; annealing, 1 min at 35°C; extension, 2 min at 72°C; number of cycles: 45. The program started with a thermal treatment of 3 min at 94°C and finished with an extension of 5 min at 72°C. The reaction was carried out in a MJ Research thermocycler.

3.3.3. Electrophoresis

Amplification products were separated by electrophoresis at 4 V/cm in a 1.5% agarose gel in 0.5X TBE running buffer. Gels were stained with 0.5 µg/mL ethidium bromide for 30 min, washed three times with distilled water, and photographed with a Polaroid camera (Polaroid film 667). RAPD experiments were repeated three times to avoid false results and to assure reproducibility. Negative controls, without DNA, were included in each experiment. In **Fig. 1** we show RAPD profiles of the strain Sp7 of *Azospirillum brasilense* when using the decanucleotide OPJ20 (Operon) as random primer and with DNA templates obtained according to the methods mentioned above. As seen in **Fig. 1**, there is no visible difference between the size and number of bands obtained using DNA prepared with different degrees of purification, and the profile obtained using DNA from thermal cell disruption (*see* **Note 5**). Although we show here a single example, we have tested this procedure may times using different primers and azospirilla isolates.

3.4. Results

The modification introduced in the original protocol ***(14)*** allowed us to evaluate the diversity of *Azospirilla* genus in soils of the sugarcane crop area of the region by using a simple and fast procedure. RAPD analysis and genome fingerprinting with this procedure showed no significant differences with con-

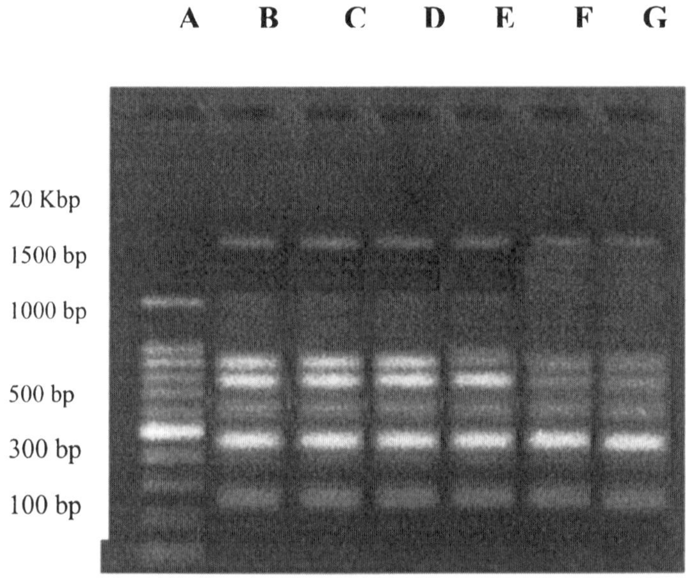

Fig. 1. RAPD profiles of the strain Sp7 of *A. brasilense* (ATCC 29145) when using DNA obtained with different methods. (A) Molecular weight marker (Ladder 100 bp, Promega); (B) DNA extracted and purified according to Dellaporta et al. ***(17)***; (C) DNA obtained according to Walsh et al. ***(18)***, with Chelex 100 resin (Bio-Rad); (D) DNA obtained by using the Wizard™ Genomic DNA Purificadion kit from Promega; (E) DNA obtained by thermal cell disruption from solid media; (F) DNA obtained by thermal cell disruption from semisolid media; (G) DNA obtained by thermal cell disruption from liquid media.

trols carried out with DNA obtained and purified with standard methods (*see* **Fig. 1**).

4. Notes

1. To obtain reliable results with the cellular thermal disruption method described here, it is important to control the purity of all cultures coming from liquid or semisolid media before extracting the DNA; owing to the conspicuous shape and motility features of these bacteria, the latter can be done microscopically.
2. Instead of the boiling water procedure, cell disruption can be carried out in a thermocycler using a one-cycle program of 95°C for 10 min.
3. The success of the amplification and reproducibility by RAPD, relies on many factors: quality and amount of DNA template, mix composition (i.e., Mg^{2+} concentration), primers used, and adjustment of the amplification program. Therefore, these factors need to be optimized before starting the RAPD analysis. Particular attention must be paid to the selection of the RAPD primers, as not all of them would yield a similar number and size of polymorphic bands. There are

primers that would not amplify any band at all. An ideal primer should yield highly reproducible DNA profiles, containing monomorphic and polymorphic bands, with a sufficient number to permit discrimination of samples.

4. As mentioned in **Note 3**, another important aspect to consider in RAPD analysis is the amount of DNA used in the amplification. An excess of DNA will produce a high-molecular-weight band (approx 20 Kbp) regardless the primer used (**Fig. 1**). The latter can be eliminated from the gel by reducing the concentration of DNA in the amplification reaction.
5. The procedure described here to obtain DNA in quality and quantity suitable for RAPD analysis has several advantages over previously mentioned techniques:
   a. It avoids the cumbersome procedure of DNA extraction and purification included in other protocols.
   b. The quality of DNA renders identical results to those from DNA obtained with more sophisticated and time-consuming procedures.
   c. The procedure allows analysis of DNA from different sources and culture media without any apparent loss of reproducibility of the fingerprint profile.
   d. The thermal cell disruption followed by PCR thermal cycling seems to eliminate contamination and other technical artifacts that usually appear as serious nuisances in standard RAPD procedures.
   e. This protocol is rather insensitive to changes in DNA load, in contrast to regular protocols that usually require thorough optimization to avoid negative effects on fingerprint reproducibility.
   f. This procedure eliminates the use of chemicals used in standard DNA extraction and purification protocols, therefore reducing the chance of DNA or any other unknown contaminants that often annoy RAPD users.
   g. This protocol also proved to be useful for ARDRA analysis ***(19)***, which requires DNA digestion with specific endonucleases after PCR amplification. The latter confirms that after the thermal treatment the DNA quality is still good enough to be used as a substrate for further DNA manipulations.

References

1. Williams, J. G. K., Kubelik, A. R., Livak, K. J., Rafalski, J. A., and Tingey, S. V. (1990) DNA polymorphisms amplified by arbitrary primers are useful as genetic markers. *Nucl. Acids Res.* **18,** 6531–6535.
2. Mazurier, S. I. and Wernars, K. (1992) Typing of *Listeria* strains by random amplification of polymorphic DNA. *Res. Microbiol.* **143,** 499–505.
3. Bazzicalupo, M. and Fani, R. (1995) The use of RAPD for generating specific DNA probes for microorganisms, in *Methods in Molecular Biology* (Clapp, J. P., ed.), Humana, Totowa, NJ, pp. 155–175.
4. Fani, R., Bandi, C., Bazzicalupo, M., Damiani, G., Di Cello, F., Fancelli, S., et al. (1995) Phylogenetic studies of the genus *Azospirillum.*, in *Azospirillum VI and Related Microorganisms* (Fendrik, I., del Gallo, M., Vanderleyden, J., and de Zamaroczy, M., eds.), Springer-Verlag, Berlin, pp. 59–75.

5. Schierwater, B. (1995) Arbitrarily amplified DNA in systematics and phylogenetics. *Electrophoresis* **16,** 1643–1647.
6. MacPherson, J. M., Eckstein, P. E., Scoles, G. J., and Gajadhar, A. A. (1993) Variability of the random amplified polymorphic DNA assay among thermal cyclers, and effects of primer and DNA concentration. *Mol. Cell. Probes* **7,** 293–299.
7. McEwan, N. R., Bakht, J., Cecchini, E., Coultart, A., Geri, C., McDonald, F. M., et al. (1998) Examination of the pipetting variation between individuals performing random amplified polymorphic DNA (RAPD) analysis. *J. Microbiol. Methods* **32,** 213–215.
8. Meunier, J. R. and Grimont, P. A. D. (1993) Factors affecting reproducibility or random amplified polymorphic DNA fingerprinting. *Res. Microbiol.* **144,** 373–379.
9. Fani, R., Bazzicalupo, M., Gallori, E., Giovannetti, L., Ventura, S., and Polsinelli, M. (1991) Restriction fragment length polymorphism of *Azospirillum* strains. *FEMS Microbiol. Lett.* **83,** 225–230.
10. Vos, P., Hogers, R., Bleeker, M., Reijans, M., van de Lee, T., Hornes, M., et al. (1995) AFLP: a new technique for DNA fingerprinting. *Nucl. Acids Res.* **23,** 4407–4414.
11. Kloepper, J. W., Lifshitz, R., and Zablotowicz, R. M. (1989) Free-living bacterial inocula for enhancing crop productivity. *Trends Biotechnol.* **7,** 39–44.
12. Steenhoud, O. and Vanderleyden, J. (2000) *Azospirillum,* a free-living nitrogen-fixing bacterium closely associated with grasses: genetic, biochemical and ecological aspects. *FEMS Microbiol. Rev.* **24,** 487–506.
13. Okon, Y. and Vanderleyden, J. (1997) Root associated *Azospirillum* species can stimulate plants. *ASM News* **63,** 366–370.
14. Joshi, A. K., Baichwal, V., and Ames, G. F. (1991) Rapid polymerase chain reaction amplification using intact bacterial cells. *BioTechniques* **10,** 42–44.
15. Sambrook, J., Fritsch, E. F., and Maniatis, T. (1989) *Molecular Cloning. A Laboratory Manual,* 2nd ed. Cold Spring Harbor Laboratory Press, NY.
16. Döbereiner, J., Baldani, V. L. D., and Baldani, J. I. (1995) Como isolar e identificar bacterias diazotróficas de plantas não-leguminosas. EMBRAPA-SPI, Brasilia-DF.
17. Dellaporta, S. L., Wood, J., and Hicks, J. B. (1983) A plant DNA minipreparation: version II. *Plant Mol. Biol. Rep.* **1,** 19–21.
18. Walsh, P. S., Metzger, D. A., and Higuchi, R. (1991) Chelex 100 as a medium for simple extraction of DNA for PCR-based typing from forensic material. *BioTechniques* **10,** 506–513.
19. Pedraza, R. O. and Díaz Ricci, J. C. (2000) *Azospirillum amazonense:* su presencia en el área cañera de la provincia de Tucumán. *Rev. Arg. Microbiol.* **32,** 199–201.

16

PCR Fingerprinting of rRNA Intergenic Spacer Regions for Molecular Characterization of Environmental Bacteria Isolates

M. Alejandra Martínez and Faustino Siñeriz

1. Introduction

The analysis of DNA or RNA molecules has been used in a large number of studies on bacterial taxonomy and typing and for diversity studies. Undoubtedly, these methods presently dominate modern taxonomic studies as a consequence of technological progress, but primarily because the present view on classification is that it should reflect the natural relationships as encoded in the DNA. Some of these methods are not fully sequence-based, although in a number of cases they have provided considerably more information than traditional phenotypic methods *(1)*.

Among genotyping methods, analyses of multigene families such as *rrn* operons and tRNA genes have been demonstrated to be highly useful *(2)*. The *rrn* operon is a multigene family that frequently presents more than one copy in most bacteria. With some exceptions *(3)*, the rDNA genetic loci in eubacteria include, in 5' to 3' order: 16S, 23S, and 5S rRNA genes, which are separated by intergenic transcribed spacer (ITS) regions. Sequence evaluations of the 16S rRNA have been used frequently as a powerful and accurate method for determining inter- and intraspecific relationships *(1)*. However, owing to their highly conserved nature, closely related organisms are often found to have nearly identical 16S rDNA sequences, limiting its power in resolving close relationships. Indeed, as evolutionary distances decrease, the diversity found in the 16S rRNA gene is often insufficient to define genetic relationships of closely related species *(4)*.

From: *Methods in Biotechnology: Environmental Microbiology: Methods and Protocols*
Edited by: J. F. T. Spencer and A. L. Ragout de Spencer

The 16S–23S rDNA intergenic regions were denoted as internal transcribed spacers (ITSs; ***5,6***), as intergenic spacer regions (ISRs; ***7***), and even long intergenic spacer regions (LISRs; ***8***). To avoid confusions and also to follow Gürtler and Mayall ***(2)***, we will call the 16S–23S rDNA spacers ISRs.

ISRs are less conserved than adjacent regions (16S and 23 S rDNA genes) as a result of a lower selective pressure during evolution, thus exhibiting a great deal of sequence and length variation at the genus and species level. The latter is partly because of the variations in the number and type of sequences with functional roles (tRNA genes) that are also found within the ISR ***(9)***. As a consequence, analysis of 16S–23S ISR length polymorphism and sequence variability has been shown to be an important supplement to 16S rDNA sequencing for differentiating bacterial species and even strains of prokaryotes, now being more frequently used for microbial typing and evolutionary and diversity studies of many groups of bacteria. Sequences of the 16S–23S regions of several species have become available for comparison.

16S–23S ISRs could be easily targeted by PCR (ISR-PCR) to show spacer length polymorphism in the different rDNA gene operons of the genome, based on their relative electrophoretic mobility on agarose gels. Length varies considerably between species (200–1500 bp), usually showing significant variations at the genus or species level and in some cases intra-species variations ***(8,10–16)***.

In screening programs to isolate strains of biotechnological interest or in diversity studies of cultivable bacteria, it is necessary to use a fast and easy procedure to characterize and differentiate them, especially when a large number of isolates is to be considered. Among PCR fingerprinting methods, ISR-PCR has been successfully used in our laboratory as a fast and reproducible procedure to cluster isolated bacteria according to different band patterns.

After clustering different isolates, they could be analyzed and identified according to morphological characteristics, biochemical properties, and several genotyping methods. It is important to note that the use of intergenic spacers is included in genomic approaches to typing, taxonomy, and evolution of bacterial isolates, and also in studies of prokaryotic diversity (most of the references cited). Nevertheless, it is important to note that further work needs to be done to determine the precise role of *rrn* recombination in phylogenetics, evolution, and typing, since concerted evolution of the *rrn* multigene family by recombination and mutation in 16S–23S rDNA spacer rearrangements has been described ***(2,17)***.

This chapter focuses on PCR fingerprinting of rRNA spacers in environmental isolates belonging to the genera *Bacillus*. We also obtained good results with actinomycetes strains. In our laboratory, different groups work with each bacterial genus, and particular characteristics were used to isolate them.

Because several copies of the rDNA operon are present in the bacterial genome, a simple and fast chromosomal DNA isolation procedure for all groups of bacteria has been shown to be adequate to ensure target sequences suitable for PCR amplification.

2. Materials

2.1. Culture Media

For *Bacillus* strains: nutrient broth composed of 5.0 g/L peptone, 3.0 g/L meat extract, 15 g/L agar, if necessary; pH 7.0.

To grow alkaliphilic strains, pH of the medium was raised to approx 10.0 after sterilization by the addition of sterile 1% $NaHCO_3$ (1 mL in 10 mL) from a stock solution, or 1 *M* Na-sesquicarbonate solution prepared with 4.2 g $NaHCO_3$ and 5.3 g Na_2CO_3 anhydrous and distilled water to 100 mL.

2.2. DNA Isolation

1. Lysis buffer TEC-SDS: 10 m*M* Tris-HCl, pH 8.0; 10 m*M* EDTA; 100 m*M* NaCl, and 2% (w/v) SDS.
2. 20 mg/mL Proteinase K in distilled water.
3. 3 *M* Sodium acetate, pH 5.2.
4. TE-buffer-saturated phenol ***(18)***.
5. Chloroform:isoamyl alcohol, 24:1 (v/v) ***(18)***.
6. Isopropanol and 70% ethanol
7. RNAase A solution (10 mg/mL stock) in distilled water ***(18)***.

2.3. PCR Amplification

1. *Taq* polymerase and 10X STR buffer (Promega).
2. Thermal cycler (PE 9700, Applied Biosystems, CA).
3. 100 μM Stock solution of primers ISR-1494 (5'–GTCGTAACAAGG TAGCCGTA–3') and ISR-35 (5'–CAAGGCATCCACCGT–3') ***(12)*** (*see* **Note 1**).

2.4. Agarose Gel Electrophoresis

1. TAE buffer (1X): 0.04 *M* Tris-acetate, 0.001 *M* EDTA, pH 8.0.
2. 2% Agarose in 1X TAE buffer.
3. Ethidium bromide staining solution ***(18)***.
4. Molecular weight markers: 1 kb and 100 bp DNA Ladders (Promega).
5. Loading buffer (Promega) or prepared according to Sambrook et al. ***(18)***.

3. Methods

3.1. DNA Isolation

1. Transfer 1.5 mL of an overnight culture to a microcentrifuge tube and spin down 2 min. Discard the supernatant.
2. Resuspend the pellet in 750 µL of lysis buffer by repeated up and down pipetting. Add 15 µL of 20 mg/mL proteinase K, mix by inverting the tube carefully, and incubate 30 min to 2 h at 55°C (*see* **Notes 2** and **3**).
3. Add an equal volume of TE-buffer-saturated phenol and mix by inverting the tube several times.
4. Centrifuge (10,000*g*, 5 min). Transfer the aqueous phase (upper) to a new tube and repeat phenol extraction once.
5. Transfer the upper aqueous phase to a clean tube and add an equal volume of chloroform:isoamyl alcohol. Again mix well and centrifuge (10,000*g*, 5 min). Repeat this extraction three times.
6. Transfer the aqueous phase to a new tube. Add 1/10 vol of 3 *M* sodium acetate.
7. Add 0.6 to 1 vol of 2-propanol and mix gently until the DNA precipitates. If a low amount was recovered, precipitation could be favored by centrifugation (15,000*g*, 10 min).
8. Discard the supernatant and wash DNA with 500 µL of 70% ethanol to remove residual salts and isopropanol. Centrifuge (15,000*g*, 5 min), carefully discard the ethanol, and dry until ethanol has been removed (*see* **Notes 4** and **5**).
9. Resuspend DNA in 100–200 µL double-distilled sterile water and 1–2 µL RNAase A. Allow to dissolve at 37°C for several hours or overnight at 4°C (*see* **Note 6**).

3.2. PCR Amplification

1. Reaction mixture. Prepare 20 µL for each sample in one mixture considering total number of samples plus an additional control tube that will not include template DNA. Each tube should contain 0.5 µL of isolated DNA, 2 µL of 10X STR buffer, 0.2 µL of *Taq* Blue polymerase (1 U), and 0.1 µL of each primer (0.5 µ*M* final concentration). Double-distilled sterile water to 20 µL. Dispense 19.5 µL for all samples tested into PCR tubes and add samples (*see* **Notes 7** and **8**).
2. Amplification conditions: initial denaturation at 94°C for 4 min followed by 30 cycles each consisting of 94°C for 1 min, 55°C for 2 min, and 72°C for 2 min, with a final extension step at 72°C for 7 min (*see* **Note 9**).
3. Evaluation of PCR fingerprint obtained: runs should be made in 2% agarose gels using appropriate markers in the range of 100 bp to 2000 bp. Electrophoresis should be performed at 50 V for 2 h (*see* **Note 10**). Stain the gel in ethidium bromide and observe under UV light (*see* **Notes 11** and **12**).

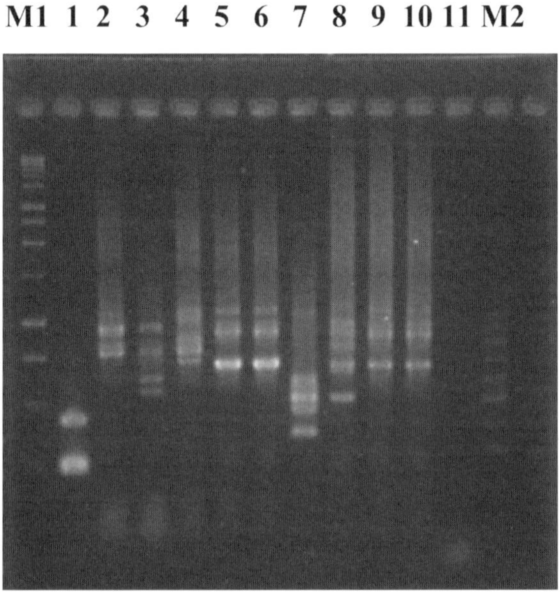

Fig. 1. ISR-PCR from *Bacillus* spp. strains. Lanes: M1, 1 kb DNA Ladder; 1, *B. subtilis* 1 A1 (obtained from BGSC[a]); 2, *B. halodurans* DSM497[T] (purchased as DSMZ[b]). From lanes 3 to 10, *Bacillus* sp. alkaliphilic strains[c]: 3, MIR32; 4, MRL1; 5, MRL2; 6, MRL22; 7, MRL33; 8, MRL4; 9, MRL5; 10, MRL5; 11, control reaction without DNA template; M2, 100 bp DNA Ladder. Band between 200 and 600 bp are obtained for members of *subtilis* and *cereus* groups (Martinez, unpublished data), and patterns including long ISR spacers, generally between 600 and 1200 bp are typical for most alkaliphilic *Bacillus* spp. tested (*see* **Note 13**). a = *Bacillus* Genentic Stock Center (BGSC); b = German Collection of Microorganisms (DSMZ); c = Isolated in soda lakes in Kenya, Africa (Breccia, personal communication).

4. Notes

1. Concerning primer sequences, the ones given were useful in *Bacillus* and even in actinomycetes strains, but they could be used to check other possibilities such as L1 5'–CAAGGCATCCACCGT–3' and G1 5'–GAAGTCGTAACAACG–3' ***(6,19)***, which mainly differ in the position of the primer complementary to 23S sequence (***3,7,9,20–22***, and others).
2. When setting up DNA isolation and PCR reactions, wear gloves to minimize the risks of DNAase contamination. In the case of PCR preparation, gloves should be powder-free because powder inhibits DNA polymerases. Precautions must be taken when handling dangerous solutions such as phenol, chloroform, and ethidium bromide.
3. Information on this topic can be found at Material Safety Data Sheets (www.sigma-aldrich.com).

4. DNA isolation. It could be necessary to optimize lysis conditions when handling wild-type isolates. Spore-forming bacterial strains might need to be processed before the end of the exponential growth phase. If poor clarification is achieved after lytic treatment, proteinase K treatment could be extended. Nevertheless, further phenol treatment ensures enough DNA suitable for PCR.
5. After two ethanol wash steps, liquid can be discarded by inverting tubes, followed by a spin down of 1 min to allow residual liquid in the walls to be collected by carefully pipeting from the bottom of the tubes. Do not overdry DNA pellets; this can make resuspension difficult.
6. DNA samples can be stored for several weeks at 4°C. Aliquots of samples can be kept at –20°C, taking into account that repeated freezing and thawing of the samples can damage DNA.
7. DNA samples obtained by this procedure are usually suitable for PCR amplification. Concentration and quality can be tested by gel electrophoresis and absorbance measurements ***(18)***. Nevertheless, OD measurements tend to overestimate DNA concentration. After electrophoretic evaluation, concentration of the samples can be equalized by repeating **steps 5–8**.
8. Total PCR reaction volume should be loaded to allow detection of all bands, even the less intense ones. Inclusion of controls without template is important because possible PCR artifacts should be considered when analyzing bands of low size (100–300 bp).
9. The use of longer annealing times and temperature ramps has been described ***(11)***, although we obtained good results as indicated previously.
10. Electrophoresis runs of 2–3% agarose gels should be relatively slow (40–50 V). Migrations of approx 10–11 cm of the front marker should be done to allow a discrete band pattern to develop. 100 bp Ladder and 1 kb Ladder (Promega) could be loaded in a separate well or in the same well. It is important to consider that bromophenol blue contained in gel-loading buffer migrates at approximately the same rate as DNA 300 bp in length.
11. The use of capture-image systems and software are useful when several bands are obtained to precisely determine the number and size of bands.
12. Absence of amplification could be caused by residual phenol or ethanol in the final preparations. To test reagents, a positive (amplifiable) control should be included in all assays. Samples can be recovered by repeating **steps 5–9**.
13. Frequently, *Bacillus* species yield several bands, and intra-species variations are sometimes observed. For example, alkaliphilic strains present larger spacers (**Fig. 1**). Evaluation of intergenic length polymorphism of hypervariable parts of conserved genomic regions revealed us typically large 16S–23S intergenic regions in alkaliphilic *Bacillus* strains in comparison with those described for other *Bacillus* species and lane 3 ***(11,2,3)***.

References

1. Vandamme, P., Pot, B., Gillis, M., De Vos, P., Kersters, K. and Swings, J. (1996) Polyphasic taxonomy, a consensus approach to bacterial systematics. *Microbiol. Rev.* **60,** 407–438.
2. Gürtler, V. and Mayall, C. (2001) Genomic approaches to typing, taxonomy and evolution of bacterial isolates. *Int. J. Syst. Evol. Microbiol.* **51,** 3–16.
3. Garcia-Martinez, J., Acinas, S. G., Anton, A. I., and Rodriguez-Valera, F. (1999) Use of the 16S–23S ribosomal genes spacer region in studies of prokaryotic diversity. *J. Microbiol. Methods* **36,** 55–64.
4. Rosselló-Mora, R. and Amann, R. (2001) The species concept for prokaryotes. *FEMS Microbiol. Rev.* **25,** 39–67.
5. Yoon, J.-H., Lee, S. T., Kim, S.-B., Goodfellow, M., and Park, Y.-H. (1997) Inter and intraspecific genetic analysis of the genus *Saccharomonospora* with 16S to 23S ribosomal DNA (rDNA) and 23S to 5S rDNA internally transcribed spacer sequences. *Int. J. Syst. Bacteriol.* **47,** 661–669.
6. Zhang, Z., Kudo, T., Nakajima, Y., and Wang, Y. (2001) Clarification of the relationship between the members of the family *Thermomonosporaceae* on the basis of 16S rDNA, 16S–23S rRNA internal transcribed spacer and 23S rDNA sequences and chemotaxonomic analyses. *Int. J. Syst. Evol. Microbiol.* **51,** 373–383.
7. Aakra, A., Utaker, J. B., and Nes, I. F. (1999) RFLP of rRNA genes and sequencing of the 16S–23S rDNA intergenic spacer region of ammonia-oxidizing bacteria: a phylogenetic approach. *Int. J. Syst. Bacteriol.* **49,** 123–130.
8. Gürtler, V., Rao, Y., Pearson, S. R., Bates, S. M., and Mayall, B. C. (1999) DNA sequence heterogeneity in the three copies of the long 16S-23S rDNA spacer of *Enterococcus faecalis* isolates. *Microbiology* **145,** 1785–1796.
9. Gürtler, V. and Stanisich, V.A. (1996) New approaches to typing and identification of bacteria using the 16S–23S rDNA spacer region. *Microbiology* **142,** 3–16.
10. Mendoza, M., Meugnier, H., Bes, M., Etienne, J., and Freney, J. (1998) Identification of Staphylococcus species by 16S–23S rDNA intergenic spacer PCR analysis. *Int. J. Syst. Bacteriol.* **48,** 1049–1055.
11. Daffonchio, D., Borin, S., Frova, G., Manachini, P., and Sorlini, C. (1998) PCR fingerprinting of whole genomes: the spacers between the 16S and 23S rRNA genes and of intergenic tRNA gene regions reveal a different intraspecific genomic variability of *Bacillus cereus* and *Bacillus licheniformis. Int. J. Syst. Bacteriol.* **48,** 107–116.
12. Senesi, S., Celandroni, F., Tavanti, A., and Ghelardi, E. (2001) Molecular characterization and identification of *Bacillus clausii* strains marketed for use in bacteriotherapy. *Appl. Environ. Microbiol.* **67,** 834–839.
13. Nakasone, K., Masui, N., Takaki, Y., Sasaki, R., Maeno, G., Sakiyama, Y., et al. (2000) Characterization and comparative study of the *rrn* operons of alkaliphilic *Bacillus halodurans* C-125. *Extremophiles* **4,** 209–214.
14. Mehta, A. and Rosato, Y. B. (2001) Phylogenetic relationships of *Xylella fastidiosa* strains from different hosts, based on 16S rDNA and 16S–23S intergenic spacer sequences. *Int. J. Syst. Evol. Microbiol.* **51,** 311–318.

15. Martínez, M. A., Delgado, O. D., Breccia, J., Baigor, M. D., and Siñeriz, F. (2002) Revision on taxonomic position of the xylanolytic *Bacillus* sp. MIR32 strain reidentified as *Bacillus halodurans* and plasmid-mediated transformation of *B. halodurans* species. *Extremophiles*, in press.
16. Hain, T., Ward-Rainey, N., Kroppenstedt, R. M., Stackebrandt, E., and Rainey, F. A. (1997) Discrimination of *Streptomyces albidoflavus* strains based on the size and number of 16S–23S ribosomal DNA intergenic spacers. *Int. J. Syst. Bacteriol.* **47,** 202–206.
17. Gürtler, V. (1999) The role of recombination and mutation in 16S–23S rDNA spacer rearrangements. *Gene* **238,** 241–252.
18. Sambrook, J., Fritsch, E. F., and Maniatis, T. (1989) *Molecular Cloning: A Laboratory Manual*, 2nd ed. Cold Spring Harbor Laboratory, Cold Spring Harbor, NY.
19. Jensen, M. A., Webster, J. A., and Straus, N. (1993) Rapid identification of bacteria on the basis of polymerase chain reaction-amplified ribosomal DNA polymorphism. *Appl. Environ. Microbiol.* **59,** 945–952.
20. Nour, M. (1998) 16S–23S and 23S–5S intergenic spacer regions of lactobacilli: nucleotide sequence, secondary structure and comparative analysis. *Res. Microbiol.* **149,** 433–448.
21. Christensen, H., Jorgensen, K., and Olsen, J. E. (1999) Differentiation of *Campylobacter coli* and *C. jejuni* by length and DNA sequence of the 16S–23S rRNA internal spacer region. *Microbiology* **145,** 99–105.
22. Salazar, O., Morón, R., and Genilloud, O. (2000) New genus-specific primers for the PCR identification of members of the genus *Saccharomonospora* and evaluation of the microbial diversity of wild-type isolates of *Saccharomonospora* detected from soil DNAs. *Intl J. Syst. Evol. Microbiol.* **50,** 2043–2055.
23. Wunschel, D., Fox, K. F., Black, G. E., and Fox, A. (1994) Discrimination among *B. cereus* group, in comparison to *B. subtilis*, by structural carbohydrate profiles and ribosomal RNA spacer region PCR. *Syst. Appl. Microbiol.* **17,** 625–635.

17

A DNA–DNA Hybridization Method for the Detection and Quantification of Specific Bacterial Taxa in Natural Environments

Christopher E. Bagwell and Charles R. Lovell

1. Introduction

Reverse sample genome probing (RSGP) is a quantitative DNA–DNA hybridization approach that can be employed for (1) assessment of genomic similarities among different reference culture organisms *(1)* and (2) for determination of the presence and relative abundances of selected microorganisms in environmental samples *(2–6)*. The environmental application of this method will be the focus of this chapter. Briefly, RSGP utilizes whole community DNA extracted from environmental samples to probe against dot-blotted DNA samples from characterized reference cultures. Under stringent hybridization conditions and using an appropriate internal standard, the quantity of probe bound in each dot reflects the abundance of each reference culture organism in the environmental sample. A major application of this method is the quantification of many different bacterial taxa in an environmental sample simultaneously. The performance of this procedure as described was optimized for the use of nylon membranes and ^{35}S radiolabel; however, alternative supports (i.e., nitrocellulose membranes, glass slides) and detection methods (colorimetric, fluorescence) are also applicable.

2. Materials

2.1. Probe Preparation

1. Probe DNA (from environmental sample, 1 μg).
2. Internal standard DNA (Lambda Bacteriophage, 0.1 ng).

From: *Methods in Biotechnology: Environmental Microbiology: Methods and Protocols*
Edited by: J. F. T. Spencer and A. L. Ragout de Spencer

3. A restriction endonuclease recognizing a tetrameric sequence (*Sau*3A1).
4. Random hexamers.
5. DNA polymerase I, large (Klenow) fragment (2 U/μL).
6. dNTPs (100 μ*M* each dNTP except dATP in 250 m*M* Tris-HCl, pH 8.0, 25 m*M* $MgCl_2$, 50 m*M* 2-mercaptoethanol).
7. [α-^{35}S]dATP.
8. Boehringer Mannheim G-25 Sephadex Quick Spin columns (or comparable product) for purification of radiolabeled probe.

Storage: Once labeled and purified, probe should be stored frozen at –20°C until used.

2.2. Master Blot Preparation

1. Genomic DNA (30 ng) from each reference culture organism.
2. Internal standard DNA (bacteriophage lambda, a linear range of concentrations).
3. Denaturing solution (0.4 *M* NaOH).
4. Neutralization buffer (2 *M* ammonium acetate).
5. Nylon hybridization membrane (Duralose-UV™ membranes; Stratagene, La Jolla, CA), cut to 10 cm × 13.5 cm size—same dimensions as the dot-blot manifold.
6. 20X SSC (1X SSC = 150 m*M* NaCl, 15 m*M* sodium citrate).
7. 4X SSC.
8. India ink (1:150 dilution in dH_2O).
9. Wash buffer (1 *M* ammonium acetate).

Storage: Prepared membranes should be stored in a desiccator at 4°C until used (*see* **Notes** section).

2.3. Prehybridization Buffer (Final Concentrations Provided)

Prehybridization buffer should be made up fresh from concentrated stock solutions of all reagents prior to use.

1. 10X NET: 1X = 150 m*M* NaCl, 15 m*M* Tris-HCl, pH 7.5, 1 m*M* disodium EDTA.
2. 0.1% Sodium dodecyl sulfate (SDS).
3. 3X Denhardt's solution: 1X = 0.02% bovine serum albumin, 0.02% Ficoll™, and 0.02% polyvinylpyrrolidone.
4. 50 μg/mL Denatured salmon sperm DNA.

Storage: NET buffer and SDS stocks can be stored at room temperature. Denhardt's solution and salmon sperm DNA should be stored frozen at –20°C until used.

2.4. Hybridization Buffer (Final Concentrations Provided)

Hybridization buffer should be made up fresh from concentrated stock solutions of all reagents prior to use.

1. 50% Deionized formamide.

2. Bio-Rad AG 501-X8 Mixed Bed Resin (for deionizing formamide).
3. 5X SSC: 1X = 150 m*M* NaCl, 15 m*M* sodium citrate.
4. 25 m*M* Potassium phosphate buffer, pH 7.4.
5. 5X Denhardt's solution.
6. 50 µg/mL Denatured salmon sperm DNA.
7. Denatured labeled probe (environmental DNA—internal standard mix).

Storage: Formamide should be deionized by stirring with 5 g/100 mL of the mixed bed resin for 1 h, filtered through a 0.2-µm-pore filter, aliquoted, and stored frozen at –20°C until use. Potassium phosphate and SSC stocks can be stored at room temperature.

2.5. Posthybridization Wash Buffers

1. 2X SSC, 0.1% SDS.
2. 0.5X SSC, 0.1% SDS.

2.6. Equipment

1. Dot blot manifold (Schleicher & Schuell Minifold or comparable product).
2. Vacuum oven or UV crosslinker.
3. Shaking water bath.
4. Liquid scintillation counter.

3. Methods

3.1. Probe Preparation

1. Add 0.1 ng of internal standard DNA to 1 µg environmental DNA to prepare the DNA probe (*see* **Note 1**).
2. Restriction digest the probe DNA (*see* **Note 2**) with *Sau*3A1 at 37°C for 30 min. Inactivate the *Sau*3A1 by incubating the digested probe at 65°C for 10 min.
3. Random prime labeling: Mix probe DNA and random hexamers (155.4 µg/mL final concentration) (*see* **Note 3**) together and denature by incubating in a boiling water bath for 10 min, and cool to room temperature. Add dNTPs (16.7 m*M* final concentration), Klenow fragment (1 µL/30 µL reaction volume), and [α-^{35}S]dATP (10 µCi), and incubate at 37°C overnight. Inactivate the DNA polymerase by heating at 65°C for 10 min.
4. Purify products by gel filtration using G-25 Sephadex® spun columns.
5. Determine radioactivity by liquid scintillation counting. Calculate specific activity.

3.2. Master Blot Preparation

1. Dispense 30 ng of DNA from each reference organism and a series of internal standards (5-, 15-, 20-, 25-, and 30-ng standards), each preset at the appropriate concentration in deionized H_2O or TE (10 m*M* Tris-HCl, pH 8.0, 1 m*M* disodium EDTA), into assigned wells of a 96-well microtiter plate.
2. Designate several control wells that will each receive a solution but no target DNA.

3. Add 50 µL denaturing solution and incubate for 30 min at room temperature.
4. Add 50 µL of neutralization solution and incubate for 30 min on ice.
5. Prewet the nylon membrane evenly with deionized H_2O, followed by 20X SSC, and assemble dot-blot manifold.
6. Add 100 µL of 20X SSC to each well and apply vacuum.
7. Mark each dot by adding 5 µL of India ink to each well (*see* **Note 4**) and applying vacuum.
8. Load samples (101 µL final volume) under vacuum.
9. Add 100 µL of wash buffer to each well and apply vacuum.
10. Add 100 µL 4X SSC to each well and apply vacuum.
11. Disassemble dot-blot manifold and allow membrane to air-dry completely (DNA side up).
12. Bake membrane at 80°C under vacuum for 2 h and store in a desiccator at 4°C until used (*see* **Notes 5** and **6**).

3.3. Prehybridization

1. Prepare 30 mL fresh prehybridization buffer.
2. Denature salmon-sperm DNA by incubating in a boiling water bath for 10 min and immediately add the salmon-sperm DNA to the prehybridization buffer (*see* **Note 7**).
3. Prehybridize in a shaking water bath at 40°C for 6 h.

3.4. Hybridization

1. Prepare 30 mL fresh hybridization buffer.
2. Denature salmon-sperm DNA and labeled probe standard mixture by incubating in a boiling water bath for 10 min, and immediately add both solutions to the hybridization buffer.
3. Hybridize in a shaking water bath at 40°C overnight.

3.5. Posthybridization

1. Wash membrane twice in 2X SSC, 0.1% SDS at room temperature for 5 min each wash without shaking.
2. Wash membrane twice in 0.5X SSC, 0.1% SDS at 45°C for 20 min each wash in a shaking water bath.
3. Using a hole punch, cut control, standard, and experimental dots from the membrane and assay the radioactivity in each by liquid-scintillation counting.

3.6. Quantification

1. Subtract background radioactivity in the control dots from all standard and experimental dots.
2. Determine DNA quantity in experimental dots by regression analysis using the internal standards method. Briefly, the quantity (ng) of bound environmental DNA per experimental dot is calculated from a linear regression of the internal

standards (i.e., $X = (Y - b)/m$, where X = quantity (ng) of environmental DNA per experimental dot, Y = radioactivity measured per experimental dot, b = Y-intercept of the internal standards regression line, and m = slope of the internal standards line). The population size of each reference culture organism in the environmental sample (1 µg) can be approximated from the quantity of bound environmental DNA (ng) per experimental dot, X, by assuming a constant genome size (i.e., 5 fg DNA per cell for *Escherichia coli*) and 1 genome copy per cell.

4. Notes

1. For best results, probe DNA should be free of contaminating polysaccharides, detergents, and humic materials. Contaminating materials will result in decreased labeling efficiency.
2. Increased hybridization specificity and sensitivity was achieved by enzymatic digestion (specifically with a restriction endonuclease recognizing 4-bp sites, such as *Sau*3AI) of the probe (internal standard DNA and environmental DNA mixture) prior to labeling. Enzymatic digestion permitted species-level resolution, while hybridizations conducted with intact genomic DNA under the same hybridization conditions could not consistently differentiate closely related species.
3. Random prime labeling using random hexamers, as compared to longer primers, results in more efficient and consistent labeling of different probe DNA templates.
4. India ink is a suitable and nonreactive method for marking dots. The dye is not washed away during overnight hybridizations or posthybridization wash steps. In addition, this method permits visual assessment of the consistency and evenness of all dots. It should be noted that India ink should be applied to all dots prior to the application of DNA to the membrane. Otherwise, decreased hybridization efficiency results.
5. Fixation of template DNAs onto nylon membranes by UV crosslinking generally resulted in inconsistencies on and between membranes. Performance of membranes baked under vacuum was much more consistent and reproducible.
6. The performance of prepared membranes stored at 4°C in a desiccator was consistent for up to 3 wk. Beyond 3 wk, background intensity tended to be higher and hybridization results unsatisfactory.
7. Efficient blocking is extremely important for obtaining low background levels. Addition of salmon-sperm DNA to the prehybridization buffer was essential.
8. Careful temperature control and the use of a shaking-water-bath incubator were essential for the removal of nonspecific hybrids and for achieving low background.

References

1. Bagwell, C. E. and Lovell, C. R. (2000) Persistence of selected *Spartina alterniflora* rhizoplane diazotrophs exposed to natural and manipulated environmental variability. *Appl. Environ. Microbiol.* **66,** 4625–4633.

2. Bagwell, C. E. and Lovell, C. R. (2000) Microdiversity of culturable diazotrophs from the rhizoplanes of the salt marsh grasses *Spartina alterniflora* and *Juncus roemerianus. Microb. Ecol.* **39,** 128–136.
3. Shen, Y., Stehmeier, L. G., and Voordouw, G. (1998) Identification of hydrocarbon-degrading bacteria in soil by reverse sample genome probing. *Appl. Environ. Microbiol.* **64,** 637–645.
4. Voordouw, G., Voordouw, J. K., Karkhoff-Schweizer, R. R., Fedorak, P. M., and Westlake, D. W. S. (1991) Reverse sample genome probing, a new technique for identification of bacteria in environmental samples by DNA hybridization, and its application to the identification of sulfate-reducing bacteria in oil field samples. *Appl. Environ. Microbiol.* **57,** 3070–3078.
5. Voordouw, G., Voordouw, J. K., Jack, T. R., Fought, J., Fedorak, P. M., and Westlake, D. W. S. (1992) Identification of distinct communities of sulfate-reducing bacteria in oil fields by reverse sample genome probing. *Appl. Environ. Microbiol.* **58,** 3542–3552.
6. Voordouw, G., Shen, Y., Harrington, C. S., Telang, A. J., Jack, T. R., and Westlake, D. W. S. (1993) Quantitative reverse sample genome probing of microbial communities and its application to oil field production waters. *Appl. Environ. Microbiol.* **59,** 4101–4114.

18

^{13}C and ^{1}H NMR Study of the Glycogen Futile Cycle in *Fibrobacter succinogenes*

Anne-Marie Delort, Genevieve Gaudet, and Evelyne Forano

1. Introduction

Fibrobacter succinogenes, a strictly anaerobic bacterium, is a major fibrolytic species found in the rumen of cattle and sheep. It degrades cellulose into glucose and cellobiose, which are further metabolized in the cells into essentially succinate and acetate *(1–7)*. A part of the metabolized carbohydrates can be stored as glycogen accounting for up to 70% of the cells dry weight. This storage is continuous throughout the various growth phases of the bacterial cultures, with the glycogen–protein ratio remaining constant in the cells *(1)*. This unusual feature suggests that glycogen synthesis is not subject to stimulation by nutrient limitations or carbon excess, as previously stated for bacteria. Indeed, NMR experiments have allowed us to show a continuous turnover of glycogen in this bacterium, suggesting the occurrence of a "futile cycle" of glycogen. Such a futile cycle process can be defined as the simultaneous synthesis and breakdown of storage carbohydrates, in the presence of exogenous soluble carbohydrate (mainly glucose). We have used different approaches to prove the occurrence of this futile cycle: in vivo ^{13}C NMR which gives direct evidence in living microorganisms; ^{1}H NMR performed on acellular extracts or extracellular medium, which quantifies the involvement of these cycles in the isotopic dilution of final metabolites (^{12}C–^{13}C ratio). These different strategies are illustrated in detail as follows:

1.1. Description of F. succinogenes *Metabolism*

In *F. succinogenes*, glucose is metabolized via glucose-6-phosphate through the Embden-Meyerhof-Parnas pathway. Succinate and acetate are subsequently

From: *Methods in Biotechnology: Environmental Microbiology: Methods and Protocols*
Edited by: J. F. T. Spencer and A. L. Ragout de Spencer

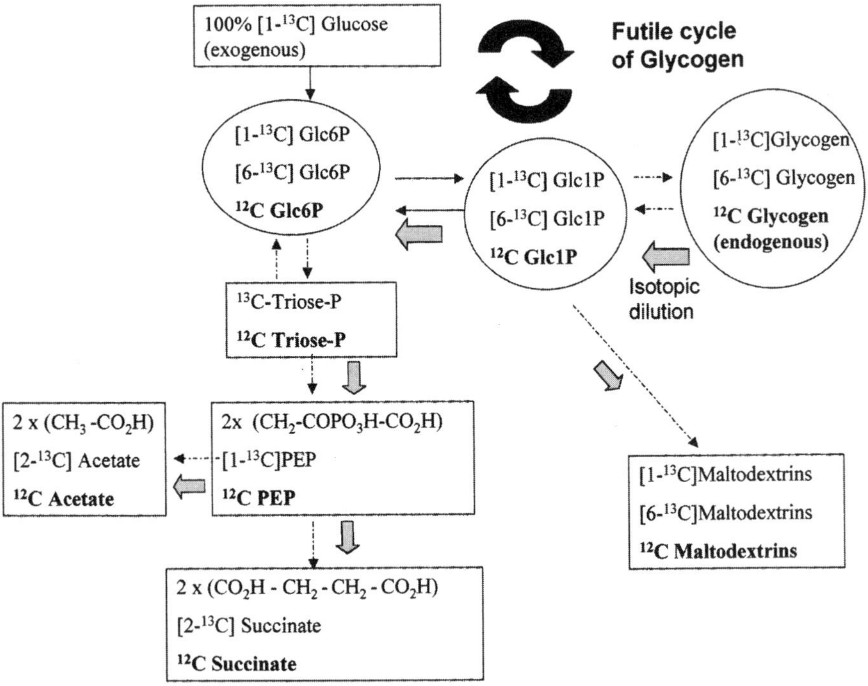

Fig. 1. Pathway of formation and labeling of succinate, acetate, glycogen, and maltodextrins from [1-^{13}C]glucose in *F. succinogenes*. ⟶ step involving one enzyme; ---→ step involving several enzymes; ⇨ isotopic dilution.

synthesized from phosphoenolpyruvate (PEP), as previously described *(3)*. A part of the glucose-6-phosphate (Glc6-P) is converted to glucose-1-phosphate (Glc1-P) through a phosphoglucomutase *(2,5)* giving rise to glycogen and, under some conditions, maltodextrins *(8)*. **Figure 1** presents the expected ^{13}C labeling of the metabolites when the cells are incubated in the presence of [1-^{13}C]glucose.

1.2. In Vivo ^{13}C NMR Experiments

In vivo NMR, a noninvasive method, allows kinetic monitoring of polysaccharide synthesis/breakdown on whole cells. Although the molecular mass of glycogen is very high it can be easily detected by NMR because of its large mobility (T_1 about 0.1–0.2 s) *(9)*. Monitoring glycogen synthesis when cells are incubated with labeled glucose is straightforward; however to visualize simultaneous degradation of glycogen, specific strategies had to be set up to impose different labeling of endogenous glycogen and exogenous glucose. We used "sequential incubations" experiments. Basically, *F. succinogenes* resting

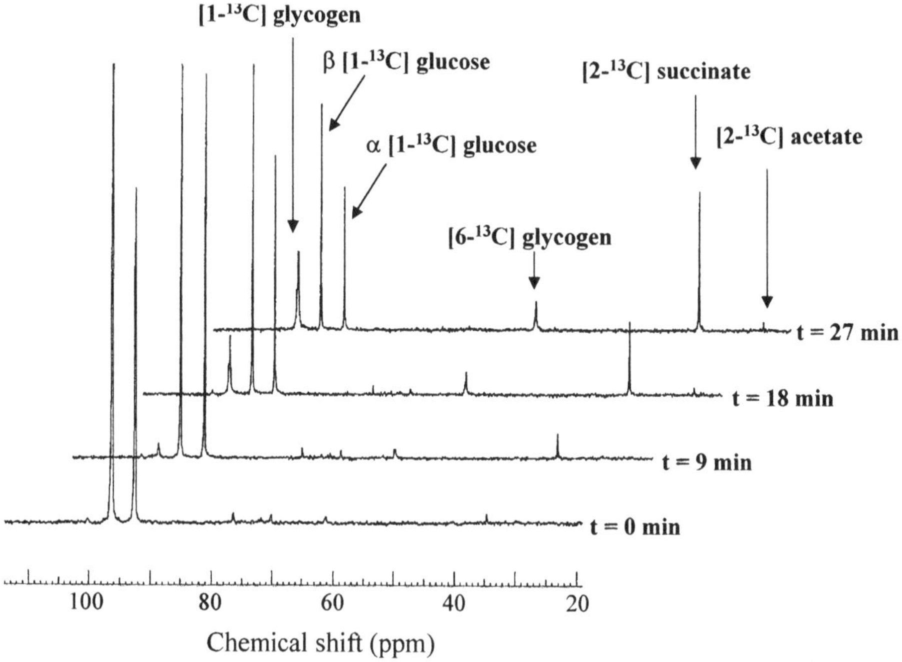

Fig. 2. Spectra of in vivo kinetics of the first incubation of *F. succinogenes* cells with [1-^{13}C]glucose. 32 m*M* [1-^{13}C]glucose is added at zero time and spectra were collected at the time indicated.

cells were first incubated with [1-^{13}C]glucose; under these conditions [1-^{13}C] and [6-^{13}C] glycogen was stored in the cells. **Figure 2** presents ^{13}C NMR spectra registered during this first incubation, which show the synthesis of [1-^{13}C] and [6-^{13}C] glycogen, [2-^{13}C]succinate and [2-^{13}C]acetate from [1-^{13}C]glucose. Then bacteria were washed anaerobically and resuspended in a medium containing [2-^{13}C]glucose and the degradation of labeled glycogen in the presence of [2-^{13}C]glucose was monitored by in vivo ^{13}C NMR. A scheme explaining the sequential incubations is presented in **Fig. 3**. During the second incubation, while the integrals of prestored [1-^{13}C] and [6-^{13}C] glycogen decreased with time, [2-^{13}C] and [5-^{13}C] glycogen integrals increased as a result of *de novo* synthesis ***(1)***. A variation of this strategy was to incubate the cells successively with [1-^{13}C]glucose and unlabeled glucose ***(1,4)***. In this case it was possible to monitor the degradation of prestored [1-^{13}C] and [6-^{13}C] glycogen and the synthesis of [2-^{13}C]succinate and [2-^{13}C]acetate resulting from this degradation. In parallel, a control experiment was carried out: the cells were incubated first with unlabeled glucose, washed, and then incubated with [1-^{13}C]glucose; this

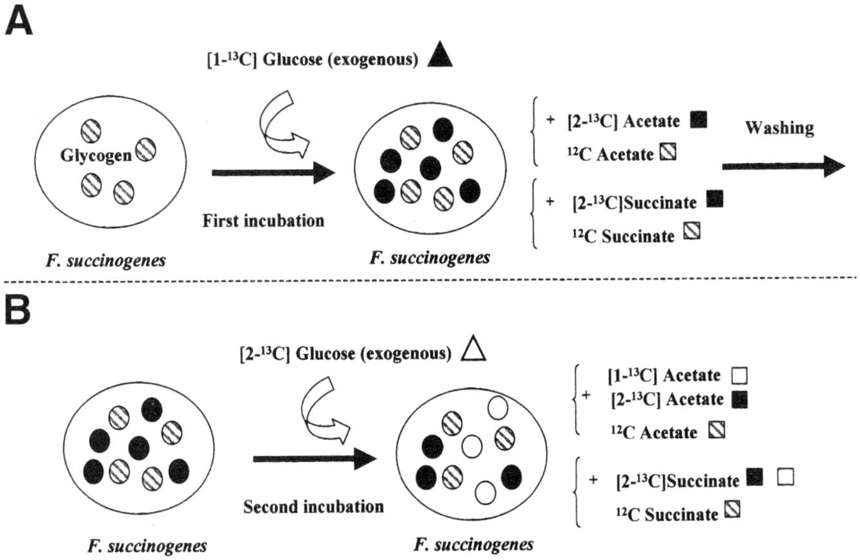

Fig. 3. Evidence of glycogen futile cycling by sequential incubations. **(A)** *First incubation*: Incubation of *F. succinogenes* resting cells with [1-^{13}C]glucose results in the synthesis of [1-^{13}C] and [6-^{13}C] glycogen (black circles) , [2-^{13}C]succinate and [2-^{13}C]acetate. **(B)** *Second incubation*: After washing, the cells are incubated with [2-^{13}C]glucose, which is converted into [2-^{13}C]succinate, [1-^{13}C]acetate, and [2-^{13}C] and [5-^{13}C] glycogen (white circles). Prestored [1-^{13}C] and [6-^{13}C] glycogen (black circles) is degraded to [2-^{13}C]succinate and [2-^{13}C]acetate. In incubations **A** and **B**, endogenous ^{12}C-glycogen (hatched circles) is converted to ^{12}C-succinate and ^{12}C-acetate.

experiment showed that *F. succinogenes* was still able to store glycogen during the second incubation.

By using these "sequential incubation" experiments, direct evidence of futile cycling of glycogen was found for different strains of *Fibrobacter* ***(4)***, indicating that this phenomenon was a common feature to this genus, even though some of the *Fibrobacter* strains tested were not phylogenetically close. However, this approach presents some drawbacks: (1) bacteria can be affected by repeated washings; indeed, a slower metabolism was observed for three strains of *Fibrobacter* (S85, 95, NR9), and one of them (HM2) did not tolerate this treatment at all ***(4)***. (2) Only qualitative observations can be made or, at best, a roughly quantitative estimation.

To obtain more precise data, indirect analyses were carried out by ^{1}H NMR experiments.

1.3. In Situ *^1H NMR Experiments*

When prestored endogenous ^{12}C-glycogen is degraded in parallel to exogenous [1-^{13}C]glucose, ^{12}C-Glc1-P and ^{12}C-Glc6P are generated simultaneously with enriched ^{13}C-Glc1-P and ^{13}C-Glc6-P. These unlabeled substrates ultimately induce an isotopic dilution in the final metabolites such as acetate, succinate, or oligosaccharides such as maltodextins (**Fig. 1**). To assess the isotopic dilution induced by the futile cycle mechanism, ^1H NMR spectrometry can be used to quantify ^{13}C–^{12}C ratios. Indeed, ^{13}C- directly linked proton signals are split with a one bond coupling constant $^1J_{13C\text{-}1H}$ giving rise to ^{13}C satellites that differ from those of ^{12}C-linked protons. This is illustrated by **Fig. 4**.

1.3.1. Analysis of ^{13}C Enrichment of Succinate by a 1D ^{13}C-Filtered Spin Echo Difference (XFSED) Sequence

From **Fig. 1**, it can be deduced that one [1-^{13}C]glucose unit is cleaved into two triose-phosphates that produce two molecules of succinate, only one of which half is labeled. In addition, succinate is a symmetrical molecule with two CH_2, thus only 25% of C2 succinate atoms are ^{13}C-labeled. Hence if ^{12}C-Glc6-P resulting from the degradation of endogenous ^{12}C-glycogen enters the glycolytic pathway, the maximum percent of labeling of [2-^{13}C]succinate should be inferior to 25% when exogenous glucose is 100% labeled. This implies that the expected isotopic dilution is low, about a few percent. It was thus necessary to develop a specific NMR sequence in order to measure very precisely the percent enrichment of the metabolites. By using a ^{13}C-Filtered Spin Echo Difference (XFSED) sequence *(4)* the spectra of exclusively ^{12}C-linked or ^{13}C-linked protons can be specifically edited (**Fig. 4**). This technique allowed us to measure very precisely (1% error) the ^{13}C enrichment of C2 atoms of succinate.

When various strains of *Fibrobacter* (S85, 095, NR9, HM2) were incubated in the presence of [1-^{13}C]glucose without ammonia, the percent of labeling of C2-succinate was about 21%, thus the labeling deficit corresponding to the degradation of endogenous glycogen was about 4% *(4)*. This means that 16% of ^{12}C-glucose units were provided by unlabeled glycogen. Although other metabolic fluxes such as glucose consumption, acetate or glycogen synthesis were modified in parallel, the contribution of endogenous glycogen to glycolysis through its degradation was shown to be fairly constant under various conditions, for instance when resting cells were incubated with [1-^{13}C]glucose in the presence of NH_4^+ *(6)* or when bacterial cellulose was added as substrate *(7)*.

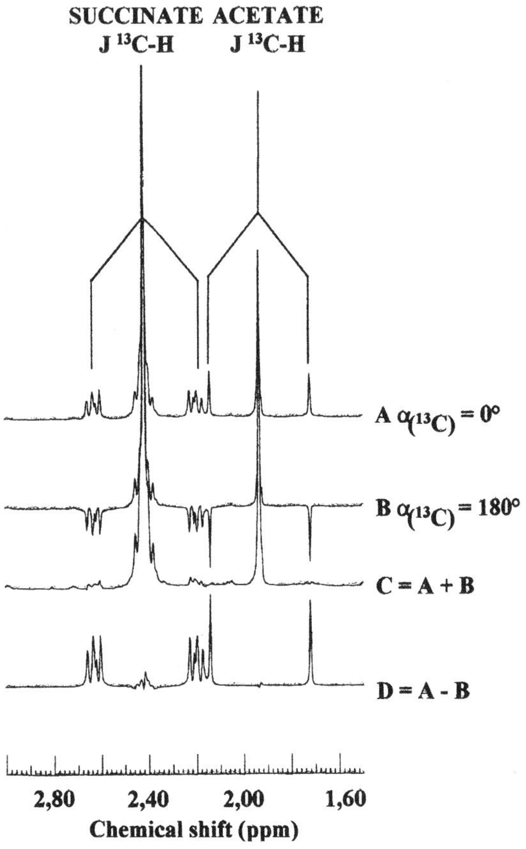

Fig. 4. ^{1}H NMR spectra acquired with ^{13}C-Filtered Spin-Echo pulse sequence. Spectra **A** and **B** represent the output of the experiment with and without 180° ^{13}C-pulse, respectively, in the middle of the Spin Echo period. Spectra **C** and **D** were obtained after addition or subtraction of the spectra **A** and **B**, respectively. Spectrum **C** corresponds to ^{12}C-linked and spectrum **D** to ^{13}C-linked protons.

1.3.2. Analysis of ^{13}C Enrichment of Maltodextrins by a 2D ^{1}H Double Quantum Filtered Correlation Spectroscopy (DQF-COSY) Sequence

Very recently, we showed the excretion of short linear maltodextrins in the extracellular medium by resting cells of *F. succinogenes* S85 incubated with [1-^{13}C]glucose and NH_4^+ *(8)*. The analysis of ^{13}C enrichment of maltodextrins was not possible using a 1D NMR experiment since many signals of glucose derivatives overlap in the same region (*see* **Fig. 5**, top trace spectrum). Homocorrelated DQF COSY spectra, which splits the information into two spectral dimensions, allowed distinction between the H2 chemical shifts of different

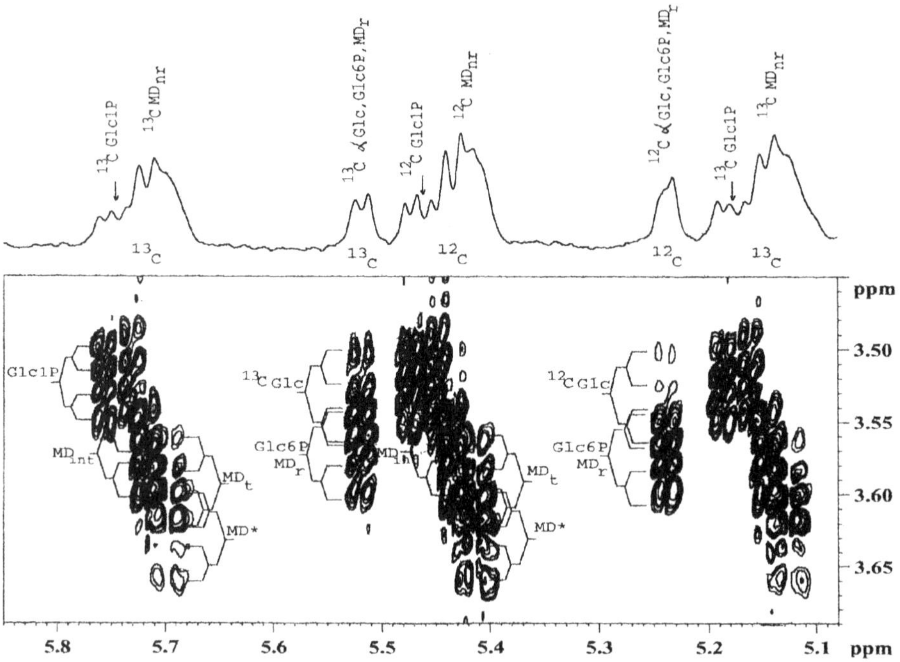

Fig. 5. Region of α anomeric signals of the DQF COSY spectra of extracellular medium obtained after incubation of *F. succinogenes* with [1-^{13}C]glucose after 18 min. Top trace represent the ^{1}H spectrum.

Glc, glucose; Glc1P, glucose-1-phosphate; Glc6P, glucose-6-phosphate; MD_rGlcα, maltodextrins reducing end Glcα unit; MDn_r Glcα,ß, maltodextrins nonreducing end Glcα,ß unit.

metabolites, including their ^{13}C satellites **(Fig. 5)**. The H1/H2 cross-peaks of three principal unlabeled ^{12}C-metabolites could be distinguished as well as that of α glucose (αGlc). The crosspeak at δ 5.459/3.517 ppm was assigned to glucose1-phosphate (Glc1-P) and those at δ 5.435–5.414/3.568–3.628 ppm to the nonreducing glucose units of maltodextrins (nr MD).

The presence of these ^{13}C-satellites is the consequence of metabolism of the exogenous [1-^{13}C]glucose, whereas the huge crosspeaks owing to ^{12}C-linked protons are because of the degradation of glycogen. This proves the involvement of the futile cycle of glycogen in this bacterium.

1.4. Concluding Remarks

These different NMR-based approaches, used by our group to evidence and quantify the continuous glycogen flux in the rumen bacterium *F. succinogenes,* could be applied to other bacterial species storing glycogen or synthesizing other types of polysaccharide in or out the cell. Indeed, polysaccharide storage has been shown to have an important role in several species from the environment: intracellular polysaccharides such as glycogen or cyclic β-glucans may play specific roles as material reserves ***(10)*** or osmoregulators ***(11)***, respectively, while exopolysaccharides, e.g., cellulose or succinoglycans, may be involved in different functions like cell protection ***(12)*** or the establishment of symbiosis ***(11)***, respectively. For example, the glycogen futile cycle was recently suggested, but not quantified, in the cellulolytic soil species *Clostridium cellulolyticum* ***(13,14)***. In addition, the use of ^{13}C-NMR to study the sugar metabolism of another soil bacteria, *Sinorhizobium meliloti*, allowed quantification of the specific enrichment of both intracellular and exocellular polysaccharides, suggesting a cyclic organization of the carbohydrate metabolism in this species ***(11)***.

2. Materials

2.1. Microbiology

F. succinogenes is a strictly anaerobic bacterium, thus all the techniques, media and buffers used for cultivation, harvesting, washing or incubating the cells must avoid any contact of the cells with oxygen. For this purpose, the methodology used is mainly based on the strictly anaerobic techniques first described by Hungate ***(15)***.

1. The strain of *F. succinogenes* used is the strain S85 (ATCC, 19169), the type strain of this species isolated from a bovine rumen.
2. Culture media: The Hungate anaerobic technique ***(15)*** is used for the preparation of medium. The culture medium is the mineral medium of Bryant and Burkey ***(16)***: 0.45 g/L K_2HPO_4, 0.45 g/L KH_2PO_4, 0.9 g/L $(NH_4)_2SO_4$, 0.9 g/L NaCl, 0.09 g/L $MgSO_4$, 0.09 g/L $CaCl_2$ with the following additions:
   a. Volatile fatty acids (m*M*, final concentration) 29 acetic acid, 7.9 propionic acid, 4.3 *n*-butyric acid, 1.3 isovaleric acid, 0.9 DL-α-methylbutyric acid, 1.0 isobutyric acid, 1.3 *n*-valeric acid.
   b. Vitamins: 0.1 mg/mL *p*-aminobenzoic acid; 0.05 mg/mL biotine.
   c. 1 mg/L Hemin.
   d. 0.5 g/L Cysteine/HCl, as a reducing agent.
   e. 1 mg/L Resazurin, the redox potential indicator.
   f. 3 g/L Cellobiose as carbohydrate.

The culture medium (usually 3 L) is prepared in round-bottom boiling flasks. The medium is prepared without cysteine or Na_2CO_3, and is autoclaved in the flasks. After autoclaving (120°C, 20 min), the flasks are opened at the flame (sterile conditions) and placed under a stream of 100% oxygen-free CO_2 directly flushed into the container in order to maintain conditions as anaerobic as possible. A 3-inch no. 19-gauge hypodermic needle bent so that about 2 inches extend down into the container is used for introducing the gas. The needle is connected to a 0.2-µm filtration unit (type Versapor ref. 6254185, GelmanSciences) that is connected on the other side to the tube delivering the gas. Na_2CO_3 and cysteine are then added successively to the medium which is then left to cool down under the gas stream until the pink resazurin reduces to a colorless compound. The flasks are then sealed with sterile butyl rubber stoppers to maintain anaerobiosis. The final pH of the medium is between 6.5 and 6.7.

When 10 mL medium tubes are needed, the medium is first prepared as explained previously without cysteine and Na_2CO_3. It is then degassed by boiling at 100°C, placed under a stream of 100% oxygen-free CO_2, and Na_2CO_3 and cysteine are added successively. After reduction of resazurin, the medium is dispensed into culture tubes ("Hungate" type anaerobic culture tubes, complete—including stoppers—16 × 125 mm, ref. 204716125, Bellco) that are also flushed by 100% oxygen-free CO_2 (before, during and after filling). The tubes are then sealed by their specific stoppers, then autoclaved.

3. Incubation buffer: The buffer used for washing or incubating the cells is an anaerobic phosphate buffer: 0.1% of resazurin was added to a solution of K_2HPO_4/KH_2PO_4 at 50 m*M*, pH 7.1 (*see* **Note 1**). The solution (in a bottle) was then degassed by boiling at 100°C, placed under a stream of 100% oxygen-free CO_2, and Na_2CO_3 (0.4%) and cysteine (0.05%) were added successively. The solution was then cooled under the gas stream until reduction of resazurin, and the bottle was sealed by a butyl rubber stopper.

2.2. NMR Spectroscopy

1. NMR spectrometer: NMR spectra are recorded on a Bruker Avance DSX300 spectrometer operating at 300.13 MHz for ^1H and 75.46 MHz for ^{13}C. Temperature is regulated at 38°C. XWIN NMR Bruker software is used for signal treatments and signal integrations. The ^2H resonance of D_2O is used to lock the field and for shimming.
2. NMR probes: A 10-mm broad band probe is used for in vivo NMR, whereas a 5 mm TXI ^1H,^{13}C,^{15}N is used for *in situ* (supernatant) experiments.
3. Chemicals: [1-^{13}C]glucose, [2-^{13}C]glucose (99% labeled), D_2O and TSP-*d4* are purchased from Eurisotop (France). Benzene and ^{12}C-glucose are from Sigma-Aldrich (France).
4. NMR tubes are from Wilmad Glass Company: 10-mm diameter tubes (WG 10M) are used for in vivo NMR, 5-mm tubes (WG 5M) are for *in situ* NMR.
5. Reference capillary: (*see* **Fig. 6**): A 1-mm diameter capillary is filled with pure benzene using a 0.5-mL syringe up to half length. The capillary is sealed under a flame (*see* **Note 2**) and a silicon tube is attached to this capillary simply by melting it to the capillary glass. This silicon tube is very useful for handling the capillary during in vivo NMR experiments. To keep it centered in the 10-mm tube, it is inserted in a cap previously bored to the right diameter. After each use, this capillary can be rinsed carefully with water and stored for many months in the fridge.

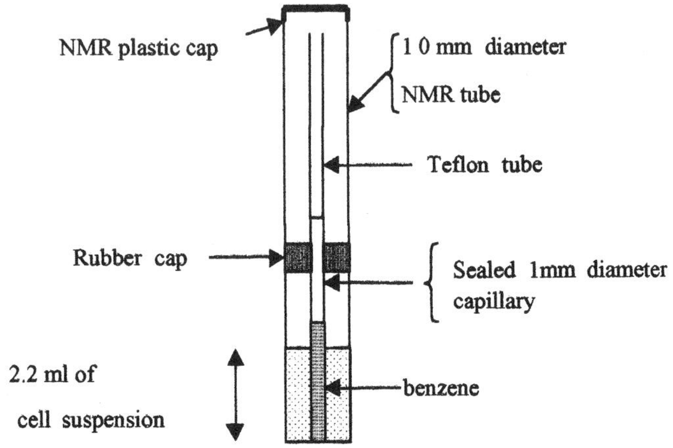

Fig. 6. ^{13}C in vivo NMR experiments: 10-mm diameter NMR tube containing a 1-mm diameter sealed capillary filled with pure benzene. Bacterial cells are under CO_2 atmosphere.

6. Pipets: 5 Gilson automatic pipets are used: one for pure bacterial cells (1 mL), for anaerobic buffer (1 mL), one for D_2O (200 µL), for glucose (100 µL), and one for the whole sample (2 mL). To introduce the various samples at the bottom of the NMR tube or to take the samples out of the NMR tube easily the tips used for the pipets are equipped with PTFE tubing (PolyLabo-Merck-Eurolab, 0.9-mm diameter, ref. 92404) of about 20 cm length (*see* **Note 3**).

3. Methods

3.1. F. succinogenes *Culture and Preparation of the Cells for NMR Experiments*

1. The strain was stored at –80°C by freezing directly a 15-h culture in Hungate tubes.
2. The bacteria were cultivated for 15 h in the anaerobic medium in Hungate tubes. The Hungate anaerobic technique was used *(15)*. A 10% inoculum was introduced by a pipet or hypodermic syringe into a culture tube under anaerobic (stream of CO_2 while the tube is open) and sterile (flame) conditions. The tubes are then incubated at 39°C in an incubator for 15 h. For large cultures, one to three 15-h cultures were inoculated under sterile and anaerobic conditions, as described previously, into a flask containing 800 mL medium. The flask was then sealed by a butyl rubber stopper and incubated for 15 h at 39°C in a waterbath.
3. At the end of the culture, the cells were harvested by transfer under anaerobic conditions to 250 mL polycarbonate centrifugation bottles (Beckman 62X122, ref. 356013) fitted with a hermetic screw cap. The cells were then centrifuged at 4500*g* for 15 min, at 4°C. The cells were washed twice in an anaerobic phosphate buffer (*see* above) and finally resuspended at about 5–10 mg protein/mL.

3.2. Sequential Incubations and ^{13}C *In Vivo NMR Experiments*

As previously stated, all the manipulations are carried out under a CO_2 stream. However, when bacteria are placed in the NMR tube sealed with a rubber stopper (*see* **Fig. 6**), anaerobiosis is maintained during the NMR experiments without any further CO_2 addition (*see* **Note 4**). Because glucose degradation is rather quick (about 45 min), the NMR apparatus must be ready; preliminary settings must be therefore made in advance, such as fixing the temperature at 38°C, shimming using a test tube and preparing NMR data files and parameters.

1. *Step 1:* Bacterial cells (the volume depends on the protein concentration of the cell suspension at the end of the washings) are transferred in a 10-mm tube under CO_2 flux, 200 µL of D_2O (10% for shimming) are added as well as an adequate volume of anaerobic buffer to reach a total volume of 2.2 mL. The final protein concentration is 5 mg/mL (*see* **Note 5**). The whole sample is mixed by pumping in and out with the pipet. The benzene capillary is then centered in the tube (*see* **Fig. 6**) and the tube is closed and quickly inserted in the NMR magnet using an air lift system. A first ^{13}C NMR spectrum is collected after shimming of the tube with the following conditions: protons are decoupled with Waltz-16 program, 360 scans are collected (acquisition time 0.256 s, relaxation time 0.5 s, 8K data points) with a pulse length of 45°. The total time of the experiment is 4.5 mn. This first spectrum constitute a reference spectrum (*see* **Note 6**), before the addition of [1-^{13}C]glucose, and it generally contains only the resonance (*see* **Note 7**) of benzene at δ 128.6 ppm which is used as an external reference for chemical shift measurements and for normalization of the values of metabolite integrals.
2. *Step 2:* The NMR tube is taken out of the magnet by the air lift system and 35 µL of 2 *M* glucose is transferred as quickly as possible into the tube under CO_2 flux. The whole sample is mixed by pumping in and out with the pipet. Then the tube is transferred very quickly back in the magnet, quickly shimmed with Z1 and Z2 coils and the monitoring of the kinetics by ^{13}C in vivo NMR is started (*see* **Note 9**). Ten spectra are collected under the same conditions as **step 1**. The time course of [1-^{13}C]glucose consumption and [2-^{13}C]acetate, [2-^{13}C]succinate and [1-^{13}C], and [6-^{13}C]glycogen formation can be followed in real time on the computer screen (*see* **Fig. 3**) so that the progress of the experiment can be checked. The experiment is stopped when [1-^{13}C]glucose is exhausted (usually 45 min). FID are Fourier transformed after application of a line broadening of 3Hz, relative integrals are measured compared to the benzene capillary area.
3. *Step 3:* The third step consists of washing the cells with the phosphate anaerobic buffer containing 32 m*M* [2-^{13}C]glucose to avoid a slight degradation of intracellular glycogen, which can happen even at 4°C. Bacterial cells are centrifuged for 15 min at 5000*g* in 10 mL Oak-ridge tubes (Nalgene ref. 3118-0010) fitted with adequate screw caps (Nalgene, 3131-0013) and resuspended in the anaerobic buffer to obtain the same cell concentration as in the first incubation.

4. The washed cells are incubated with 32 m*M* [2-^{13}C]glucose, and the kinetic of glucose and [1-^{13}C] and [6-^{13}C] glycogen degradation as well as of formation of [1-^{13}C] and [2-^{13}C] acetate, [2-^{13}C]succinate, [2-^{13}C] and [5-^{13}C] glycogen (*see* **Fig. 2**) are monitored by in vivo ^{13}C NMR as previously described in **step 2** (*see* **Note 9**).

3.3. In Situ ^{1}H NMR Experiments

1. The first step consists of incubating resting cells of *F. succinogenes* with 32 m*M* [1-^{13}C]glucose as described in **Subheading 3.2.**, **steps 1** and **2** (*see* **Note 10**) except that twice the volume was used (4.4 mL). After **step 1**, 2.2 mL is taken out to be used as a reference to check how much ^{12}C succinate was already present at zero time (resulting from the degradation of intracellular glycogen of the cells stored in ice).
2. The cell suspensions, sampled at time zero and after incubation with glucose, are centrifuged (13,000*g*, 10 min, 4°C), and the supernatants are kept as extracellular media. 400 µL of cell-free supernatants are mixed with 100 µL of 25 m*M* solution of deuterated 3-trimethylsilylpropionate sodium salt (TSP-*d4*) in D_2O and transferred to a 5-mm diameter NMR tube. To measure the percentage enrichment ($^{13}C/^{12}C$), ^{1}H NMR spectra are acquired using internal TSP-*d4* (δ 0.0) as a reference standard for chemical shifts in a 5-mm inverse probe ($^1H/^{13}C/^{15}N$) with the ^{13}C-Filtered Spin-Echo Difference (^{13}C FSED) pulse sequence presented below:

$$\text{Preparation}–(90°)_H\text{-}\tau/2–(180°)_H/(\alpha)_X–\tau/2–(90°)_X–\text{FID}(^1H)$$

where the subscripts denote the nucleus experiencing the pulse (H-proton; X-^{13}C). The last carbon pulse is a purging pulse used to remove all spurious signals originating from pulse imperfections. Spectra with $\alpha = 0°$ or 180° are acquired in subsequent scans and are stored independently in two blocks of memory. Extensive phase cycling is used to compensate quadrature detection artifacts (CYCLOPS) *(17)* and 180° pulse imperfections (EXORCYCLE) *(18)*. In a preparation period, the solvent resonance was presaturated by irradiation for 3.5 s at 60 dB. The evolution interval τ was adjusted according to the one-bond C, H coupling constant for succinate ($\tau = 1/^1J(C,H) = 7.8$ ms). After 8 dummy scans, 128 scans were accumulated into 8K of memory with an acquisition time of 3.5 s, a spectral width of 4000 Hz and a relaxation delay of 3.0 s. This repetition time is not sufficient for full relaxation of all protons, so for quantitative analysis, a calibration measurement is made with an additional 10 s relaxation delay. Both time domain spectra ($\alpha = 0°$, $\alpha = 180°$) are proceeded under identical conditions (same window function, same phase correction). After Fourier transformation the subspectra are edited. The sum of the two spectra yields the subspectrum of protons directly bonded to ^{12}C nuclei or another heteronucleus (O or N), and their difference gives the subspectrum of protons directly bonded to ^{13}C nuclei (*see* **Fig. 4**). Before quantitative analysis, both spectra are corrected for the presence of succinate before the incubation with glucose (*see* **step 1**). The zero incubation time sample also containing TSP-*d4* is used for quantitative comparison of the integrals. After the zero incubation time correction, relative integral intensities are determined for all the relevant signals in both spectra by using a standard integration procedure. Normalized intensities (TSP-*d4* = 100) are then corrected for the differences in T1 relaxation times by using correction factors determined from the calibration experiment. In addition, ^{13}C integrals are corrected for the natural abundance contribution (1.1%). Isotopic enrichment (*e*) at the C2 position of succinate is determined from the corrected integral intensities [I (^{13}C) and I(^{12}C)] of CH_2 signals in ^{13}C and ^{12}C subspectra with the following equation:

$$e = I(^{13}C)/[I(^{13}C) + I(^{12}C)]$$

The systematic error of the spin-echo method for the determination of the CH_2 groups is less than 1% and negative (i.e, the spin-echo slightly underestimates the enrichment of succinate.

3. To analyze the labeling of maltodextrins in the supernatant obtained after incubation with glucose (*see* **step 1**), the pH of the solutions is corrected to 7.40 by adding a few microliters of NaOH (0.1 *M*), the solution is freeze-dried, and the residue dissolved in 500 μL 99.98% D_2O and transferred to a 5-mm diameter NMR tube (*see* **Note 11**). The 2D NMR COSY-DQF experiment is performed with the homo-nuclear double quantum filtered DQF COSY sequence published by Davies et al. ***(19)*** with pulse field gradient, using a 5-mm inverse probe ($^{1}H/^{13}C/^{15}N$). The NMR parameters are the following: spectral width 2.8 ppm, data points 1024 and 256 for F2 and F1 dimensions respectively, acquisition time 0.61 s, relaxation delay 1.5 s, number of scans 48, 90° pulse 8 μs at 6 dB, delay for homospoil/gradient recovery 100 μs, homospoil/gradient pulse 1 ms, GPZ1:GPZ2 = 1:2. Sine window function was applied in both dimension before 2D Fourier transformation.

4. Notes

1. The incubation buffer has to be adapted to the microbial species studied: it has to have about the same osmotic pressure as the culture medium, and a pH value and other physiochemical conditions adapted to the physiological conditions of the cells.
2. Working under a fume hood is recommended as benzene is volatile and highly carcinogenic. Melted glass must be manipulated quickly with a grip, taking care of bubble formation owing to the evaporation of benzene and thus carefully checking that the capillary is perfectly sealed.
3. The number of pipets and the tips equipment are key points for correct and rapid preparation the samples.
4. Anaerobiosis is maintained because the rubber cap is highly impermeable to oxygen. In addition, the cell density is high and *F. succinogenes* metabolism also generates CO_2 continuously in the tube.
5. Some slight metabolic variations can be observed from one culture to the other, mainly because glycogen content can be variable. Therefore, to compare various

kinetics corresponding to different bacterial cultures we do not consider strict protein concentrations but we use bacterial concentrations corresponding to the same kinetics of glucose (32 m*M*) degradation (complete degradation after 45 min). To do so, a first incubation is performed with a bacterial suspension at 5 mg/mL protein, and according to this first result, the bacterial concentration is further adjusted (dilution or concentration). Note, however, that the final bacterial concentration should be around 4–6 mg/mL. If a larger concentration is needed it means that bacteria are not in a correct metabolic state and experiments are not continued.

6. This first experiment also allows to shim correctly on the bacterial sample and to adjust the temperature at 38°C (it was previously stored in ice). Therefore the setting up will be extremely brief in the second step when kinetics are monitored (no time is lost for these adjustments).
7. In some cases, when intracellular glycogen concentration is high, natural abundance signals of glycogen can be detected ($\delta = 61.0$ to 100.1 ppm).
8. The time necessary to carry out all these operations is 30 s; however, no time gap is observed on the kinetics as these 30 s compensate more or less the temperature decrease of the sample when it is out of the magnet (from 38°C to room temperature).
9. After washing, a slight decrease of the metabolism rate is usually observed; in some cases, specific strains cannot cope with this "sequential incubation protocol" and thus cannot be used.
10. ^{13}C NMR is very useful to check the metabolic state of the cells and stop the incubation precisely at the end of glucose consumption; however, the same experiment can be carried out on the bench in a water bath and the experiment can be followed by assaying glucose or succinate by enzymatic methods.
11. This freeze-drying step is important for removal of H_2O, otherwise a huge signal will overlap the interesting anomeric ^1H signals of maltodextrins. Note that this operation will modify the concentration of acetate in the medium, so that any assay of acetate should be done, if necessary, before freeze drying. On the contrary, succinate concentration is not changed.

References

1. Gaudet, G., Forano, E., Dauphin, G., and Delort, A.-M. (1992) Futile cycling of glycogen in *Fibrobacter succinogenes* as shown by *in situ* ^{13}C- and ^1H-NMR investigation. *Eur. J. Biochem.* **207,** 155–162.
2. Matheron, C., Delort, A.-M., Gaudet, G., and Forano, E. (1996) Simultaneous but differential metabolism of glucose and cellobiose in cells, studied by *in vivo* ^{13}C-NMR. Evidence of glucose 6-phosphate accumulation. *Can. J. Microbiol.* **42,** 1091–1099.
3. Matheron, C., Delort, A.-M., Gaudet, G., and Forano, E. (1997) Re-investigation of glucose metabolism in *Fibrobacter succinogenes* using NMR and enzymatic

assays. Evidence of pentose phosphate phosphoketolase and pyruvate formate lyase activity. *Biochim. Biophys. Acta* **1335,** 50–60.

4. Matheron, C., Delort, A.-M., Gaudet, G., Forano, E., and Liptaj, T. (1998a) ^{13}C- and ^{1}H-NMR study of glycogen futile cycling in strains of the genus *Fibrobacter. Appl. Environ. Microbiol.* **64,** 74–81.
5. Matheron, C., Delort, A.-M., Gaudet, G., and Forano, E. (1998b) *In vivo* ^{13}C NMR study of glucose and cellobiose metabolism by four cellulolytic strains of the genus *Fibrobacter. Biodegradation* **9,** 451–461.
6. Matheron, C., Delort, A.-M., Gaudet, G., Liptaj, T., and Forano, E. (1999) Interaction between carbon and nitrogen metabolism in *Fibrobacter succinogenes S85*: a ^{13}C- and ^{1}H nuclear magnetic resonance and enzymatic study. *Appl. Environ. Microbiol.* **65,** 1941–1948.
7. Bibollet, X., Bosc, N., Matulova, M., Delort, A.-M., Gaudet, G., and Forano, E. (2000) ^{13}C- and ^{1}H-NMR study of cellulose metabolism by *Fibrobacter succinogenes* S85. *J. Biotechnol.* **77,** 37–47.
8. Matulova, M., Delort, A.-M., Nouaille, R., Gaudet, G., and Forano, E. (2001) Concurrent maltodextrin and cellodextrin synthesis by *Fibrobacter succinogenes* S85 as identified by 2D NMR spectroscopy. *Eur. J. Biochem.* **268,** 3907–3915.
9. Sillerud, L. O. and Shulman, R. G. (1983) Structure and metabolism of mammalian liver glycogen monitored by carbon-13 nuclear magnetic resonance. *Biochemistry* **22,** 1087–1094.
10. Preiss, J. (1984) Bacterial glycogen synthesis and its regulation. *Annu. Rev. Microbiol.* **38,** 419–458.
11. Portais, J.-C., Tavernier, P., Gosselin, I., and Barbotin, J.-N. (1999) Cyclic organization of the carbohydrate metabolism in *Sinorhizobium meliloti. Eur. J. Biochem.* **265,** 473–480.
12. Ross, P., Mayer, R., and Benziman, M. (1991) Cellulose biosynthesis and function in bacteria. *Microbiol. Rev.* **55,** 35–58.
13. Guedon E., Desvaux M., and Petitdemange H. (2000) Kinetic analysis of *Clostridium cellulolyticum* carbohydrate metabolism: importance of glucose 1-phosphate and glucose 6-phosphate branch points for distribution of carbon fluxes inside and outside cells as revealed by steady-state continuous culture. *J. Bacteriol.* **182,** 2010–2017.
14. Desvaux, M., Guedon, E., and Petitdemange, H. (2001) Kinetic and metabolism of cellulose degradation at high substrate concentrations in steady-state continuous culture of *Clostridium cellulolyticum* on a chemically defined medium. *Appl. Environ. Microbiol.* **67,** 3837–3845.
15. Hungate, R. E. (1966) The rumen bacteria, in *The Rumen and Its Microbes*, Academic Press, New York and London, pp. 8–81.
16. Bryant, M. P. and Burkey, L. A. (1953) Cultural methods and some characteristics of the more numerous groups of bacteria in the bovine rumen. *J. Dairy Sci.* **36,** 205–217.

17. Hoult, D. I. and Richards, R. E. (1975) Critical factors in the design of sensitive high resolution magnetic field resonance spectrometers. *Proc. R. Soc. London A* **344,** 311–340.
18. Morris, G. A. and Freeman, R. (1978) Selective excitation in Fourier transform nuclear magnetic resonance. *J. Magn. Res.* **29,** 433–462.
19. Davies, L. A., Laue, E. A., Keeler, J., Moskau, D., and Lohman, J. (1991) Absorption mode 2D NMR spectra recorded using pulse field gradients *J. Magn. Res.* **94,** 637–644.

19

Chemical Analysis and Biological Removal of Wood Lipids Forming Pitch Deposits in Paper Pulp Manufacturing

Ana Gutiérrez, José C. del Río, and Ángel T. Martínez

1. Introduction

Wood extractives cause production and environmental problems in pulp and paper manufacturing. The lipophilic fraction is the most problematic, and it includes free fatty acids, resin acids, waxes, fatty alcohols, sterols, sterol esters, glycerides, and other oxidized compounds *(1–5)*. During wood pulping and refining of paper pulp, the lipophilic extractives in the parenchyma cells and softwood resin canals are released, forming colloidal pitch *(6)*. These colloidal particles can coalesce into larger droplets that deposit in pulp or equipment, forming the so-called pitch deposits, or remain suspended in the process waters. Pitch deposition results in low-quality pulp, and can cause the shutdown of mill operations *(7)*. Economic losses associated with pitch problems in kraft mills often amount to 1% of sales. In addition, some wood extractives could have a detrimental environmental impact when released into wastewaters. This is especially important in modern environmentally-sound pulp manufacturing processes, where chlorine bleaching has been replaced by elemental chlorine-free (ECF) or totally chlorine-free (TCF) bleaching *(8,9)*.

Traditionally, pitch deposition has been reduced by debarking and seasoning logs and wood chips, and by adding physicochemical control agents *(10–13)*. However, the cost is high and often the results are far from satisfactory. As alternatives to the above, biological removal of wood extractives by treatment with enzymes *(14–16)* or microorganisms *(17,18)* has been suggested in recent years for pitch control. The screening for the most adequate organisms to

From: *Methods in Biotechnology: Environmental Microbiology: Methods and Protocols*
Edited by: J. F. T. Spencer and A. L. Ragout de Spencer

degrade lipophilic extractives responsible for pitch deposition during pulping of different types of wood requires the use of simple and sensitive analytical methodologies. The broad range of molecular masses of lipophilic extractives and their structural diversity represent two important difficulties for their chemical analysis. In the present chapter, we describe an optimized methodology for the chromatographic separation and chemical identification of complex mixtures of wood lipids, without previous fractionation. Gas chromatography (GC) and gas chromatography–mass spectrometry (GC-MS) are employed, using short- and medium-length high-temperature capillary columns (with thin films), respectively, that enable elution and separation of high-molecular-weight lipids ***(19–24)***. This methodology was successfully applied for the study of the biological removal of free and esterified wood sterols by selected fungal strains ***(25,26)***. The results show that some of the new fungi efficiently degrade steroids involved in pitch deposit formation in chlorine-free pulps, which are not removed by the other microbial preparations currently available. In this chapter, the GC-MS profiles of lipophilic extractives from eucalypt wood after solid-state fermentation (SSF) treatment with four selected fungi is used to illustrate this methodology.

For more detailed characterization, a simple solid-phase extraction (SPE) method using aminopropyl phase cartridges is described. This enables fractionation of the complex mixture of lipids isolated from woods and pitch deposits into major lipid classes, which can be subsequently characterized and quantified by GC and GC-MS. The advantages of SPE over traditional sample preparation methods include increased speed and simplicity, reduced solvent usage, and improved selectivity ***(27)***. The method outlined in this chapter is based on that previously developed by Kaluzny et al. ***(28)*** for separation of animal lipids. However, a different fractionation scheme with a reduced number of steps and cartridges, as well as a different solvent system, has been used for the fractionation of wood lipids.

2. Materials

1. GC equipment: Hewlett-Packard HP 5890 gas chromatograph equipped with a split-splitless injector and a flame ionization detection (FID) system (Hewlett-Packard, Hoofddorp, Netherlands).
2. GC-MS equipment: Varian Star 3400 gas chromatograph (Varian, Walnut Creek, CA) with an ion-trap detector (Varian Saturn 2000).
3. High-temperature capillary column (DB-5HT, 0.25 mm id, 0.1 μm film thickness) from J&W Scientific (Folsom, CA) especially processed for an extended temperature of 400°C (15 m for GC-MS analysis and 5 m for GC-FID analysis).
4. Accessories used for GC injector: Thermogreen LB-2 with extremely low bleeding (from 100°C to 350°C) and Therm-O-Ring™ seals processed for using at temperatures up to 375°C without being degraded (Supelco, Bellefonte, PA).

5. Standard compounds (palmitic acid, dehydroabietic acid, sitosterol, cholesteryl oleate, and triheptadecanoin).
6. Aminopropyl phase cartridges (500 mg) from Waters (Division of Millipore, Milford, MA).
7. Solvents: acetone, hexane, chloroform, diethyl ether, and acetic acid.
8. Wood chips from *Eucalyptus globulus* (e.g., 2–4 × 20–40 mm).
9. Fungal strains: *Bjerkandera adusta* CBS 230.93, *Phlebia radiata* CBS 184.83, *Pleurotus pulmonarius* CBS 507.85, and *Ceriporiopsis subvermispora* CBS 347.63 conserved at the Centraalbureau voor Schimmelcultures fungal culture collection (Utrecht, The Netherlands).
10. SSF equipment: Horizontal rotary fermentor including six 2-L bottles with a capacity for 350 g of wood chips (28 cm rotating diameter), which were individually sterilized at 120°C for 30 min and inoculated under sterile conditions (as described below) before being assembled into the fermentor, where they were flushed with sterilized wet air (165 mL/min), rotated 1 h/d at 1 rpm, and maintained at 28°C *(29)*.

3. Methods

The methods described below outline (1) the GC and GC-MS conditions to analyze lipophilic compounds within a wide molecular-mass range in the same chromatographic analysis in a short period of time, (2) the fractionation of the wood extracts using a SPE procedure, and (3) the fungal treatment of wood chips to remove lipophilic extractives, the efficiency of which is analyzed by the above chromatographic methods.

3.1. Chromatographic Analyses

High-temperature, short- and medium-length capillary columns with thin films have been used for the rapid identification and quantification of lipophilic wood extractives with no prior derivatization nor fractionation (*see* **Note 1**). This analytical method is being used routinely in our laboratory for the evaluation of biological removal of extractives from eucalypt and pine woods, which implies the analysis of a great number of samples.

3.1.1. GC

The capillary column used for GC analyses was of 5-m length because it enables simultaneous analysis and quantification of the main classes of lipophilic extractives (fatty acids, steroid hydrocarbons, sterols, sterol esters, and triglycerides) in the same chromatographic run (in a short period of time) as shown in **Fig. 1** (*see* **Note 1**). The injector and the detector temperatures were set at 300°C and 350°C, respectively. Sample volumes of 1 μL were injected in the splitless mode. Helium was used as the carrier gas. The oven was temperature-programmed from 100°C (1 min) to 350°C (3 min) at a rate of 15°C/min.

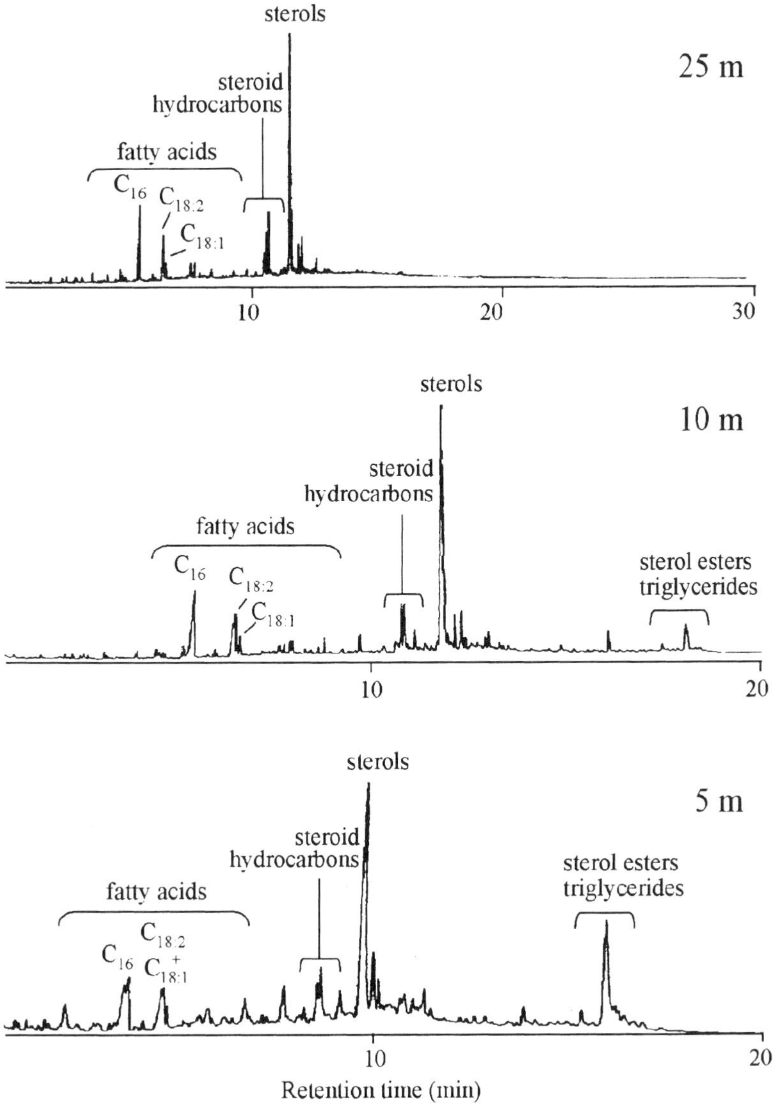

Fig. 1. GC-FID chromatograms of the lipid extract of *E. globulus* wood obtained with high-temperature capillary columns of different lengths. The identity of major compounds is shown in the chromatograms.

A mixture of standard compounds (*see* **Subheading 2.**) was used to elaborate a calibration curve for the quantification of wood extractives. All peaks were quantified by area.

3.1.2. GC-MS

The capillary column used for GC-MS analyses was of 15-m length (*see* **Note 1**). Sample volumes of 1 μL were injected with an autoinjector using a septum-equipped programmable injector (SPI) system. The temperature of the injector during the injection was 120°C, and 0.1 min after the injection was programmed to 380°C at a rate of 200°C/min and hold 10 min. Helium was used as the carrier gas. The oven was temperature programmed from 120°C (1 min) to 380°C (5 min) at 10°C/min. The temperatures of the ion trap and the transfer line were set at 200°C and 300°C, respectively. Compounds were identified by computer comparison of the mass spectra with those in the Wiley and NIST libraries, by mass fragmentography, and when possible by comparison with standard compounds.

3.2. SPE Fractionation

The SPE fractionation scheme described here has been used when a more detailed characterization of some compounds that eluted close together or were present in very minor amounts was required.

The SPE implies a physical extraction process involving a liquid and a solid phase. Process parameters include the stationary phase of the SPE cartridge, the conditioning of the phase, the volume and quantity of the sample, the choice of solvents, the volumes used, and the flow rates, as well as the suppliers of the columns. Different stationary phases are available from several suppliers *(30)*. In the SPE method described here, the lipophilic compounds of interest are retained in the column and elute from the column in order of increasing polarity. Aminopropyl phase cartridges (500 mg) were used. The column separations described here can be performed with different quantities of lipid extract (5–20 mg). The cartridge was loaded and eluted by gravity (*see* **Note 2**). It is necessary to avoid allowing the columns to become completely dry between the different elution steps. The procedure for the fractionation of eucalypt wood lipids follows (**Fig. 2**): The aminopropyl column was conditioned with hexane (4 mL). Next, the dried chloroform extract (containing free fatty acids, squalene, steroid hydrocarbons, waxes, free sterols, sterol esters, and triglycerides) was taken up in a minimal volume (0.5 mL) of hexane:chloroform (4:1) (*see* **Note 2**), and loaded into the cartridge column, leaving the entire lipid mixture in the column. The column was first eluted with 8 mL hexane. The fraction A (containing sterol esters, waxes, and hydrocarbons) was saved, and a new tube was placed below the column. Next, the column was loaded with 6 mL hexane:chloroform (5:1) and the fraction B was eluted (containing triglycerides) and saved. The column was subsequently loaded with 10 mL chloroform and the fraction C was eluted (containing free sterols) and saved. Finally,

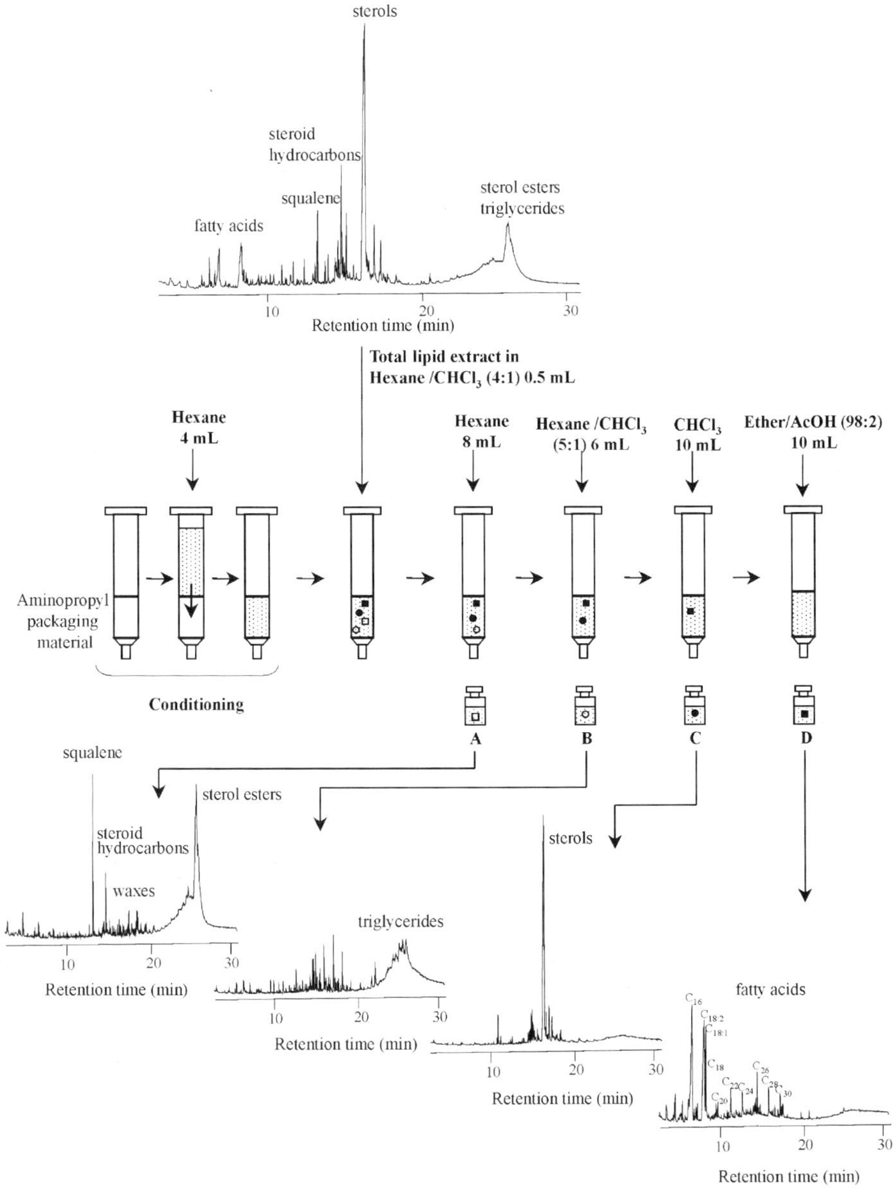

Fig. 2. Elution scheme used for the separation and isolation of lipophilic extractives from *E. globulus* wood in SPE aminopropyl columns and GC-MS (15-m column) chromatograms of the total lipid extract and the four fractions obtained (A–D).

the column was loaded with 10 mL diethyl ether:acetic acid (98:2), and the fraction D was eluted (containing free fatty acids) and saved. Each isolated fraction was dried under nitrogen and weighed. GC and GC-MS were used to monitor the purity and to determine the amount of solvent needed to elute each fraction (**Fig. 2**).

3.3. Biotechnological Control of Pitch Deposits

After a large screening of fungal strains from different origins (*see* **Note 3**), the four fungi listed above were selected for the removal of lipophilic extractives involved in pitch deposit formation from eucalypt and other woods used as raw material for chlorine-free paper pulp manufacturing. Inocula were grown in liquid media, and fungal pellets were used to inoculate wood chips. In order to remove the most recalcitrant lipophilic extractives (such as free and esterified sterols in the case of eucalypt), the inoculated wood was incubated for several weeks under SSF conditions. These are similar to those in chip piles at the pulp mill during natural seasoning of wood, which is traditionally used for decreasing pitch troubles. Therefore, the treatment described could be extrapolated to the mill scale, and can be considered as a controlled seasoning of wood using selected fungal inocula (instead of depending on the action of the wood-inhabiting microorganisms as in the case of natural seasoning).

3.3.1. Inoculum Preparation

Flasks (1 L) with 100 mL of a medium containing (per L) 10 g glucose, 2 g ammonium tartrate, 1 g KH_2PO_4, 1 g yeast extract, 0.5 g $MgSO_4 \cdot 7\ H_2O$, 0.5 g KCl, and 1 mL of a mineral solution (containing 10 mg $B_4O_7Na_2 \cdot 10\ H_2O$, 7 mg $ZnSO_4 \cdot 7\ H_2O$, 5 mg $FeSO_4 \cdot 7\ H_2O$, 1 mg $CuSO_4 \cdot 5\ H_2O$, 1 mg $MnSO_4 \cdot 4\ H_2O$, 1 mg $(NH_4)_6Mo_7O_{24} \cdot 4\ H_2O$, and 100 mL water) were sterilized by autoclaving at 120°C for 20 min *(31)*. Each flask was inoculated with two plugs from a fungal colony grown on 2% malt-extract agar slants, and was maintained under stationary incubation at 28°C for 15 d. After this time period the mycelia were washed and homogenized under sterile conditions. The washed and homogenized mycelia (preinoculum) were grown in 1-L flasks with 200 mL of the above glucose-ammonium-yeast extract medium for 5 d, and the pellets produced were washed and used to inoculate the eucalypt chips.

3.3.2. Fungal Treatment of Wood

The sterilized wood chips inoculated with fungal pellets (0.5 mg/kg wood) were incubated under SSF conditions (corresponding to the water-holding capacity) in the rotary fermentor described previously. After an adequate incubation period (typically 21 d), the treated chips were dried in an aerated oven at 60°C, ground to sawdust using a knife mill, extracted with acetone in a Soxhlet

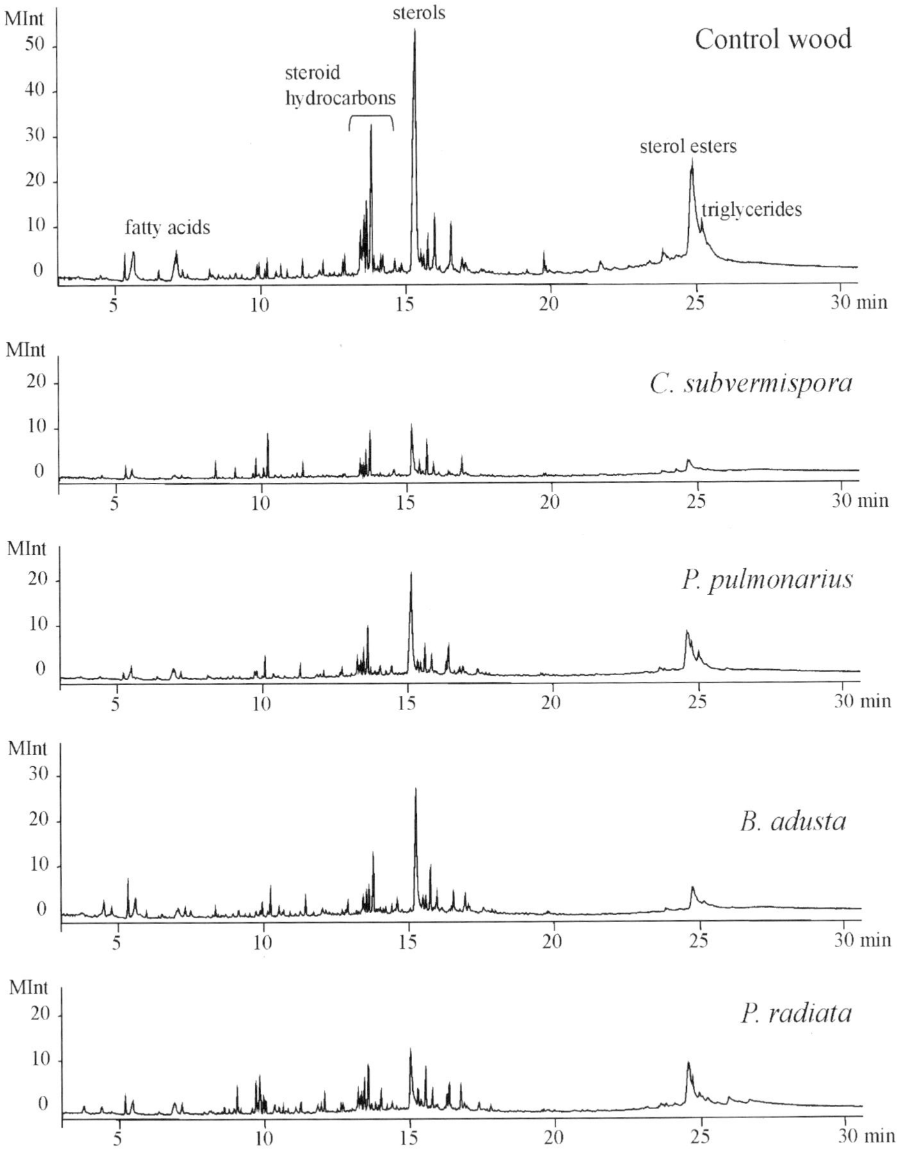

Fig. 3. Total-ion GC-MS (15-m column) chromatograms of the lipophilic fraction of the acetone extracts from the *E. globulus* wood treated with the selected fungal strains (21 d).

apparatus for 8 h, and the lipophilic compounds dissolved in chloroform and analyzed by GC-MS as previously described. After pretreating wood with these fungi, the abundance of most pitch-forming compounds, such as free and esterified sterols, was strongly decreased, as illustrated in **Fig. 3**, which shows the GC-MS profiles of lipophilic extractives from eucalypt wood after SSF treatment with the four selected fungi and the noninoculated control. These results show that the four fungi selected are promising organisms for the biological control of pitch in paper-pulp manufacturing, when applied as a SSF pretreatment of the raw material (*see* **Note 3**).

4. Notes

1. Several authors have used short GC capillary columns, which allow the elution and separation of high-molecular-mass lipids, for the analysis of wood lipophilic extractives, although they do not allow the best resolution ***(20,21)***. In the present study, the range of temperature of the analysis has been extended by using high-temperature capillary columns. On the other hand, capillary columns with thin films, which are necessary for an optimal analysis of high-molecular-mass lipids such as waxes, sterol esters, and triglycerides ***(32)*** are preferred for this study. Therefore, the column finally selected for the chromatographic analyses was a DB5-HT capillary column of 0.25 mm id with a film thickness of 0.1 µm. Previous procedures for the analysis of wood extractives used conventional temperature columns and different film thickness or internal diameters ***(20,21)***. Different column lengths (from 25 to 2 m) and different temperature programs were investigated for the analysis of the lipid extract from eucalypt wood by GC-FID. As the main objective was to obtain short elution times, high-temperature programming rates, which enhance the speed of the analysis, were preferred. **Figure 1** shows the chromatograms of eucalypt wood lipophilic extractives obtained with different column lengths and program temperature rates. It can be observed that the high-molecular-mass lipids (sterol esters and triglycerides) start to elute as the column length decreases. Shorter lengths (2 m) were also attempted, but the resolution was not good enough for quantitative purposes. The length of the column used should be a compromise between the optimum, in terms of resolution, and the need to limit the exposure time of the sample to high temperatures to the minimum ***(32)***. After this comparative study, a 5-m capillary column was selected for the rapid analysis of wood extractives, since it enables elution and separation of compounds with a wide range of molecular masses (from fatty and resin acids to sterol esters and triglycerides) in the same chromatographic analysis, in a short period of time (20 min), and with enough resolution for quantification of the different component groups. The chromatograms obtained by GC-FID using the 5-m capillary column had to be reproducible in GC-MS in order to identify the different compounds. However, GC-MS systems cannot support very short columns. The minimum column length of a capillary column of 0.25 mm id suggested by the manufacturers of the Varian Saturn 2000 GC-MS system was limited to 15 m, although shorter columns (10–12 m) have also been successfully used. The utilization of a high-temperature capillary column made it possible to increase the final temperature up to the 380°C necessary for the detection of

sterol esters and triglycerides in a 15-m capillary column in a short period of time (30 min). The good reproducibility of chromatograms of the eucalypt wood extractives obtained by GC-FID with a 5-m column and by GC-MS with a 15-m column, can be observed by comparing the chromatograms of **Figs. 1** and **2**. Although some triglycerides and sterol esters from eucalypt wood may elute closely, their differentiation is possible in the 15-m column by their mass spectra. On the other hand, this has not been a problem for the quantification in GC-FID when the purpose was categorizing the extractives into chemical classes as in the case of the quantification of wood extractives degradation after fungal treatment. However, when a more accurate characterization of some compounds was required, the extract was fractionated by the simple SPE procedure described in this chapter.

2. The separation and isolation of neutral and polar lipid classes for subsequent use or analysis has been the subject of many reports in the literature. The advantages of SPE include smaller sample and solvent requirements and ease of use compared to conventional solvent extraction techniques. The principles of SPE and the methods of isolation and fractionation of lipids in biological and food matrices have been extensively reviewed *(27,30,33)*. A basic method for the separation of individual neutral and polar lipid classes using aminopropyl bonded SPE cartridges was developed by Kaluzny et al. *(28)* for the separation of lipids from bovine adipose tissue. Chen et al. *(34)* used a SPE method for the fractionation of wood extractives and described a procedure that involved the use of three different SPE cartridges in a multi-step fractionation that resulted in seven fractions. However, both the GC elution order described and the purity of some of the SPE fractions obtained is questionable. Using the SPE method outlined in **Fig. 2,** lipophilic wood extractives are also fractionated in aminopropyl-bonded SPE cartridges; however, only one cartridge is used per sample and some elution steps have been eliminated, resulting in only four final fractions. The present method has been optimized for eucalypt wood extractives *(35)*. The column was loaded and eluted by gravity. Gravity flow is essential when the full resolving power of the sorbent is required *(30)*, and it demands little additional equipment. However, when a major flow-rate is needed, positive displacement or vacuum can be applied. The recovery rate was approx 95%, and the purity of each isolated fraction was confirmed by GC and GC-MS using the chromatographic methods described above. In the SPE procedure outlined here, lipids elute from the aminopropyl column in order of increasing polarity. Polar lipids such as fatty acids or those having a polar group such as sterols are likely to interact more strongly with the aminopropyl group on the columns through hydrogen bonding to the primary amine group *(27)*. On the other hand, it must be reemphasized that the relative compositions of the solvent mixtures (shown in the methods section) should not be changed, since the physical environment of the columns may be altered. This may result in less than optimal separation, recovery, and/or purity of the fractions obtained *(28)*. For example, we found that when the sample is taken up in pure chloroform and the column subsequently eluted with

chloroform:hexane (1:5), as proposed by Chen et al. ***(34)***, triglycerides and sterols may coelute in fraction A. This does not occur if the sample is taken up in pure hexane and subsequently eluted with hexane. However, since the eucalypt wood extractives were not totally soluble in pure hexane, it was necessary to use the mixture hexane:chloroform (4:1) described here, in which these samples were completely soluble. The proportion of chloroform should not be increased, since higher amounts of the more polar solvent may result in some triglyceride elution in fraction A. The SPE fractionation method described here has also been successfully applied to the fractionation of the extracts of pitch deposits occurring during the kraft pulping of eucalypt wood.

3. The results obtained after GC and GC-MS analysis of wood chips treated with a variety of fungi from several taxonomic groups showed different patterns of degradation of compounds responsible for pitch deposit formation ***(25,26)***. Commercial preparations for pitch biocontrol during manufacturing of mechanical pulps from pine and other softwood are currently available ***(1)***. These are based on the use of the so-called sap-stain fungi (from the group of ascomycetes) being able to hydrolyze triglycerides and degrade the fatty acids. The four fungi described here are among the most promising strains for pitch biocontrol in those cases where more recalcitrant wood extractives (including free and esterified sterols) are at the origin of deposits, as shown in **Fig. 3**. The experiments described here were at the laboratory scale, and the wood samples (chips of different sizes) were inoculated with plugs from fungal cultures (fungal screening experiments in flasks) or pellets from liquid cultures (SSF bioreactor experiments). Scaling-up the fungal treatment implies optimization of inocula for large-scale use (considering supports, stabilizers, and enhancers of fungal growth). After pretreating wood with these fungi, which belong to the group of basidiomycetes, the abundance of free and esterified sterols strongly decreased. Laboratory pulping of pretreated wood has shown that the biological removal of sterols from the raw material results in pulps with reduced amounts of these compounds ***(36)***, as well as in a decrease in the potential toxicity of the process effluents. This is because natural extractives (which were degraded by chlorine-containing bleaching reagents but survive peroxide bleaching of pulp) are among the most toxic compounds in effluents from manufacturing chlorine-free paper pulps ***(8,9)***. The pulp yield after biological pretreatment (47–50%) could be further optimized by modifying the duration of the biological pretreatment (which always results in some loss of wood weight) or by improving the fungal strains used to treat wood (with the purpose of increasing lipid degradation and decreasing carbohydrate hydrolysis). The final selection of the most promising fungal treatment for industrial control of pitch deposition should be a compromise between the optimum in terms of extractive removal, the lowest decrease of pulp yield, and the conservation of those pulp properties of interest for the different types of paper to be produced ***(36,37)***. The ultimate goal is a sustainable production of paper pulp—i.e., a pulp mill in ecological balance with nature by the use of biotechnological tools enabling a higher closure of circuits, and a better performance of environmentally sound TCF bleaching processes.

Acknowledgments

The authors appreciate financial support from the European contracts FAIR-CT95-560 and QLK5-99-1357 and the Spanish CICYT (projects FEDER 1FD97-0742 and AGL-2002-00393 AGR-FOR). A. G. is thankful for a "Ramón y Cajal" contract of the Spanish MCYT.

References

1. Gutiérrez, A., del Río, J. C., Martínez, M. J., and Martínez, A. T. (2001) The biotechnological control of pitch in paper pulp manufacturing. *Trends Biotechnol.* **19,** 340–348.
2. Hillis, W. E., ed. (1962) *Wood Extractives.* Academic Press, London.
3. Fengel, D. and Wegener, G., eds. (1984) *Wood: Chemistry, Ultrastructure, Reactions.* De Gruyter, Berlin, Germany.
4. Rowe, J. W., ed. (1989) *Natural Products of Woody Plants. I and II. Chemicals Extraneous to the Lignocellulosic Cell Wall.* Springer-Verlag, Berlin, Germany.
5. Gutiérrez, A., del Río, J. C., González-Vila, F. J., and Martín, F. (1999) Chemical composition of lipophilic extractives from *Eucalyptus globulus* Labill. wood. *Holzforschung* **53,** 481–486.
6. Sjöström, E., ed. (1993) *Wood Chemistry. Fundamentals and Applications.* Academic Press, San Diego, CA.
7. Hillis, W. E. and Sumimoto, M. (1989) Effect of extractives on pulping, in *Natural Products of Woody Plants* (Rowe, J. W., ed.), Springer-Verlag, Berlin, Germany, pp. 880–920.
8. Leach, J. M. and Thakore, A. N. (1976) Toxic constituents in mechanical pulping effluents. *Tappi J.* **59,** 129–132.
9. Liss, S. N., Bicho, P. A., and Saddler, J. N. (1997) Microbiology and biodegradation of resin acids in pulp mill effluents: a minireview. *Can. J. Microbiol.* **43,** 599–611.
10. Allen, L. H. (1980) Mechanisms and control of pitch deposition in newsprint mills. *Tappi J.* **63,** 81–87.
11. Hamilton, K. A. and Lloyd, J. A. (1984) Measuring the effectiveness of talc for pitch control. *Appita* **37,** 733–740.
12. Allen, L. H. (1988) Pitch control during production of aspen Kraft pulp. *Pulp Paper Can.* **89,** 87–91.
13. Allen, J., Sitholé, B. B., MacLeod, J. M., Lapointe, C., and McPhee, F. J. (1991) The importance of seasoning and debarking in the Kraft pulping of aspen. *J. Pulp Paper Sci.* **17,** J85–J91.
14. Fujita, Y., Awaji, H., Taneda, H., Matsukura, M., Hata, K., Shimoto, H., et al. (1992) Recent advances in enzymic pitch control. *Tappi J.* **75,** 117–122.
15. Fischer, K. and Messner, K. (1992) Reducing troublesome pitch in pulp mills by lipolytic enzymes. *Tappi J.* **75,** 130–135.
16. Fischer, K., Puchinger, L., and Schloffer, K. (1993) Enzymatic pitch control of sulfite pulp on pilot scale. *J. Biotechnol.* **27,** 341–348.

17. Farrell, R. L., Blanchette, R. A., Brush, T. S., Hadar, Y., Iverson, S., Krisa, K., et al. (1993) Cartapip™: a biopulping product for control of pitch and resin acid problems in pulp mills. *J. Biotechnol.* **30,** 115–122.
18. Chen, T., Wang, Z., Gao, Y., Breuil, C., and Hatton, J. V. (1994) Wood extractives and pitch problems: analysis and partial removal by biological treatment. *Appita* **47,** 463–466.
19. Evershed, R. P., Prescott, M. C., Spooner, N., and Goad, L. J. (1989) Negative ion ammonia chemical ionization and electron impact ionization mass spectrometric analysis of steryl fatty acyl esters. *Steroids* **53,** 285–309.
20. Sitholé, B. B., Sullivan, J. L., and Allen, L. H. (1992) Identification and quantitation of acetone extractives of wood and bark by ion exchange and capillary GC with a spreadsheet program. *Holzforschung* **46,** 409–416.
21. Örså, F. and Holmbom, B. (1994) A convenient method for the determination of wood extractives in papermaking process waters and effluents. *J. Pulp Paper Sci.* **20,** J361–J366.
22. Lusby, W. R., Thompson, M. J., and Kochansky, J. (1984) Analysis of sterol esters by capillary gas chromatography-electron impact and chemical ionization-mass spectrometry. *Lipids* **19,** 888–901.
23. Wakeham, S. G. and Frew, N. M. (1982) Glass capillary gas chromatography-mass spectrometry of wax esters, steryl esters and triacylglycerols. *Lipids* **17,** 831–843.
24. Adler, E. (1977) Lignin chemistry—past, present and future. *Wood Sci. Technol.* **11,** 169–218.
25. Martínez, M. J., Barrasa, J. M., Gutiérrez, A., del Río, J. C., and Martínez, A. T. (1999) Fungal screening for biological removal of extractives from *Eucalyptus globulus* Labill. wood. *Can. J. Bot.* **77,** 1513–1522.
26. Gutiérrez, A., del Río, J. C., Martínez, M. J., and Martínez, A. T. (1999) Fungal degradation of lipophilic extractives in *Eucalyptus globulus* wood. *Appl. Environ. Microbiol.* **65,** 1367–1371.
27. Ebeler, S. E. and Shibamoto, T. (1994) Overview and recent developments in solid-phase extraction for separation of lipid classes, in *Lipid Chromatographic Analysis* (Shibamoto, T., ed.), Marcel Dekker, Inc., New York, pp. 1–49.
28. Kaluzny, M. A., Duncan, L. A., Merritt, M. V., and Epps, D. E. (1985) Rapid separation of lipid classes in high yield and purity using bonded phase columns. *J. Lipid Res.* **26,** 135–140.
29. Valmaseda, M., Almendros, G., and Martínez, A. T. (1991) Chemical transformation of wheat straw constituents after solid-state fermentation with selected lignocellulose-degrading fungi. *Biomass Bioenerg.* **1,** 261–266.
30. Christie, W. W. (1992) Solid-phase extraction columns in the analysis of lipids, in *Advances in Lipid Methodology—One* (Christie, W. W., ed.), The Oily Press, Dundee, Scotland, pp. 1–18.
31. Guillén, F., Martínez, A. T., and Martínez, M. J. (1992) Substrate specificity and properties of the aryl-alcohol oxidase from the ligninolytic fungus *Pleurotus eryngii*. *Eur. J. Biochem.* **209,** 603–611.
32. Christie, W. W. (1994) *Gas Chromatography and Lipids: A Practical Guide.* The Oily Press, Dundee, United Kingdom.

33. Ruiz-Gutiérrez, V. and Pérez-Camino, M. C. (2000) Update on solid-phase extraction for the analysis of lipid classes and related compounds. *J. Chromatogr.* **A 885,** 321–341.
34. Chen, T. C. (1994) Solid-phase extraction can rapidly separate lipid classes from acetone extracts of wood and pulp. *Tappi J.* **77,** 235–240.
35. Gutiérrez, A., del Río, J. C., González-Vila, F. J., and Martín, F. (1998) Analysis of lipophilic extractives from wood and pitch deposits by solid-phase extraction and gas chromatography. *J. Chromatogr.* **823,** 449–455.
36. Gutiérrez, A., Martínez, M. J., del Río, J. C., Romero, J., Canaval, J., Lenon, G., et al. (2000) Fungal pretreatment of *Eucalyptus* wood can strongly decreases the amount of lipophilic extractives during chlorine-free kraft pulping. *Environ. Sci. Technol.* **34,** 3705–3709.
37. Martínez-Íñigo, M. J., Gutiérrez, A., del Río, J. C., Martínez, M. J., and Martínez, A. T. (2000) Time course of fungal removal of lipophilic extractives from *Eucalyptus globulus* Labill. wood. *J. Biotechnol.* **84,** 119–126.

20

Propolis

Chemical Micro-Heterogeneity and Bioactivity

José Domingos Fontana, Juliana Adelmann, Mauricio Passos, Marcelo Maraschin, Cristina A. de Lacerda, and Fernando Mauro Lanças

1. Introduction

Propolis, bee glue or hive dross is a waxy:resinous (30%:50%) sealer and protector mass prepared by bees from heterogeneous plant material. Essential oils and pollen are other components. Phenolics such as flavonoids and phenol-carboxylic acids are strategic components in propolis to render it bioactive against several pathogenic microorganisms, for instance as a bacteriostatic and/or bactericidal agent. The flora surrounding the hive (buds, twigs, bark, and, less importantly, flowers) are the basic source for phenolics, and it exerts an outstanding influence on the final composition of the propolis *(1)*, and hence on its physical, chemical, and biological properties, too. Although the wax component is a necessary supplement provided by the bee secretory apparatus, far less is known about the degree of influence of these hard-working insects in changing all the other natural components collected from nature, including minor ingredients like essential oils (10%), mainly responsible for the delicate and pleasant odor. The contribution of the flora to propolis plus the exact wax content may then explain physical properties such as color, taste, texture, melting point (*see* **Note 1**), and, more importantly from the health standpoint, a lot of pharmaceutical applications. The main constituents—wax and plant resins—are responsible for the sticky, nonwettable, and scented aspects of propolis. Although propolis fractions may be obtained from apolar organosolvents like hexane till water *(2)*, ethanol (or hydroethanolic mixes from 1:9 to 3:7 for extracts or balsams; *see* **Note 2**) is more often utilized as the effective solubi-

From: *Methods in Biotechnology: Environmental Microbiology: Methods and Protocols*
Edited by: J. F. T. Spencer and A. L. Ragout de Spencer

lizing tool, owing to its more safe chemical nature and the high yields (50–70 g% of dry solids). This is the usual formulation for dermal application in the case of inflammations and wound infections. Propolis concentrates are also usually incorporated in toothpastes, shampoos, toilette soaps, creams, tablets, and chocolates *(3)*. Increased water solubility of several of the bioactive components may be achieved by forming a complex of propolis with cyclodextrins *(4)*. Although propolis has limited nutritional value as compared to royal jelly and honey, about 1% of its content consists of amino acids, and arginine and proline account for about half of this amount. The outstanding bioactivity of propolis comes, in part, from complex phenolics; applications of medical interest include activity against pathogenic bacteria, yeasts, fungi, and viruses *(5,6)* (*see* **Table 1** and **Note 7**). One basic structure is shown in **Fig. 1** (pinocembrin *(7)*, a bacteriostatic flavanone). Its activity may be explained by the extensive variability experienced by the basic triple phenylpropane nucleus of flavones (luteolin) and flavonols (kempferol), which may be further reduced to form flavanones (pinocembrin), open up to form chalcones (olivin), rearranged to form isoflavones (genistein), progressively oxidized through several single phenolic groups (quercetin, a pentahydroxyflavonol), methylated (rhamnetin), prenylated (sericetin, a derivative of the 2'-hydroxylated flavonol galangin), or even mono- (vitexin) or oligo-(robinin) glycosylated, most often with glucose, rhamnose, xylose, galactose, arabinose, and glucuronic acid *(8)*. Because bees possess efficient enzymatic apparatuses, it is worth mentioning their expected capability to perform complex phenolic modifications, since for some of them (e.g., kempferol, quercetin) up to 2 or 3 tenths glycoslylated variants are known. Further, some plants like *Petunia* may provide a strategic natural chemical combination of different pharmacologically active compounds such as flavonol (kempferol) and phenol-carboxylic acid (ferulic acid) interconnected by an acetylated sugar dissacharide bridge, as in the case of petunoside *(9)*. The degree of polarization the bees maintain for native simple or complex phenolics of the propolis may play an important role in the desired pharmacological activity, since the polar substituents improve water solubility, a property usually desired for drug formulation, blood transport, and distribution to human tissues (bioavailability). Apitherapy, the medicopharmaceutical use of bee products, is a historical and well established business in several countries, and propolis is a particularly rich chapter of it *(10)*.

2. Materials

1. Propolis samples were provided by Latsis (Curitiba-Pr) and Breyer Co. (União da Vitória, Paraná, Brazil) and kept frozen at –18°C until the laboratory work.
2. Each cold propolis mass (100-g sample) is powdered with a Waring blender at low speed for 1–2 min and twice extracted with 4–5 vol absolute ethanol and/or

Table 1
Pharmacological Activities of Propolis

Bioactivity	Target/comments	Reference
Bacteriostatic and bactericidal	*Staphylococcus aureus* (including MRSa, methycillin-resistant strains), growth inhibition, synergistic effects	***2,11,12***
	Bacillus larvae, hive assepsy	***13***
	Streptococcus mutans, anticariogeny and glusoyl-transferase inhibition, *de novo* plaque inhibition	***14–16***
	Klebsiella pneumoniae	***17***
	Helicobacter pilori, anti-ulcer	***18***
	Mycobacterium sp.	***19***
Fungistatic and fungicide	Fungicide effect, pinocembrin	***20***
	Dermatophytes	***21***
	Aspergillus sulphureus, ochratoxin A synthesis inhibition	***22***
Viriostatic and virucide	*Herpes simplex* virus, inhibition by methyl-butenyl-caffeate	***23***
	Anti-influenza virus, cinnamic acid esters	***24***
	Rhinovirus infections, flavonoids	***25***
	HIV infection chemotherapy, Newcastle Disease virus inhibition	***26,27***
Antitumoral (in vitro; cytostatic and cytotoxic)	Preferential antitumoricide, artepillin C	***28***
	Melanoma and glioblastoma inhibition	***29***
	Selective cytotoxicity, caffeic acid phenethyl ester (CAPE)	***30***
	Colon carcinogenesis prevention, phenyl-ethyl-methyl-caffeate	***31***
	Ehrlich carcinoma	***32***
	Suppression of type 5 adenovirosis	***33***
Antioxidant	Free radical scavenging	***34***
	Lypoxygenase inhibition, caffeic acid phenethyl ester (CAPE)	***35***
	Free radical scavenging, flavonoids	***36***
Wound healing and tissue repair	Dermatitis, allyl-caffeic acids	***37***
	Gastric ulcers healing	***38***
	Sunscreen agent	***39***
	Anti-inflammatory effect mechanism	***40***

(continued)

Table 1 *(Continued)*

Bioactivity	Target/comments	Reference
	Liver protection, chronic alcoholism	***41***
	Hepatic damage protection, carbon tetrachloride toxicity	***42***
	Trophic chronic varicoses treatment	***43***
Anesthesia	Local anesthesia	***44***
	Superior to procaine, attributed to conifer flavonoids like pinocembrin and phenol carboxylic acid derivatives	***45***
Immunobenefits	Tetanus toxoid, immunoadjuvant	***46***
	HIV-1 replication suppression	***47***
Cardiovascular aids	Anti-hyperlipidemia, attributed to dihydroflavonoids	***48***
	Trypanosomiasis	***49***
Blood and intestinal parasites	Giardiasis treatment	***50***
Respiratory diseases	Upper respiratory tract	***51***
Mutagenesis	Antimutagenic action of propolis	***52,53***

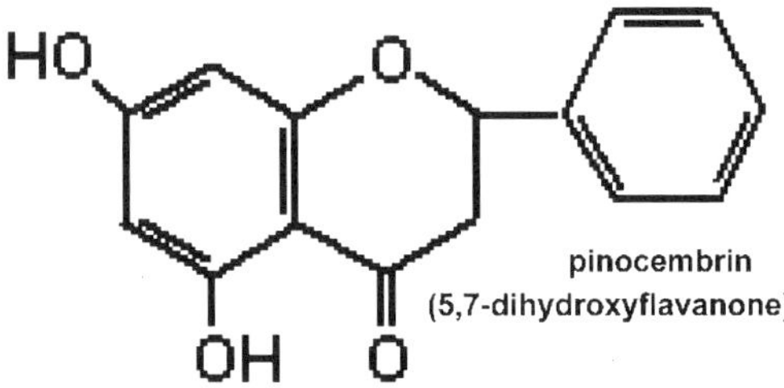

Fig. 1. Pinocembrin, a flavanone derivative of the flavone chrysin.

with ethanol:water (7:3), followed by centrifugation in a clinical apparatus (1000*g*) for 5 min. Obtained yields of total solids in the yellow to brown clear supernatants are in the range of 55 to 70% (w/w), depending on the propolis source.

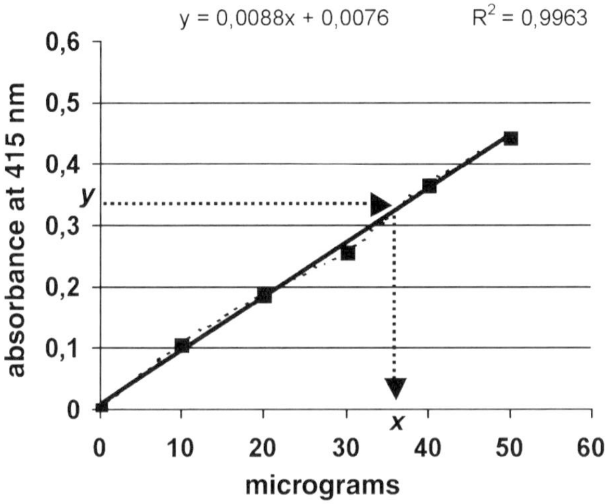

Fig. 2. Calibration curve for quercetin in the visible spectrometry using the acetic Al^{3+} nitrate reagent and total flavonoid determination in a propolis extract.

3. Methods

3.1. Extraction, Quantitation, and Characterization

3.1.1. Spectrophotometry

The total flavonoids content, a parameter often used to evaluate propolis quality for the purpose of the preparation of pharmaceutical formulations, either in crude extracts or in downstream fractions (*see* **Note 3**), is measured by direct spectrophotometry in the visible range (*see* **Note 6**) of any apparatus (1-mL glass or quartz cells), by following the color increase through the addition of the phenolics complexing reagent, aluminum nitrate. Propolis (e.g., 25 mg total solids/mL) is diluted 1:10 with ethanol, and samples containing up to 50 µg total flavonoids (e.g., 200 µL for an estimated 10% content of flavonoids in the total solids of the original extract) are adjusted to 0.5 mL with ethanol, mixed with 0.1 mL $Al(NO_3)_3$, 0.1 mL of 1 *M* CH_3COOK, and 4.3 mL of 80% ethanol; the increased yellow color is read at 415 nm (blue zone of the visible spectrum) against a blank of the solvent mix. Quantification is carried out with a calibration curve prepared from a stock solution of 1 mg/mL quercetin (flavonol or pentahydroxyflavone) using samples progressively from 10 to 50 µL (10–50 µg) (**Fig. 2**).

In the above example the 1:10 diluted sample of propolis (e.g, 150 µL; hence, 375 µg of total solids) led to an absorbance of 0.33 at 415 nm, thus corresponding in the calibration curve of quercitin to 37 µg of flavonoids and, finally, corresponding to (37/375) × 100 or (37 × 10) × (1000/150) = 2466.7; (2466.7/

25,000) × 100 = 9.87 g% of total flavonoids in the total solids of the original propolis extract. It is worth mentioning that, unlike quercitin and kaempferol (naturally yellow-colored flavonols), chrysin (a light-yellow flavone) and its hydrogenated derivative pinocembrin (an uncolored flavanone) experience minor or no color increase with aluminum nitrate, leading to underestimation of the total flavonoids content.

3.1.2. Thin-Layer Chromatography Coupled to Densitometry

Alternatively, because very often any propolis sample contains more than one flavonoid component, the undiluted propolis sample can be spotted on a silica gel 60 chromatoplate (e.g., article 1.05553 from Merck, Germany) along with application of a known amount of any standard (e.g., pinocembrin, 2 μg). The plate is run with a mobile phase consisting of hexane:ethyl acetate (3:2 v/v) and then developed with a spray of the mix anisaldehyde (0.5 mL):methanol (95 mL):sulfuric acid (5 mL) followed by progressive heating to 100–105°C for 2–3 min until full color development (e.g., the naturally colorless pinocembrin turns to yellow and then to deep orange while phenol-carboxylic acids turn to pink and then violet colors; other lipids most often stain as deep wine or violet spots). The multicolored plate is then photographed (**Fig. 3A**) and the photo used for densitometric records using a Shimadzu CS-9301PC dual wavelength flying spot scanning spectrophotometer or similar apparatus for densitometry in the visible range. The light beam is initially focused on the flavonoid standard spot and a spectrum is obtained to establish the wavelength of maximum absorption (e.g., 440 nm [blue] in the case of orange-stained flavonoids or 660 nm [red] in the case of other propolis components). Each chromatographic lane is then scanned and the absorbances registered as illustrated in **Fig. 3B,C**. Using the maximum absorbance obtained for the known amount of standard in the range of 2–4 μg (e.g., the strong and slower spot at Rf = 0.56 in lane "pc-me/pc" in **Fig. 3A** and the dashed line for "pc-std" in **Fig. 3B**; $A_{440\ nm}$ = 1.05 at 52 mm of the TLC pathway); calculations of the same component in any sample is an immediate result. The whole results of **Fig. 3A–C**) thus indicated that poplar (*Populus* sp.) bud as well as its propolis product PP-4 are enriched in pinocembrin (>10%) whereas propolis PP-2, collected from a nearby geographic region devoid of poplar trees, has a trace pinocembrin content.

3.1.3. Gas Liquid Chromatography

The profile of "total volatile derivatives" of propolis is obtained through capillary gas chromatography (cGC). A moisture-free sample (e.g., 1 mg of whole native propolis or any of its derived fractions) is dissolved in 50 μL of dry pyridine and 50 μL of BSTFA (bis-trimethylsilyl-trifluoroacetamide) as per-silylating agent for 30 min at 70°C. Any phenolic, alcoholic, and (undisso-

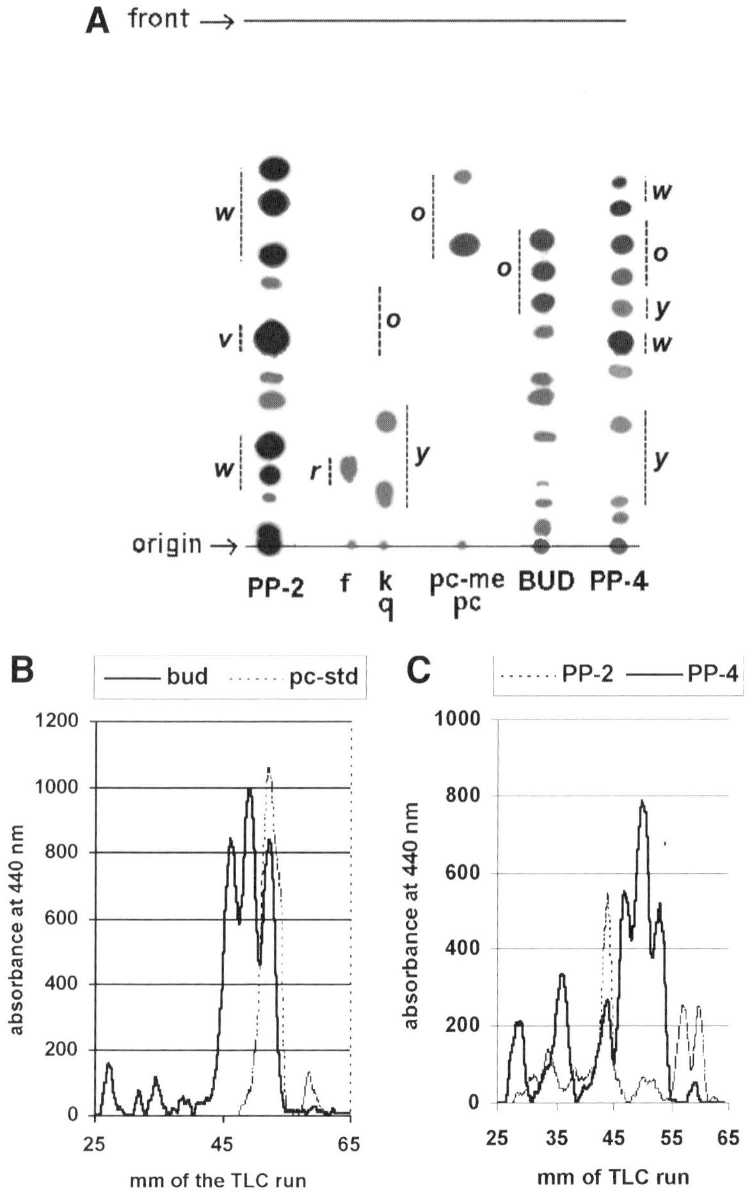

Fig. 3. (**A**) TL chromatogram of propolis (PP 2 and PP-4) and poplar bud (BUD) samples (fractions from the eluent of silica gel columns with ethyl acetate) and selected standards of phenol-carboxylic acid (f – ferulic acid and flavonoids [k + q = kempferol and quercetin; pc = pinocembrin, and pc-me, its 7-methyl ether as the faster migrating and weaker spot]). The letters *w*, *v*, *r*, *o*, and *y* denote the wine, violet, rose, orange, and yellow colors arising after spraying with acidified *p*-anisaldehyde and heating. (**B** and **C**) Densitometric profiles of the same samples and lanes from **Fig. 3A**, except for the f and k + q standards.

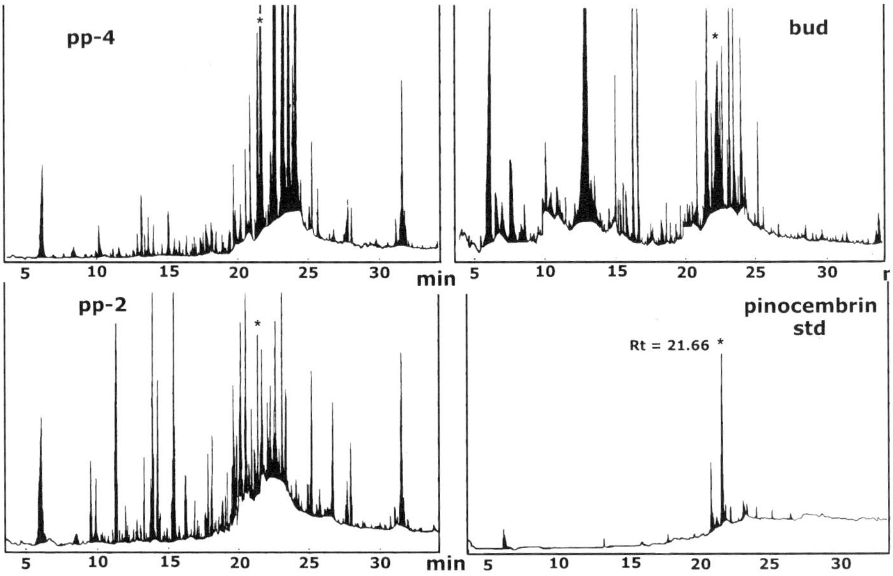

Fig. 4. Gas-liquid chromatograms for ethanol extracts of propolis samples PP-4 and PP-2 or poplar buds, and for the pinocembrin standard. *Note:* For the detection of the volatile components, a flame ionization detector (FID) was used.

ciated) carboxyl group hydrogen is thus replaced by a trimethylsylanol group (–Si–$[CH_3]_3$), producing a volatile derivative from the parent compound. A suitable stationary phase is, for example, a capillary column of 5% phenyl–95% dimethylsiloxane (e.g., Agilent HP-5; ≥25 m length, 0.25 mm internal diameter, 0.50 µ of chromatographic film) programmed at 5°C/min from 100°C to 300°C. One µL injections are enough to obtain the multicomponent profiles seen in **Fig. 4** (*see also* **Note 4**). The apparatus will quantify any desired peak as percentage of the total volatiles eluted, or as µg per volume if compared to a peak obtained from any standard (the bottom right chromatogram).

3.1.4. Capillary Electrophoresis

Analysis of propolis regarding particular components such as flavonoids may also be carried out by capillary zone electrophoresis (CZE), specifically the capillary micellar electrokinetic chromatography (cMEKC) variant. Flavonoid phenolic groups require strong alkalization for complete dissociation to generate anions; this limitation can be overcome if they are used in the form of a complex with an anionic detergent like sodium dodecylsulfate (SDS), using a gentle basic pH with a pH 8.0 buffer (e.g., borate). Flavonoids will then generate micells of different molecular size and electrophoretic behavior, allowing

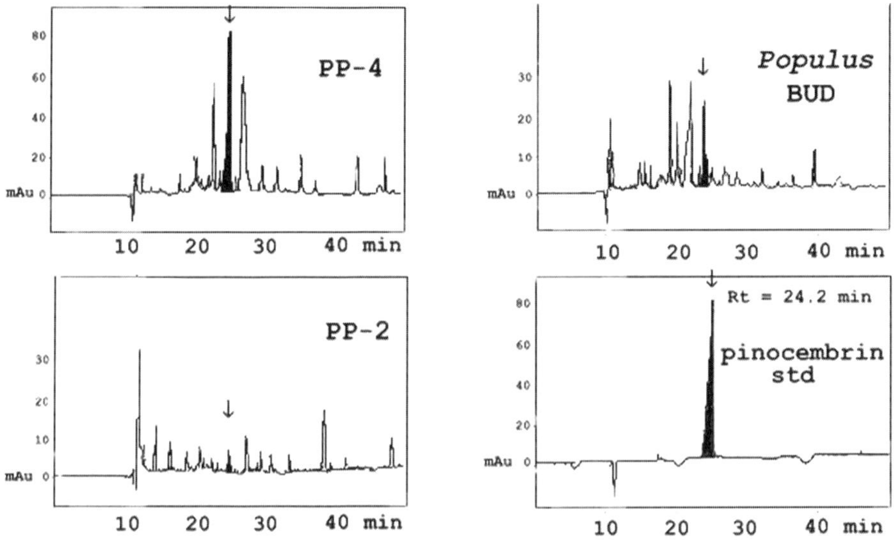

Fig. 5. Capillary electrophoretograms (MEKC mode) for propolis samples, poplar bud, and pinocembrin standard. *Note:* The wavelength used for differential diode array (DAD) monitoring for phenolics was 275 nm.

their resolution. Inclusion of 10% of methanol in the buffer system may help dissolve the compounds and medium polarity modification. A typical result obtained on a Agilent CE^3 apparatus is shown in **Fig. 5**. For phenol-carboxylic acids CZE, *see* **Note 5**.

3.2. Bioactivity Against Pathogenic Bacteria

Antibiosis as bacteriostatic or bactericidal activity is more properly assessed in liquid media and then expressed as minimum inhibitory concentration (MIC), a function of the resulting bacterial cell mass turbidity reduction ***(54)***, but this methodology applies to individual bioassays. Conversely, the disk diffusion method in solid phase (agar-based nutrient broth), although limited, for instance, by the solubility of the compound(s) being assayed, allows for the detection and the semiquantification of several samples in just one Petri dish (**Fig. 6**). For details, *see* **Note 8**.

4. Notes

1. Propolis native masses may vary considerably in appearance. Colors may be grayish, greenish, yellowish, brown, and, less often, even reddish; hence, the corresponding colors of the alcohol extracts. Native propolis is usually a hard, brittle material at 15°C, which turns ductile at 30°C and experiences liquefaction from 60°C.

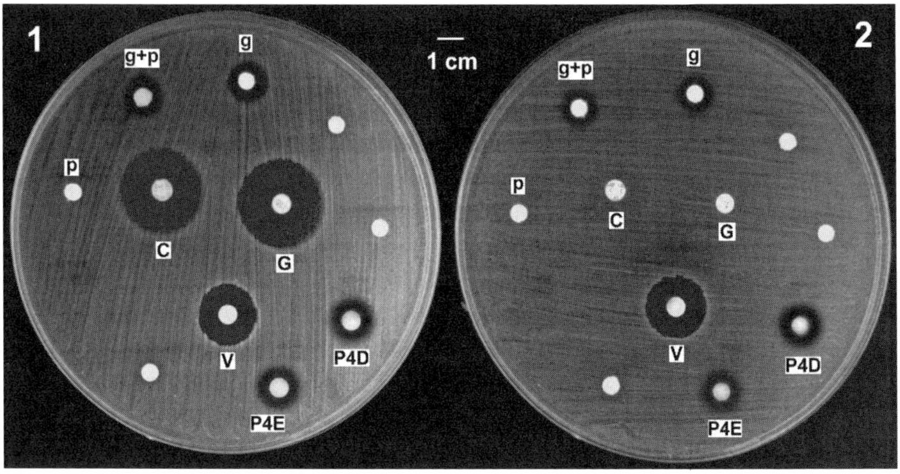

Fig. 6. Antibiogram according to the disk diffusion method in agar-solidified medium. (1) *Staphylococcus aureus*, wild-type strain; (2) *S. aureus*, methycilin-resistant strain (MRSa). *Note:* Clindamycin (C), vancomycin (V), gentamycin (G) disks with respectively 2, 30, and 10 μg each, pinocembrin (p), gallic acid (g), pinocembrin + gallic acid (p+g) disks with 200 μg amounts of the flavonoid or phenol-carboxylic acid, and propolis PP-4 ethanol (P4E) or DMSO (P4D) extracts with 2.2 or 2.7 mg of total solids/disk, respectively. Unlabeled disks contained ferulic or chlorogenic acid (200 μg) or simply pure DMSO as a control for the P4D disk. It is evident from the above disk antibiogram that bacteriostatic or bactericidal action of pure pinocembrin (flavanone) would require a higher dose. The absence of inhibition zone (halo) was also confirmed with its parent compound, the flavone chrysin (200 μg).

2. Although 70–80% ethanol is the default safe solvent for most of the pharmaceutical liquid, viscous, clean and yellow-to-brown preparations from propolis, the addition of a higher content of water in these extracts causes the progressive decrease in solubility of waxes and/or less hydrophylic components, and this results in preparations with a whitish or milky but stable turbidity. This formulation is more convenient for some applications where higher organosolvent content is to be avoided (e.g., sprays for sore throat). Starting from 1 g of each native propolis mass (e.g., the current samples) and 4 vol of extractant (one single extraction after gentle warming), the total solids yield for each independent extraction procedure depends on polarity and increases with the ethanol content in the extractant, as shown in **Table 2**.
3. If enrichment in flavonoids is desired, an absolute ethanol extract of propolis can be eluted from silica gel or silicic acid columns. Warm hexane or petrol ether will wash out most of the hydrocarbons, waxes, and more apolar components. Next elution with ethyl acetate yields most of the flavonoids family, usually as deep yellow-colored fractions.

Table 2
Native Propolis Extractability as Function of the Ethanol Concentration in the Aqueous Extractant

	Extractant				
Propolis sample	H_2O	25% EtOH	50% EtOH	75% EtOH*	Absolute EtOH**
PP-2**	1.3%	1.6%	5.9%	14.0%	25.5%**
PP-4*	0.4%	0.7%	14.8%	47.1%*	41.3%

Note: A single extraction step after 2 min heating at 70°C followed by quick centrifugation at 2000*g*.

4. In the GC of TMS derivatives of propolis, even when the major peak is recorded as the maximum of the electrometric signal (full scale), it is not uncommon to detected 100–200 signals for the secondary and minor components. Considering the chemical microheterogeneity of propolis, GC is the superior technique for the wide-spectrum detection of multi-components. One additional advantage of this technique is its combination with mass spectrometry (GC-MS): any component is then fully characterized in a single chromatographic run.
5. If CE is the analytical tool, it may encompass the detection and resolution of phenol-carboxylic acids (e.g., ferulic acid) with no need of a micell-generating detergent in the electrophoretic buffer. Since these phenolics bear a formal negative charge at pH ≥7.0, it is sufficient to use an alkaline buffer (e.g., 25–50 m*M* phosphate or borate in the pH 8.0–9.0 range), with the addition of 10–20% of methanol as cosolvent and polarity modifier.
6. For spectroscopic monitoring of propolis components, the indicated wavelengths of maximum absorbance for each class of flavonoids in **Table 3** may be useful.
7. Allergenic effects are seldom attributed to propolis except for a few mentions in the literature *(55)*.
8. Each commercial antibiotic (e.g., vancomycin) or antimicrobial heterogeneous mix (e.g., propolis) is dissolved in the most appropriate volatile solvent (acetone, ethanol, methanol, or even sterile water) to attain a concentration from 1 to 50 mg/mL. DMSO may substitute for any of these solvents, but in this case, because of its high boiling point, control disks need to be prepared omitting the antibiosis agent. Successive 5 µL volumes are then loaded in each 4-mm diameter filter paper circle or disk prepared from no. 1 or preferably 3 MM chromatographic paper from Whatman (previously rendered sterile by washing with acetone) under a sterile air stream for solvent removal until the desired amount of solid(s) for each disk is reached (e.g., 4 × 5 µL = 20 µL; 20–1000 µg/disk). Mueller-Hinton agar plates are prepared. A loop of bacterial mass from a slant of a wild type (ATCC, cat. No. 25923) or a methycilin-resistant (LACEN) strain of *Staphylococcus aureus* (MRSa) is cultivated in the Nutrient Broth liquid medium for about 2 h in

Table 3
Wavelengths of Absorbance for Flavonoids

Flavonoid class	Example	Solvent	Wide ranges of absorption (nm) for the flavonoid class	Main and (secondary) λ_{max} (nm) for the given example
Anthocyanins	Cyanidin-3-rutinoside	0.01% HCl in methanol	269–289 310–333 495–538	523 (290)
Flavones	Luteolin	Ethanol	248–286 332–356	350 (255, 268)
Flavonols	Kaempferol	Ethanol	252–268 345–379	368 (268)
Flavanones	Naringenin	Ethanol	215–233 278–290 312–335	325 (290, 224)
Isoflavones	Genistin	Ethanol	241–275 296–302 320–335	262 (330)
Chalcones	Neoplathymenin	95% Ethanol	235–266 320–385	393 (268, 320)
Aurones	Hispidol	95% Ethanol	234–272 254–355 388–413	388 (234, 254)
Xanthones	Mangiferin	95% Ethanol	230–245 250–265 305–330 340–400	258 (242, 316, 364)

order to attain a cell density equal to 0.5 on the McFarland scale. This corresponds to a turbidity equivalent to barium sulfate generated by a mixture of 0.05 mL of 1.175% of barium chloride dihydrate with 9.95 mL of 1% sulfuric acid ***(54)*** or to 1.5×10^8 cells/mL. A cotton swab is wetted in the bacterial suspension and exhaustively but gently spread in the agar plate to obtain a uniform film of bacterial cells. The previously prepared disks (up to 12) are then each laid on a 14-cm-diameter Petri dish Mueller-Hinton medium surface with the help of a small pair of pliers, providing the best distribution possible. The plate is then incubated at 36°C for 24 h. Around any disk containing an antibiosis agent able to diffuse to the medium as a circle, a clear inhibition zone is generated due to the contrast with the thick, milky, and less transparent layer where the bacterium experienced free growth. The larger the diameter of a circular inhibition zone (transparent halo), the higher the antibiosis power (e.g., 3 cm for gentamycin in plate 1; **Fig. 6**).

Acknowledgments

The authors are thankful for the collaboration of undergraduates M. H. R. Santos (UFPR) and D. Breyer (PUC-PR), and of graduate student C. Diagone (USP-São Carlos-SP); financial support from CNPq, Fundação Araucária-SETI-PR, and FAPESP; and also the propolis samples provided by Breyer Co. (União da Vitória-PR). The MRSa strain of *S. aureus* was kindly provided by C. K. Fontana from State Central Laboratory (LACEN) in Curitiba-PR-Brazil.

References

1. Bankova, V., Boudourova-Krasteva, G., Sforcin, M. J., Frete, X., Kujumgiev, A., Maimoni-Rodella, R., et al. (1999) Phytochemical evidence for the plant origin of Brazilian propolis from Sao Paulo state. *Z. Naturforsch.* **54,** 401–405.
2. Fontana, J. D., Passos, M., Santos, M. H. R., Fontana, C. K., Oliveira, B. H., Schause, L., et al. (2000). Profiling propolis flavonoids and phenol carboxylic acids by means of micellar electrokinetic chromatography and biological activity. *Chromatographia* **52,** 147–151.
3. Ackermann, T. (1991). Fast chromatography study of propolis crudes. *Food Chem.* **42,** 135–138.
4. Szente, L. and Szejtli, J. (1987) Formulation of propolis with beta-cyclodextrin. *Acta Pharm. Technol.* **33,** 218–221.
5. Marcucci, M. C. (1995) Propolis—Chemical composition, biological properties and therapeutic activity. *Apidologie* **26,** 83–99.
6. Dobrowolski, J. W., Vohora, S. B., Sharma, K., Shah, S. A., Naqvi, S. A. H., and Dandiya, P. C. (1991) Antibacterial, antifungal, antiamoebic, antiinflammatory and antipyretic studies on propolis bee products. *J. Ethnopharmacol.* **35,** 77–82.
7. Harborne, J. B. (1967) *Comparative Biochemistry of the Flavonoids*. Academic Press, London.
8. Birkofer, L., Kaiser C., and Kosmol, H. (1965) Konstitution des Petunosids. *Z. Naturf.* **20,** 605.
9. Villanueva, V. R., Barbier, M., Gonnet, M., and Lavie, P. (1970) Propolis flavonoids: isolation of a new bacteriostatic agent, pinocembrine (5,7-dihydroxy flavanone). *Ann. Inst. Pasteur (Paris)* **118,** 84–87.
10. Stangaciu, S. (1998) A guide to the composition and properties of propolis. *Apiacta* **33,** 71–77.
11. Grange, J. M. and Davey, R. W. (1990) Antibacterial properties of propolis (bee glue). *J. Roy. Soc. Med.* **83,** 159–160.
12. Nakano, M., Shibuya, T., Sugimoto, T., Fukuda, S., and Kurimoto, M. (1995) Anti-MLRSA (anti-methicillin resistant *Staphylococcus aureus*) compounds in Brazilian propolis. *Honey Bee Science* **16,** 175–177.

13. Lindenfelser, L. A. (1968) In vivo activity of propolis against *Bacillus larvae*. *J. Invertebr. Pathol.* **12,** 129–131.
14. Koo, H., Cury, J. A., Rosalen, P. L., Park Y. K., and Bowen W. H. (2000) Effects of *Apis mellifera* propolis on glucosyltransferases. *J. Dent. Res.* **79,** 1130–1136.
15. Ikeno, K., Ikeno, T., and Miyazawa, C. (1991) Effects of propolis on dental caries in rats. *Caries Res.* **25,** 347–351.
16. Murray, M. C., Worthington, H. V., and Blinkhorn, A. S. (1997) A study to investigate the effect of a propolis-containing mouthrinse on the inhibition of de novo plaque formation. *J. Clin. Periodontol.* **24,** 796–798.
17. Dimov, V., Ivanovska, N., and Manolova, N. (1991) Immunomodulatory action of propolis. Influence on anti-infectious protection and macrophage function. *Apidologie* **22,** 155–162.
18. Hashimoto, T., Aga, H., Tabuchi, A., Shibuya, T., Chaen, H., Fukuda, S., et al. (1998) Anti-*Helicobacter pylori* compounds in Brazilian propolis. *Natural Medicins* **2,** 518–520.
19. Jozwik, Z. and Baraniecka-Wloszycka (1976) The effect of propolis on *Mycobacterium* sp. *Annales Universitatis Mariae Curie, Sklodowska C.* **31,** 143–150.
20. Metzner, J., Schneidewind, E. M., and Friederich, E. (1977) Effect of propolis and pinocembrin on fungi. *Pharmazie* **32,** 730–732.
21. Cizmárik, J. and Trupl, J. (1976) Effects of propolis on dermatophytic fungi. *Pharmazie* **31,** 55–55.
22. Pepeljnjak, S., Jalsenjak, I., and Maysinger, D. (1982) Inhibition of growth and biosynthesis of ochratoxin A in *Aspergillus sulphureus* NRRL 4077 by propolis extract. *Pharmazie* **37,** 439–440.
23. Amoros, M., Lurton, E., Boustie, J., Girre, L., Sauvager, F., and Cormier, M. (1994) Comparison of the anti-herpes simplex virus activities of propolis and 3-methyl-but-2-enyl caffeate. *J. Nat. Prod.* **57,** 644–647.
24. Serkedjieva J. and Manolova, N. (1992) Anti-influenza virus effect of some propolis constituents and their analogues (esters of substituted cinnamic acids). *J. Nat. Prod.* **55,** 294–302.
25. Szmeja, Z., Kulczynski, B., Sosnowski, Z., and Konopacki, K. (1989) Therapeutic value of flavonoids in *Rhinovirus* infections. *Otolaryngol. Pol.* **43,** 180–184.
26. Vlietinck, A. J., De Bruyne, T., Apers, S., and Pieters, L. A. (1998) Plant-derived leading compounds for chemotherapy of human immunodeficiency virus (HIV) infection. *Planta Med.* **64,** 97–109.
27. Hegazi, A. G., El Berdiny, F., El Assily, S., Khashabah, E., Hassan, N., and Popov, S. (1993) Studies on some aspects of antiviral activity. 1- Influence of propolis on Newcastle disease virus. *Vet. Med. J. Giza* **41,** 53–56.
28. Matsuno, T., Jung S. K., Matsumoto, Y, Saito, M., and Morikawa, J. (1997) Preferential cytotoxicity to tumor cells of 3,5-diprenyl-4-hydroxycinnamic acid (artepillin C) isolated from propolis. *Anticancer Res.* **17,** 3565–3568.
29. Guarini, L., Su, Z. Z., Zucker, S., Lin, J., Grunberger, D., and Fisher, P. B. (1992) Growth inhibition and modulation of antigenic phenotype in human melanoma and glioblastoma multiforme cells by caffeic acid phenethyl ester (CAPE). *Cell. Mol. Biol.* **38,** 513–527.

30. Grunberger, D., Banerjee, R., Eisinger, K., Oltz, E. M., Efros, L., Caldwell, M., et al. (1988) Preferential cytotoxicity on tumor cells by caffeic acid phenethyl ester isolated from propolis. *Experientia* **44,** 230–232.
31. Rao, C. V., Desai, D., Rivenson, A., Simi, B., Amin, S., and Reddy, B. S. (1995) Chemoprevention of colon carcinogenesis by phenylethyl-3-methylcaffeate. *Cancer Res.* **55,** 2310–2315.
32. Scheller, S., Krol, W., Swiacik, J., Owczarek, S., Gabrys, J., and Shani, J. (1989) Antitumoral property of ethanolic extract of propolis in mice-bearing Ehrlich carcinoma, as compared to bleomycin. *Z. Naturforsch. [C]* **44,** 1063–1065.
33. Su, Z. Z., Lin, J., Grunberger, D., and Fisher, P. B. (1994) Growth suppression and toxicity induced by caffeic acid phenethyl ester (CAPE) in type-5 adenovirus-transformed rat embryo cells correlate directly with transformation progression. *Cancer Res.* **54,** 1865–1870.
34. Basnet, P., Matsuno, T., and Neidlein, R. (1997) Potent free radical scavenging activity of propel isolated from Brazilian propolis. *Z. Naturforsch. [C]* **52,** 828–833.
35. Sud'ina, G. F., Mirzoeva, O. K., Pushkareva, M. A., Korshunova, G. A., Sumbatyan, N. V., and Varfolomeev, S. D. (1993) Caffeic acid phenethyl ester as a lipoxygenase inhibitor with antioxidant properties. *FEBS Lett.* **329,** 21–24.
36. Pascual, C., Gonzalez, R., and Torricella R. G. (1994) Scavenging action of propolis extract against oxygen radicals. *J. Ethnopharmacol.* **41,** 9–13.
37. Ginanneschi, M., Acciai, M. C., Sertoli, A., and Bracci, S. (1989) Propolis allergy: synthesis and patch testing of gamma,gamma-dimethylallyl caffeic acid ester and its o-methyl derivatives. *Contact Dermatitis* **21,** 267–269.
38. Cristea, A., Oitã, N., and Dobrescu, D. (1979) The efficacy of propolis in cicatrization of gastric ulcers. *Apicultura* **7,** 19–20.
39. Hermann, G. R. and Bracamonte, J. A. (1990) Propolis extract as a sunscreen agent. *Drug and Cosmetic Industry* **147,** 38.
40. Khayyal, M. T., el-Ghazaly, M. A., and el-Khatib, A. S. (1993) Mechanisms involved in the antiinflammatory effect of propolis extract. *Drugs Exp. Clin. Res.* **19,** 197–203.
41. Lin, S. C., Lin, Y. H., Chen, C. F., Chung, C. Y., and Hsu, S. H. (1997) The hepatoprotective and therapeutic effects of propolis ethanol extract on chronic alcohol-induced liver injuries. *Am. J. Chin. Med.* **25(3-4),** 325–332.
42. Mahran, L. G., el-Khatib, A. S., Agha, A. M., and Khayyal, M. T. (1996) The protective effect of aqueous propolis extract on isolated rat hepatocytes against carbon tetrachloride toxicity. *Drugs Exp. Clin. Res.* **22,** 309–316.
43. Philipov, I. A., Cherkasova, A .I., Dudov, I. A., and Rodionova, V. V. (1987) A clinical study on the therapeutic effect of bee products on trophic chronic varicoses ulcerations of the inferior limbs. *Apiacta* **3,** 81–82.
44. Ibricevic, H. (1981) The examination of propolis as a potential local anaesthetic. *Stomatol-Vjesn.* **10,** 125–128.
45. Paintz, M. and Metzner, J. (1979) Local-anesthetic action of propolis and some of its constituents. *Pharmazie* **34,** 839–841.

46. Kivalkina, V. P. (1971) The adjuvant effect of propolis in tetanus toxoid immunization. *Veterinary* **11,** 45–46.
47. Harish, Z., Rubinstein, A., Golodner, M., Elmaliah, M., and Mizrachi, Y. (1997) Suppression of HIV-1 replication by propolis and its immunoregulatory effect. *Drugs Exp. Clin. Res.* **23,** 89–96.
48. Choi, J. S., Yokozawa, T., and Oura, H. (1991) Antihyperlipidemic effect of flavonoids from *Prunus davidiana. J. Nat. Prod.* **54,** 218–224.
49. Castro, S. L. and Higashi, K. O. (1995) Effect of different formulations of propolis on mice infected with *Trypanosoma cruzi. J. Ethnopharmacol.* **46,** 55–58.
50. Miyares, C., Hollands, I., Cataneda, C., Gonzalez, T., Fragoso, T., Curras, R., et al. (1988) Clinical trial with a preparation based on propolis "propolisina" in human giardiasis. *Acta Gastroenterol. Latinoam.* **18,** 195–201.
51. Focht, J., Hansen, S. H., Nielsen, J. V., van den Berg-Segers, A., and Riezler, R. (1993) Bactericidal effect of propolis in vitro against agents causing upper respiratory tract infections. *Arzneimittelforschung* **43,** 921–923.
52. Varanda, E. A., Monti, R., and Tavares, D. C. (1999) Inhibitory effect of propolis and bee venom on the mutagenicity of some direct- and indirect-acting mutagens. *Teratog. Carcinog. Mutagen* **19,** 403–413.
53. Yamada, J. and Tomita, Y. (1996) Antimutagenic activity of caffeic acid and related compounds. *Biosci. Biotechnol. Biochem.* **60,** 328–329.
54. Baron, E. J. , Peterson, L. R., and Finegold, S. M. (1994) *Bailey and Scott's Diagnostic Microbiology* (9th ed) (Shanahan, J. F., ed.), Mosby-Year Book, Inc., pp. 168–188.
55. Bellegrandi, S., D'Offizi, G., Ansotegui, I. J., Ferrara, R., Scala, E., and Paganelli, R. (1996) Propolis allergy in an HIV-positive patient. *J. Am. Acad. Dermatol.* **35,** 644.

21

Enzymatic Saccharification of Cellulosic Materials

Luiz Pereira Ramos and José Domingos Fontana

1. Introduction

Lignocellulosic materials are renewable resources that can be directly or indirectly used for the production of biomolecules and commodity chemicals *(1,2)*. However, the industrial utilization of these renewable materials has been compromised by factors such as the close association that exists among the three main components of the plant cell wall—cellulose (*see* **Note 1**), hemicellulose (*see* **Note 2**) and lignin (*see* **Note 3**)—and the low efficiency by which lignocellulosic substrates are converted through biological processes such as enzymatic hydrolysis and fermentation.

Several models have been proposed to explain the internal structure of the plant cell wall. Owing to the linearity of the cellulose backbone, adjacent chains form a framework of water-insoluble aggregates of varying length and width *(3)*, and these aggregates or elementary fibrils consist of both ordered (crystalline) and less ordered (amorphous) regions *(3,4)*. The lattice forces that are responsible for maintaining the crystalline regions are basically the result of an extensive inter- and intramolecular hydrogen bonding. According to Fengel and Wegener *(4)*, microfibrils are formed by an association of several elementary fibrils with an average diameter of 3.5 nm, and the occurrence of fibrillar structures with wider dimensions appears to be dependent on the origin and treatment of the sample. Four of these elementary fibrils are then held together by a monolayer of hemicelluloses, forming thread-like structures 25 nm wide, which are enclosed in a matrix of hemicellulose and lignin. Therefore, it is the tight association of these microfibrils that ultimately confers great chemical and physical stability to the plant cell wall.

From: *Methods in Biotechnology: Environmental Microbiology: Methods and Protocols*
Edited by: J. F. T. Spencer and A. L. Ragout de Spencer

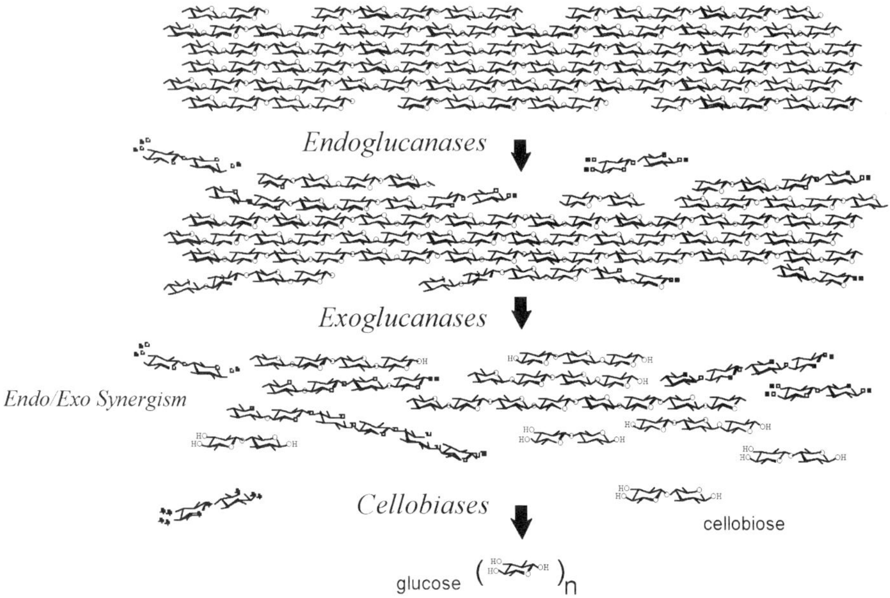

Fig. 1. Synergism among cellulases during hydrolysis of cellulosic materials.

The highly organized, crystalline structure of cellulose, added to a close interaction with hemicelluloses and lignin, has often been described as an obstacle to enzymatic hydrolysis *(5–7)*. Indeed, this tight association restricts the action of microbial hydrolases to the surface of the composite, therefore eliciting lower substrate accessibilities as compared to other water-soluble materials such as starch.

The cellulose-degrading enzymes produced by the soft-rot fungi *Trichoderma reesei,* the most widely known cellulolytic microorganism to date, are composed of three major enzyme components: endoglucanases (EG, EC 3.2.1.4), exocellobiohydrolases (CBH, EC 3.2.1.91), and β-glucosidases (EC 3.2.1.21) *(7–8)*. These enzymes are usually glycosylated, occur in multiple forms, and have distinct specific activities (*see* **Note 4**). In general, the random action of EGs gradually increases the availability of cellulose chain ends, therefore increasing the substrate availability for CBH catalysis. β-glucosidases would then catalyze the conversion of cellobiose to glucose, thus releasing the end-product inhibition of both EGs and CBHs (**Fig. 1**). At least five types of endoglucanases have been identified in the cellulase system secreted by the soft-rot fungus *T. reesei* ***(9,10)***. On the other hand, two unrelated cellobiohydrolases (CBH I and II) have been described in the culture filtrate of this fungus ***(11)***, and these enzymes have been shown to act synergistically during the enzymatic hydrolysis of cellulosic substrates ***(12)***.

Together with the aforementioned cellulases, most commercial preparations also contain a great deal of hemicellulase activity that usually accounts for the hydrolysis of the hemicellulose present within cellulosic materials ***(5)***. This activity arises from hemicellulases themselves (xylanases, mannanases, esterases) and/or from unespecific β-(1-4)-glycanases such as the endoglucanase I (EG I) of *T. reesei*. By contrast, oxidative enzymes, such as laccases, phenol oxidases, and peroxidases, are not usually present in these preparations, and any residual lignin that might be present remains insoluble after complete saccharification of the substrate ***(13)***.

Several workers ***(14–16)*** have shown that high levels of β-glucosidase activity are effective in reducing the inhibitory effects of cellobiose accumulation and consequently influence the efficiency of the overall saccharification process. Holtzapple et al. ***(17)*** showed that both exo- and endoglucanases are very sensitive to increased concentrations of cellobiose, whereas β-glucosidases are inhibited by high concentrations of glucose. However, the drop of enzyme activity in the system due to glucose accumulation was negligible when compared to that of cellobiose. Breuil et al. ***(18)*** also studied the influence of cellobiase activity on hydrolysis of both filter paper (filter paper assay) and steam-treated aspen. It was suggested that commercial cellulase preparations such as Celluclast 1.5 L (Novo Nordisk), which lack cellobiase activity for hydrolysis, should be supplemented with an excess of commercial β-glucosidase preparations such as Novozym 188 (Novo) to ensure minimal accumulation of cellobiose during hydrolysis. Other authors suggested that the accumulation of soluble sugars in the reaction mixture may also affect the adsorption profile of specific cellulase components onto the pretreated substrate ***(19)***.

During a time course hydrolysis of cellulosic materials (**Fig. 2**), the gradual decrease in the hydrolysis rate is due primarily to the end-product inhibition of the enzymes ***(5–8)***. However, this has also been associated with increases in substrate recalcitrancy during hydrolysis, therefore involving factors such as lignin accumulation and redistribution, increased cellulose crystallinity, and reduction of the substrate-available surface area owing to a partial coating of the residual substrates with irreversibly adsorbed enzymes and/or lignin ***(6,20)***. Therefore, besides enzyme-related factors, changes in substrate composition during hydrolysis have a direct implication in substrate–enzyme interaction, upon which effective hydrolysis depends.

The initial fast rate by which cellulose is hydrolyzed has been associated with the occurrence of more accessible regions at the surface of the substrate ***(8)***. At the fiber level, these more accessible regions are often associated with cracks and defects that are encountered along the fiber axis. At the molecular level, they are characterized by a larger pore volume and/or available surface

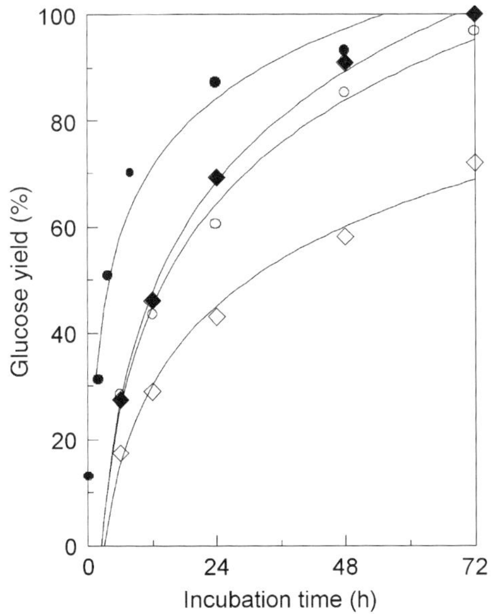

Fig. 2. Enzymatic hydrolysis of typical steam-treated hardwoods. Hydrolysis was carried out with a mixture of Celluclast 1.5 L and Novozym 188 (Novo Nordisk) at 2% (w/v) in 50 m*M* acetate buffer, pH 4.8, 150 rpm. Open losangles, water-insoluble fraction (35% lignin) plus 4.5 FPU/g cellulose. Closed losangles, water-insoluble fraction plus 7.4 FPU/g cellulose. Open circles, alkali-insoluble fraction (7% lignin) plus 8.7 FPU/g cellulose. Close circles, alkaline peroxide-treated fraction (0.5% lignin) plus 9.8 FPU/g cellulose.

area ***(21)***, higher availability of chain ends (that is, lower degree of polymerization or higher polydispersity) ***(22,23)*** and a lower crystallinity index (amorphous cellulose) ***(24)***.

Considering all of these factors, it is easy to conclude that enzymes are not so easy to handle compared to other chemicals such as mineral acids, and care must be taken to guarantee comparability among any experimental dataset. To attain good reproducibility in laboratory experiments, one has to have perfect control over the following parameters: (a) agitation and reaction temperature; (b) type of buffer and its ionic strength; (c) activity profile of the enzyme chosen; (d) substrate moisture content and its consistency in the reaction mixture; (e) enzyme loading (or the enzyme:substrate ratio); and (f) the method used for analysis and calculation of the hydrolysis yields. Measures to ensure reproducibility of these parameters are shown below.

2. Materials

2.1. Reagent

Glacial acetic acid, sodium hydroxide, a batch of commercial cellulase, chromatographic standards for glucose and cellobiose, sulfuric acid, deionized distilled water, sodium azide.

2.2. Hydrolysis Buffer

The buffer most widely used for the enzymatic hydrolysis of cellulose is the acetate buffer, pH 4.8, containing a suitable amount of a preservative such as sodium azide at 0.02%. For the purpose of this procedure, two buffer concentrations are required: 50 and 150 mmol/L. The pH at 4.8 is considered ideal for fungal cellulases such as those produced by *T. reesei* and other highly cellulolytic filamentous fungi, but other types of biocatalysts may require slightly different experimental conditions. In these cases, the citrate buffer is also commonly used.

2.3. Substrate Preparation

Successful saccharification of cellulosic substrates usually depends on whether the substrate has never been dried. There are exceptions to this, particularly when the substrate is completely delignified and presents a large surface area for enzymes to act upon. In any case, it is always advisable to use never-dried cellulosic materials for complete hydrolysis to occur. On the other hand, knowledge of the substrate moisture content is a major requirement for this experimental procedure because not only must the substrate loading be done on a dry basis, but also moisture can be a source of error when found in exceedingly high amounts. High moisture content complicates substrate homogeneity and has a negative effect on sampling. Also, too much moisture may dilute the reaction mixture beyond acceptable limits. For these reasons, the first step toward hydrolysis is the determination of the substrate moisture content (and the corresponding dry mass), a measurement that can easily be done gravimetrically.

2.4. Enzyme Solution

The enzyme solution must be freshly prepared from stock prior to saccharification. The stock must be always stored at 4°C and brought up from this to room temperature prior to sampling.

3. Method

Enzymatic hydrolyses of cellulosic materials are usually performed using commercial cellulase preparations (*see* **Note 5**), in some cases supplemented with β-glucosidase preparations such as the *Aspergillus niger* Novozym 188 (Novo Nordisk). Enzyme preparations containing a final activity of 10 to 50 filter paper units (FPU) and 20 to 40 cellobiase units (CBU) per g of cellulose are usually applied in batch suspensions of the substrate at consistencies ranging from 2 to 5% (w/v) (*see* **Note 6**). Even though filter paper is a more representative substrate to measure the actual cellulase activity used for hydrolysis (*see* **Note 7**), enzyme loading can also be evaluated in relation to the endoglucanase activity, using either carboxymethylcellulose (CMC) or hydroxyethylcellulose (HEC) as model substrates (*see* **Note 8**).

Preparation of the enzyme solution from which aliquots are transferred to the reaction mixture is one of the most critical steps toward good reproducibility of experimental data. For this purpose, we have established the following procedure in our laboratory:

1. Choose one complete cellulase preparation such as those produced by Novo Nordisk (Celluclast 1.5 L), Yakult Honsha (Onozuka), Meiji Seika Kaisha (Meicelase), Iogen, Genencor International (Multifect) and Röhm Enzyme Finland Oy (formerly Primalco Ltd. Biotec) (*see* **Note 5**).
2. Check for the availability of information on enzyme properties such as protein content and the activity level towards Whatman no. 1 filter paper, cellobiose, and, if required, CMC and/or HEC. If these are not available, they will have to be determined according to the I.U.P.A.C. recommendations described by Ghose *(25)*.
3. Based on the enzyme properties cited previously, calculate the amount of enzyme required to generate 100 mL of an enzyme solution containing 4 FPU/mL (for a delignified material, this is the enzyme loading required to generate, under the experimental conditions described in this procedure, a final activity of 50 FPU/g of substrate).
4. Using a 100-mL volumetric flask, weigh the corresponding amount of enzyme that provides the desired activity (*see* **Note 9**). Before sampling, the enzyme preparation must be brought up to room temperature and gently homogenized to avoid errors in the volume and activity taken from the stock.
5. Complete the volumetric flask up to the 100 mL volumetric label with hydrolysis buffer and close the flask with an appropriate lid (e.g., a rubber stopper). Before transferring the enzyme solution to the reaction mixture, the solution must be homogenized without foaming, and conditioned at the temperature chosen to carry out the hydrolysis experiment.
6. The enzyme solution must always be prepared in excess to allow for the determination of its protein content and activity profile against filter paper, cellobiose, and CMC or HEC, if required (*see* **Note 10**). In principle, storage of the dilute biocatalyst must be avoided, but, if this is unavoidable, refrigeration at 4°C is highly recommended.

Complete saccharification of delignified cellulosic materials can be achieved according to the following experimental procedure:

1. For a substrate consistency of 2% (m/v) (*see* **Note 6**), weigh three batches of substrate containing 2 g in relation to its dry mass (*see* **Note 11**) and place them separately into 250 mL flat-bottomed flasks.
2. Add 25 mL of 150 m*M* acetate buffer, pH 4.8, to the flat-bottomed flasks and bring the total reaction volume to 75 mL by adding distilled water; the substrate moisture content must be taken into account to ensure that the final concentration of the hydrolysis buffer is 50 m*M* (*see* **Note 12**).
3. Close the flat-bottomed flasks with rubber stoppers to avoid moisture loss while the suspensions are conditioned in a shake incubator preset at 40°C and 145 rpm (*see* **Note 13**).
4. Prepare a flask containing 75 mL of 50 m*M* acetate buffer, pH 4.8, to serve as an enzyme blank.
5. Condition, at 40°C, 100 mL of an enzyme solution containing 4 FPU/mL in 50 m*M* acetate buffer, pH 4.8; activity against filter paper and cellobiose must be measured in this enzyme solution to ensure control of the actual enzyme loading (a fixed enzyme:cellulose ratio is desirable among hydrolyses of different cellulosic materials) (*see* **Note 14**).
6. Transfer precisely 25 mL of the enzyme solution to two of the three flat-bottomed flasks described above; to the third, add 25 mL of 50 m*M* acetate buffer, pH 4.8, to serve as the substrate blank.
7. Immediately after the enzyme loading, allow complete homogenization of the reaction mixture and collect 1–2 mL of the reaction supernatant in a test tube (this aliquot must correspond as much as possible to the time zero of the enzymatic saccharification) (*see* **Note 15**); equivalent aliquots must also be withdrawn from both enzyme and substrate reaction controls.
8. Choose a minimum of six points along the hydrolysis time course. For example, aliquots can be withdrawn at 6, 12, 24, 48, and 72 h of incubation (*see* **Note 16**). After collecting an aliquot, the test tube must be covered with a marble and immediately transferred to a boiling water bath to stop the reaction by enzyme denaturation (5 min is usually enough for that).
9. Remove the test tubes from the boiling water bath and let them cool down to room temperature. The aliquot must then be transferred to Eppendorf tubes and centrifuged at 10,000 rpm for 2 min.
10. Dilute each reaction aliquot to a sugar concentration intermediate to the range of the method used for analysis. If the analysis is carried out by liquid chromatography, dilution to 1 to 2 mg/mL is usually adequate when the method of detection is differential refractometry (*see* **Note 17**). However, higher dilution rates will be required if analysis is to be based on a typical method for the determination of reducing sugars, such as the dinitrosalicylic acid (DNS) or the Nelson-Somogyi methods *(26)* (*see* **Note 18**).
11. Calculate the hydrolysis yield in relation to either the dry mass or cellulose content of the substrate (*see* **Note 19**).

4. Notes

1. Hemicelluloses represent a family of heteropolysaccharides present in all plant materials, whose chemical nature varies from tissue to tissue and from species to species. These polysaccharides are formed by a wide variety of building blocks, including pentoses (e.g., xylose, rhamnose, and arabinose), hexoses (e.g., glucose, mannose, and galactose) and uronic acids (e.g., 4-*O*-methyl-glucuronic and galacturonic acids) ***(4)***. Some hemicelluloses, particularly heteroxylans, also show a substantial degree of acetylation. Despite the complexity of these heteropolysaccharides, their structure seems to be generally rod-shaped with branches and side chains folded back to the main chain by means of hydrogen bonding. This rod-like structure facilitates their interaction with cellulose microfibrils, resulting in a tight association that gives greater stability to the aggregate.
2. In plants, the hemicelluloses are generally combined with lignin. Lignin is a phenolic macromolecule that is primarily formed by the free-radical polymerization of *p*-hydroxy cinnamyl alcohol units with varying methoxyl contents ***(4,27)***. The chemical structure of lignin is very complicated and is based on three monomeric precursors: coniferyl alcohol, sinapyl alcohol, and *p*-coumaryl alcohol. The proportions of these monomers vary among species, and this ratio has been used for taxonomic purposes. Depending on the degree of methoxylation, the aromatic group is either guaiacyl (derived from coniferyl alcohol) or syringyl (derived from sinapyl alcohol). The former has only one methoxyl group adjacent to the phenolic hydroxyl group, whereas the latter has two. The most important physical property of this organic macromolecule is its rigidity, which not only gives strength to the plant tissue but also prevents the collapse of the water-conducting elements.
3. Cellulose is a linear homopolysaccharide that consists of glucose (D-glucopyranose) units linked together by β-(1-4) glycosidic bonds (β-D-glucan). This polysaccharide is widespread in nature, occurring in both primitive and highly evolved plants. The size of the cellulose molecule is normally given in terms of its degree of polymerization (DP), i.e., the number of anhydroglucose units present in a single molecule. However, the conformational analysis of cellulose has indicated that cellobiose (4-*O*-β-D-glucopyranosyl-β-D-glucopyranose), rather than glucose, is its basic structural unit ***(4)***.
4. The mode of action of microbial cellulases has been elucidated primarily by investigating the properties of purified enzyme components acting on substrates ranging from soluble chromophores ***(28)*** to bacterial ***(29)*** and microcrystalline cellulose ***(19)***. However, the isolation of single components is difficult to achieve even through long and tedious purification procedures, and small contaminants, such as traces of endoglucanase activity in purified CBH II preparations ***(30)***, are likely to remain. This problem has been gradually overcome by genetic engineering, as many of these enzymes have been successfully cloned and are now being made available for research purposes ***(31)***. In this way, enzymes are finding their way to rather useful industrial applications, including biopolishing of textile fibers ***(32)*** and fiber modification in pulp and paper manufacturing ***(33–35)***.

5. There is a great variety of cellulase preparations available in the literature for bioconversion of lignocellulosics, but each of them has been produced for specific applications such as biopolishing of textiles, saccharification of cellulosics, dewatering of industrial fibers, biobleaching, deinking of printed office paper, and fiber modification for better fiber properties. Those that are capable of completely hydrolyzing cellulose are known as complete systems, that is, enzymes in which all of the key components of the cellulase system are found in optimum ratio. Enzymes for other applications have usually gone through some degree of modification in order to enhance the desired effect on cellulose fibers, be them from pulp, paper, or textiles.
6. Substrate consistencies ranging from 2% to 5% (m/v) are recommended. Higher consistencies of 10% are seldomly used in lab-scale experiments because, at this level, there is very little motion inside the flat-bottomed flask and less efficient mass transfer during hydrolysis.
7. The I.U.P.A.C. Commission recommends filter paper as the ideal substrate for the determination of cellulase activity in enzyme preparations. However, the method described by Ghose ***(25)*** in *Pure and Applied Chemistry* has been the subject of many modifications, and these are usually oriented to solving problems such as: (a) the required normalization of the amount of filter paper used to determine the enzyme activity (e.g., the use of paper strips with 70 ± 0.5 mg in the FPU assay); (b) the avoidance of vortex mixing of the reaction mixture after addition of the filter-paper strip; (c) the replacement of methods for reducing sugar determination by high-performance liquid chromatography to avoid interference of other reducing chemicals that may be present in the substrate hydrolysate ***(36)***; and (d) the establishment of equivalent β-glucosidase loadings on cellulase assays if one is to avoid distinct levels of end-product inhibition ***(14–16)***.
8. The endoglucanase activity of cellulase preparations is generally determined against two model substrates—CMC and HEC. The former is the classic substrate for determining endoglucanase activity, whereas the latter offers several advantages for conducting the activity assay because HEC solutions have lower viscosity than equivalent CMC solutions and the assay is carried out for shorter incubation times of 10 min ***(25)***. Using the HEC substrate has the following benefits: (a) the assay is faster and easier to handle; (b) owing to the lower viscosity of the HEC solution, an increased substrate availability is attained together with a better mass transfer during the assay; and (c) shorter residence times provide activity measurements at the initial rate of hydrolysis, avoiding undesired effects such as end-product inhibition. However, care must be taken when one attempts to compare CMCase with HECase activity because the assays are carried out under different conditions and the activities are expressed in different units (IU/mL for CMCase and nkat/mL for HECase activities).
9. Enzymes can be obtained either as a freeze-dried powder or as dark solutions containing a great deal of stabilizers to increase shelf life and avoid contamination. While the former is easier to weigh, the latter requires a flask to which the enzyme stock is transferred with the aid of a Pasteur pipet. Therefore, even liquid

cellulase preparations must be weighed prior to dilution because this allows for a better reproducibility in the overall experimental procedure.

10. No matter how accurately the calculations are made, activity assays must always be carried out before the experiments and directly on the enzyme solution used for hydrolysis. Enzyme loadings based on protein should be avoided when the stock corresponds to a liquid preparation or crude extract. By contrast, pure (or purified) enzymes can be loaded into the reaction mixture based on protein content because, in these preparations, there is little to interfere with the methods used for protein determination (e.g., Lowry, Bradford, and Biuret methods).
11. If the substrate contains a significant amount of non-cellulosic components such as lignin, substrate loadings can also be expressed in relation to cellulose content. However, this strategy has the disadvantage of leading to distinct substrate consistencies per reaction mixture when hydrolysis is performed against substrates with variable amounts of lignin. As a result, uneven rates of mass transfer would occur during hydrolysis, causing the enzymes to be unevenly distributed within substrates and compromising the comparability among the hydrolysis experiments.
12. When setting up the hydrolysis experiment, the substrate moisture content must be taken into account to avoid dilution of the reaction mixture, reduction in the buffer ion strength, and a probable underestimation of reaction yields. For example, if a substrate with a moisture content of 70% is to be hydrolyzed, 2 g in relation to dry mass will correspond to 6.7 g of wet fibers, meaning that 4.7 g of moisture is also added to the system. Hence, only 45.3 mL of water would have to be used to complete the 75 mL described in the experimental procedure.
13. Agitation, preferably orbital, must be maintained constant and at a level not high enough to produce enzyme inactivation due to shearing (literature recommends 145–150 rpm). Temperature control must also be very strict at 40 or 45°C (the lower temperature range seems to impair lower enzyme deactivation in long-term hydrolysis experiments).
14. It is highly recommended that the amount of both filter paper and cellobiase activities used for hydrolysis always be calculated in relation to the cellulose content of the substrate **(Fig. 2)**. Cellulose is the natural substrate for the enzymes, and enzyme loadings based on dry mass may lead to a misinterpretation of substrate accessibility because the enzyme-substrate ratios are bound to be unequal, leading to distinct combined effects of end-product inhibition and available surface area. On the other hand, when the substrate contains an appreciable amount of hemicelluloses, the total amount of carbohydrates works as a better reference for enzyme loading because commercial cellulases usually present high hemicellulase activity.
15. When the reaction kinetics is under evaluation, the aliquot volume retrieved from the reaction mixture must not influence the hydrolysis yield by gradually generating a concentration factor that will elicit higher substrate consistencies, overestimated carbohydrate analysis, and higher levels of glucose and/or cellobiose inhibition.
16. Depending on the substrate and experimental conditions, one hydrolysis run may

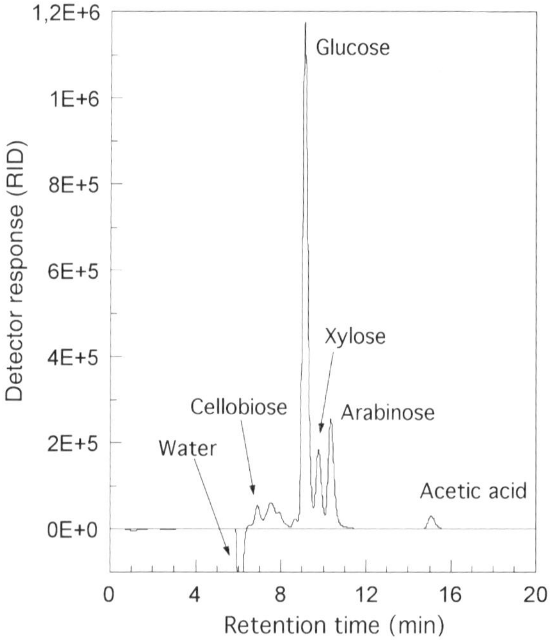

Fig. 3. HPLC analysis in Aminex HPX-87H (Bio-Rad) of a substrate hydrolysate after dilution with HPLC-grade water (1:10, v/v). Hydrolysis was carried out for 72 h at 2% (w/v) in 50 m*M* acetate buffer, pH 4.8, 150 rpm, using a mixture of Celluclast 1.5 L and Novozym 188 (Novo Nordisk) containing a final activity of 15 FPU/g cellulose.

last as long as 96 h for the complete saccharification of cellulose and, even so, not all pretreated materials are amenable to complete hydrolysis under this time frame (**Fig. 2**).

17. Several cellulosic materials contain a considerable proportion of hemicellulose, and this observation stresses the suitability of chromatographic methods to evaluate sugar release upon hydrolysis. Methods for the determination of reducing sugars cannot differentiate between glucose and hemicellulose sugars, and variations in the calibration curve for pentoses and hexoses may introduce further distortions to the calculated yields. The most widely used HPLC procedure for the determination of sugars in reaction hydrolysates utilizes ion-exchange columns such as the Aminex HPX-87H (Bio-Rad) (**Fig. 3**). Analysis in this column does not require any special sample preparation other than dilution to an acceptable concentration range and filtration through a 0.45-mm filter to avoid damage to the pipe lines, piston seals, and column-packing materials. The HPLC procedure includes conditioning of the Aminex HPX-87H column at 65°C and elution of the HPLC system with 10 m*M* H_2SO_4 at a 0.6 mL/min flow rate. Detection is usually performed by differential refractometry.

18. The DNS method is less sensitive and definitely more expensive than the Nelson-Somogyi method, but it is easier to perform and usually not as susceptible to the interference of other reducing compounds inevitably present in the reaction supernatant. Several authors have shown that pretreated materials derived from softwoods ***(37,38)*** and hardwoods ***(36,39,40)*** contain varying amounts of non-carbohydrate components that are inhibitory to cellulases and may interfere with methods such as DNS and Nelson-Somogyi.
19. Consider a delignified substrate containing 85% of cellulose and 15% of polyoses (heteroxylan) as an example: if a 1:10 dilution of a 24 h hydrolysis aliquot yields 1.35, 0.21, and 0.09 mg/mL of glucose, xylose, and arabinose (as determined by HPLC), 71.5% of the substrate cellulose content (2 g × 0.85 = 1.7 g of cellulose, which, by hydrolysis, gives 1.89 g of glucose; in 100 mL, this corresponds to 18.89 mg/mL for a total of 13.5 mg/mL measured by HPLC) and 88.2% of the substrate hemicellulose content (2 g × 0.15 = 0.3 g of hemicellulose, which, by hydrolysis, gives 0.34 g of pentoses; in 100 mL, this corresponds to 3.4 mg/mL for a total of 3.0 mg/mL measured by HPLC) would have been hydrolyzed based on the substrate theoretical yields. On the other hand, if sugar release is measured by methods such as DNS, the final hydrolysis yield would have accounted for all of the carbohydrates present in the reaction supernatant as reducing sugars, but it would not be possible to determine how much each of cellulose or hemicellulose was hydrolyzed.

Acknowledgments

The authors thank CNPq and F. Araucaria-SETI-PR for funding.

References

1. Parisi, F. (1989) Advances in lignocellulosics hydrolysis and in the utilization of the hydrolysates. In *Advances in Biochemical Engineering and Biotechnology*, vol. 38 (Fiechter, A., ed.), Springer-Verlag, Berlin, pp. 53–87.
2. Ghosh, P. and Singh, A. (1993) Physicochemical and biological treatments for enzymatic/microbial conversion of lignocellulosic biomass. *Adv. Appl. Microbiol.* **39,** 295–333.
3. Atalla, R. H. (1988) Structural transformations in cellulose. In *Steam Explosion Techniques: Fundamentals and Industrial Applications* (Focher, Marzetti, A., and Crescenzi, V., eds.), Gordon and Breach, Philadelphia, pp. 97–120.
4. Fengel, D. and Wegener, G. (1989) *Wood: Chemistry, Ultrastructure, Reactions.* Walter de Gruyter, Berlin.
5. Wood, T. M. and Garcia-Campayo, V. (1990) Enzymology of cellulose degradation. *Biodegradation* **1,** 147–161.
6. Ramos, L. P. and Saddler, J. N. (1994) Bioconversion of wood residues: mechanisms involved in pretreating and hydrolyzing lignocellulosic materials. In *Enzymatic Conversion of Biomass for Fuels Production* (Himmel, M. E., Baker, J. D., and Overend, R. P., eds.), ACS Symposium Series 566, American Chemical Society, Washington, DC, pp. 325–341.

7. Coughlan, M. P. (1985) The properties of fungal and bacterial cellulases with comment on their production and application. *Biotechnol. Gen. Eng. Rew.* **3,** 39–109.
8. Fan, L. T., Gharpuray, M. M., and Lee, Y. H. (1987) *Cellulose Hydrolysis.* Springer-Verlag, New York.
9. Saloheimo, M., Nakari-Setälä, T., Tenkanen, M., and Penttilä, M. (1997) DNA cloning of a *Trichoderma reesei* cellulase and demonstration of endoglucanase activity by expression in yeast. *Eur. J. Biochem.* **249,** 584–591.
10. Saloheimo, A., Henrissat, B., Hoffrén, A. M., Teleman, O., and Penttilä, M. (1994) A novel, small endoglucanase gene, EGl5, from *Trichoderma reesei* isolated by expression in yeast. *Molecular Microbiology* **13,** 219–228.
11. Claeyssens, M., Tomme, P., Breher, C. F., and Hehre, E. J. (1990) Stereochemical course of hydrolysis and hydration reactions catalysed by cellobiohydrolases I and II from *Trichoderma reesei. FEBS Lett.* **263,** 89–92.
12. Nidetzki, B., Steiner, W., and Claeyssens, M. (1995) Synergistic interaction of cellulases from *Trichoderma reesei* during cellulose degradation. In *Enzymatic Degradation of Insoluble Carbohydrates* (Saddler, J. N. and Penner, M. H., eds.), ACS Symposium Series 618, American Chemical Society, Washington, DC, pp. 90–112.
13. Ramos, L. P., Cotrim, A. R., Silva, F. T., and Chen, C.-L. (1999) Characterization of residual lignin after SO_2-catalyzed steam explosion and enzymatic hydrolysis of *Eucalyptus viminalis* wood chips. *J. Agric. Food Chem.* **47,** 2992–2302.
14. Chan, M., Breuil, C., Schwald, W., and. Saddler, J. N. (1989) Comparison of methods for quantifying the hydrolytic potential of cellulase enzymes. *Appl. Microbiol. Technol.* **31,** 413–418.
15. Breuil, C., Chan, M., and Saddler, J. N. (1990) Comparison of the hydrolytic activity of commercial cellulase preparations. *Appl. Microbiol. Biotechnol.* **34,** 31–35.
16. Ramos, L. P., Breuil, C., and Saddler, J. N. (1993) The use of enzyme recycle and the influence of sugar accumulation on cellulose hydrolysis by *Trichoderma cellulases. Enzyme Microb. Technol.* **15,** 19–25.
17. Holtzapple, M. T., Cognata, M., Shu, Y., and Hendrickson, C. (1990) Inhibition of *Trichoderma reesei* cellulase by sugars and solvents. *Biotechnol. Bioeng.* **36,** 275–287.
18. Breuil, C., Chan, M., Gilbert, M., and Saddler, J. N. (1992) Influence of β-glucosidase on the filter paper activity and hydrolysis of lignocellulosics substrates. *Biores. Technol.* **39,** 139–142.
19. Tanaka, M., Nakamura, H., Taniguchi, M., Morita, T., Matsuno, R., and Kamikubo, T. (1986) Elucidation of adsorption processes of cellulases during hydrolysis of crystalline cellulose. *Appl. Microbiol. Biotechnol.* **23,** 263–268.
20. Converse, A. O., Ooshima, H., and Burns, D. S. (1990) Kinetics of enzymatic hydrolysis of lignocellulosic materials based on surface area of cellulose accessible to enzyme and enzyme adsorption on lignin and cellulose. *Appl. Biochem. Biotechnol.* **24/25,** 67–73.
21. Grethlein, H. E. (1985) The effect of pore distribution on the rate of enzymatic hydrolysis of cellulosic substrates. *Bio/Technology* **3,** 155–160.

22. Ramos, L. P., Nazhad, M. M., and Saddler, J. N. (1993) The effect of enzymatic hydrolysis on the morphology and fine structure of pretreated cellulose. *Enzyme Microb. Technol.* **15,** 821–831.
23. Ramos, L. P., Zandoná Filho, A., Deschamps, F. C., and Saddler, J. N. (1999) The effect of *Trichoderma cellulase* on the fine structure of a bleached softwood kraft pulp. *Enzyme Microb. Technol.* **24,** 371–380.
24. Sinitsyn, A. P., Gusakov, A. V., and Vlasenko, E. Y. (1991) Effect of structural and physico-chemical features of cellulosic substrates on the efficiency of enzymatic hydrolysis. *Appl. Biochem. Biotechnol.* **30,** 43–59.
25. Ghose, T.K. (1986) Measurement of cellulase activities. *Pure Appl. Chem.* **59,** 257–268.
26. Chaplin, M. F. and Kennedy, J. F. (1994) *Carbohydrates Analysis: A Practical Approach.* IRL Press, Oxford, UK.
27. Higuchi, T. (1981) Lignin structure and morphological distribution in plant cell wall. In *Lignin Biodegradation: Microbiology, Chemistry, and Potential Applications* (Kirk, T. K., Higuchi, T., and Chang, H.-M., eds.), CRC Press, Boca Raton, FL, pp. 1–20.
28. Claeyssens, M. and Aerts, G. (1992) Characterization of cellulolytic activities in commercial *Trichoderma reesei* preparations: an approach using small, chromogenic substrates. *Biores. Technol.* **39,** 143–146.
29. Sridosuk, M., Kleman-Leyer, K., Keränen, S., Kirk, T. K., and Teeri, T. (1998) Mode of action on cotton and bacterial cellulose of a homologous endoglucanase-exoglucanase pair from *Trichoderma reesei. Eur. J. Biochem.* **251,** 885–992.
30. Reinikainen, T., Henriksson, K., Siika-aho, M., Teleman, O., and Poutanen, K. (1995) Low level endoglucanase contamination in a *Trichoderma reesei* cellobiohydrolase II preparation affects its enzymatic activity on β-glucan. *Enzyme Microb. Technol.* **17,** 888–892.
31. Thomas, S. R., Laymon, R. A., Chou, Y. C., Tucker, M. P., Vinzant, T. B., Adney, W. S., et al. (1995) Initial approaches to artificial cellulase systems for conversion of biomass to ethanol. In *Enzymatic Degradation of Insoluble Carbohydrates* (Saddler, J. N. and Penner, M. H., eds.), ACS Symposium Series 618, American Chemical Society, Washington, DC, pp. 208–236.
32. Cavaco-Paulo, A. (1997) Cellulases in textile processes. In *Proceedings of the Fifth Brazilian Symposium on the Chemistry of Lignin and Other Wood Components* (Ramos, L. P., ed.), Sepia, Curitiba, PR, Brazil, pp. 404–412.
33. Pommier, J. C., Goma, G., Fuentes, J. L., Rousset, C., and Jokinen, O. (1990) Using enzymes to improve the process and product quality in the recycled paper industry. Part II. Industrial applications. *Tappi J.* **73,** 197–202.
34. Pere, J., Siika-aho, M., Buchert, J., and Viikari, L. (1995) Effects of purified *Trichoderma reesei* cellulases on the fiber properties of kraft pulp. *Tappi J.* **6,** 71–78.
35. Mansfield, S. D., Wong, K. K. Y., Jong, E., and Saddler, J. N. (1996) Modification of douglas-fir mechanical and kraft pulps by enzyme treatment. *Tappi J.* **79,** 125–132.

36. Schwald, W., Chan, M., Breuil, C., and Saddler, J. N. (1988) Comparison of HPLC and colorimetric methods for measuring cellulolytic activity. *Appl. Microbiol. Biotechnol.* **28,** 398–403.
37. Larsson, S., Palmqvist, E., Hahn-Hägerdahl, B., Tengborg, C., Sternberg, K., Zacchi, G., et al. (1999) The generation of fermentation inhibitors during dilute acid hydrolysis of softwood. Enzyme Microb. Technol. **24,** 151–159.
38. Boussaid, A., Robinson, J., Cai, Y., Gregg, D. J., Saddler, J. N., and Cai, Y. G. (1999) Fermentability of the hemicellulose derived sugars from steam-exploded softwood (Douglas-fir). *Biotechnol. Bioeng.* **64,** 284–289.
39. Dekker, R. F. H. (1988) Inhibition of *Trichoderma reesei* β-glucosidase activity derived from autohydrolysis-exploded *Eucalyptus regnans. Appl. Microbiol. Biotechnol.* **29,** 593–598.
40. Sinitsyn, A. P., Clesceri, L. S., and Bungay, H. R. (1982) Inhibition of cellulases by impurities in steam-exploded wood. *Appl. Biochem. Biotechnol.* **7,** 455–458.

22

Molecular Identification of Microbial Populations in Petroleum-Contaminated Groundwater

Kazuya Watanabe, Natsuko Hamamura, and Nobuo Kaku

1. Introduction

In microbiology, morphological observation does not generally provide valuable information for species identification. Identification of microorganisms has therefore relied on physiological characterization, although this method is available only after they are isolated and cultivated in the laboratory. Because over 99% of microorganisms in the natural environment cannot be cultivated by standard techniques, alternative methods are necessary for the detection and identification of a large fraction of natural microbial populations. This is of particular relevance to microorganisms in the anaerobic ecosystem, as each of them requires a specific culture condition according to its specialized ecological niche. In addition, cultivation of anaerobes is generally laborious and time consuming. Culture-independent molecular phylogenetic approaches developed in the last two decades have enabled the detection and identification of such natural microbial populations.

In molecular phylogenetic approaches, natural microbial populations are identified by using nucleotide sequences retrieved from the environment without cultivation. Database search with the environmental sequences allows the determination of phylogenetic positions of the organisms present at the study site. Genes coding for small subunit rRNA (ssu rDNA; 16S rDNA for *Bacteria* and *Archaea*) have most frequently been used for this purpose, because numerous sequences (over 5000 sequences) stored in the databases (e.g., GenBank) enable reliable phylogenetic identification. One of the typical molecular approaches includes the following procedures: (1) DNA extraction from an

From: *Methods in Biotechnology: Environmental Microbiology: Methods and Protocols*
Edited by: J. F. T. Spencer and A. L. Ragout de Spencer

environmental sample, (2) PCR amplification of certain gene fragments, (3) cloning of PCR-amplified fragments into *Escherichia coli* cells, (4) sequencing of the cloned fragments, and (5) database search and phylogenetic analysis with the determined sequences. Other approaches may include denaturing gradient gel electrophoresis *(1)* or restriction fragment-length polymorphism *(2)*. Among them, the design of steps 1 and 2 should be particularly careful, because these steps may largely affect species of microorganisms to be detected by the molecular approach *(3)*. These two steps should be optimized for each environmental sample, while commonly applicable procedures are available for the subsequent steps.

For the design of step 1, attention should be paid to (a) avoid selective disruption of certain microbial cells, (b) prevent released DNA from fragmentation, and (c) minimize co-extraction of substances inhibitory to PCR amplification, e.g., humics. In some cases, sufficient amounts of cells need to be collected by filtration and/or centrifugation before the extraction of DNA. We recommend a combination of physical, chemical, and enzymatic treatments as described in **Subheading 2.** In step 2, PCR primers should be carefully selected. In order to detect diverse microorganisms, universal primers targeting ssu rDNA have been used (**Table 1**), which were designed from nucleotide sequences conserved among most of the organisms that had been isolated and cultivated. Domain-specific (*Bacteria, Archaea* and *Eukarya*) primers are also available (**Table 1**). However, we should realize that none of these primers is perfect; rDNA fragments of some groups of organism are not amenable to PCR amplification owing to the presence of specific nucleotide substitutions in the primer regions (primer mismatches), even when we use universal primers. In this regard, we recommend using multiple primer sets for one sample, which enables the evaluation of possible biases caused by primer-specific amplification of preferential groups of rDNA.

In this chapter, we describe a method that has proven useful for molecular identification of microbial populations in oil-contaminated groundwater. Our study has been investigating groundwater obtained from underground crude-oil storage cavities (situated at Kuji in Iwate, Japan), whose characteristics have been described elsewhere *(4)*. The method we describe here is a modified version of the previous method. Special emphases are placed on steps 1 and 2, and readers are referred to previous articles *(4,5)* for the subsequent steps.

2. Materials

2.1. DNA Extraction

1. 0.22-μm Membrane filter (e.g., type GV; Millipore) (*see* **Note 1**).
2. 15-mL Screw-capped polypropylene tube.

Table 1
Primers Used for Amplification of 16S rDNA Fragments

Primer[a]	Position[b]	Sequence (5' to 3')[c]	Specificity	Reference
8F	8–27	AGAGTTTGATYMTGGCTCAG	*Bacteria*	*6*
25F	25–42	CYGGTTGATCCTGCCRG	*Archaea*	*7*
A109F	109–125	ACKGCTCAGTAACACGT	*Archaea*	*8*
0112aR	112–128	CCACGTGTTACTSAGC	*Archaea*	*9*
0348aF	333–348	TCCAGGCCCTACGGG	*Archaea*	*9*
I-341F	341–356	CCTACGGGIGGCIGCA	*Bacteria*	*10*
341F	341–357	CCTACGGGAGGCAGCAG	*Bacteria*	*1*
Uni515F	515–533	GTGYCAGCMGCCGCGTAA	Universal	*11*
I-533R	515–533	TIACCGIIICTICTGGCAC	*Bacteria*	*10*
0517R	517–531	ACCGCGGCKGCTGGC	Universal	*9*
534R	518–534	ATTACCGCGGCTGCTGG	*Bacteria*	*1*
0690aR	704–690	TTACAGGATTTCACT	*Archaea*	*9*
907R	907–926	CCGTCAATTCMTTTRAGTTT	Universal	*12*
Ar915R	915–934	GTGCTCCCCCGCCAATTCCT	*Archaea*	*13*
Ar958R	958–976	YCCGGCGTTGAMTCCAATT	*Archaea*	*14*
968F	968–984	AACGCGAAGAACCTTAC	*Bacteria*	*15*
1068F	1053–1068	GCATGGCYGYCGTCAG	*Bacteria, Archaea*	*9*
1100AR	1100–1115	TGGGTCTCGCTCGTTG	*Archaea*	*16*
1114F	1099–1114	GCAACGAGCGCAACCC	*Bacteria, Archaea*	*9*
1401R	1385–1401	CGGTGTGTACAAGACCC	*Bacteria*	*15*
Uni1408R	1390–1408	TGACGGGCGGTGTGTRCAA	Universal	*11*
1391R	1391–1407	GACGGGCGGTGTGTRCA	Universal	*17*
1406F	1391–1406	TGYACACACCGCCCGT	*Bacteria, Archaea*	*9*
Univ1392R	1392–1406	ACGGGCGGTGTGTRC	Universal	*18*
1492R	1492–1510	GGTTACCTTGTTACGACTT	Universal	*12*
1525R	1525–1541	AAGGAGGTGATCCAGCC	*Bacteria, Archaea*	*9*
1538F	1524–1538	CGGTTGGATCACCTC	*Archaea*	*9*
1546R	1529–1546	CAKAAAGGAGGTGATCC	*Bacteria*	*6*

[a] F, forward primer; R, reverse primer.
[b] Corresponding to the numbering in the sequence of the 16R rRNA gene of *Escherichia coli*.
[c] I, inosine.

3. 1.5-mL Microcentrifuge tube.
4. Cell suspension buffer: 10 m*M* Tris-HCl, pH 8.0, 1 m*M* EDTA, 0.35 *M* sucrose.
5. Proteinase K: 100 mg dissolved per mL of TE.

6. Lysis solution: 100 m*M* Tris-HCl, pH 8.0, 0.3 *M* NaCl, 20 m*M* EDTA, 2% (w/v) sodium dodecyl sulphate.
7. Ice-cold absolute ethanol.
8. 70% Ethanol.
9. Phenol/chloroform solution: Mix an equal volume of Tris buffer (0.1 *M* Tris-HCl, pH 8.0) saturated phenol and chloroform/isoamyl-alcohol solution (24:1). Add 0.1% (w/v) 8-hydroxyquinoline, and store in a light-tight bottle at 4°C.
10. RNase A: 10 mg/mL of glycerol solution.
11. TE buffer: 10 m*M* Tris-HCl, pH 8.0, 1 m*M* EDTA.

2.2. PCR Amplification of ssu rDNA

1. *Taq* DNA polymerase (Amplitaq Gold, Applied Biosystems).
2. 10X buffer: 150 m*M* Tris-HCl, pH 8.0, 500 m*M* KCl.
3. 25 m*M* $MgCl_2$ solution.
4. Deoxynucleotide triphosphate (dNTP) mixture. 2 m*M* each dNTP (dATP, dCTP, dGTP, and dTTP) dissolved in water and pH adjusted with NaOH to 7.0.
5. PCR primers: Dissolve in TE buffer at 25 μ*M*. Some of the PCR primers widely used to amplify 16S rDNA fragments are listed in **Table 1**.
6. Template DNA.
7. 1.5% (w/v) Agarose gel. Dissolve DNA-grade agarose in TBE buffer (10.8 g Trizma® base, 5.5 g of boric acid and 4 mL of 0.5 *M* EDTA at pH 8.0, per liter) by using microwave oven. After cooling down to 50°C, the agarose solution is pored into a gel-casting tray.
8. DNA size marker: We have routinely been using the 50- to 2500-bp DNA size marker (Takara). This marker contains 50-, 100-, 200-, 300-, 400-, 500-, 525-, 700-, 1000-, 1250-, 1500-, 2000-, and 2500-bp DNA fragments.
9. Electrophoresis buffer (TBE buffer).
10. 10X Loading buffer: 50% glycerol, 0.25% bromophenol blue, 0.25% xylene cyanol FF.
11. Ethidium bromide solution (10 mg per mL) (*see* **Note 2**).

3. Methods

3.1. DNA Extraction

1. Collect microorganisms in 2 L of groundwater on a 0.22-μm membrane by vacuum filtration (*see* **Note 3**).
2. Place the membrane filter in a 15-mL tube and add 1.0 mL of the cell suspension buffer; suspend the cells in the buffer by pipeting, vortexing, and scraping with a spatula (*see* **Note 4**).
3. Add 100 μL of proteinase K solution, and incubate at 37°C for 1 h (*see* **Note 5**).
4. Add 1.5 mL of lysing solution, and subject the solution to vortexing (*see* **Note 6**).
5. Incubate the tube at 70°C for 30 min with occasional shaking.
6. Subject the solution to three cycles of freezing at –20°C and thawing at 70°C.

7. Remove the membrane filter.
8. Add 2.5 mL of phenol/chloroform and mix by vortex.
9. Centrifuge the mixture at 12,000*g* for 5 min.
10. Transfer the aqueous phase to a new 15-mL tube, and repeat the phenol/chloroform extraction as in **steps 8** and **9**.
11. Transfer the top aqueous phase to a new tube, add 2 vol of cold ethanol, and incubate at –20°C for at least 2 h.
12. Centrifuge the tube at 12,000*g* for 10 min at 4°C.
13. Discard the supernatant, wash the pellet (may not be visible) with 1 mL of 70% ethanol, and centrifuge at 12,000g for 5 min.
14. Dissolve the pellet in 100 μL of TE buffer and add 1 μL of RNase A (10 mg per mL), followed by incubation at 37°C for 1 h.
15. Add 2 volumes of absolute ethanol and incubate the solution at –20°C for 1 h.
16. Centrifuge at 12,000*g* for 10 min.
17. Discard the supernatant and wash the pellet with 70% ethanol.
18. Dissolve DNA in 100 μL of TE buffer.
19. Quantify the extracted DNA by measuring the UV absorption spectrum (*see* **Note 7**).

3.2. PCR Amplification

1. Mix the reagents at appropriate proportions in a PCR tube (*see* **Note 8**). A typical mixture (50 μL) contains 1.25 U *Taq* DNA polymerase, 15 m*M* Tris-HCl, pH 8.0, 50 m*M* KCl, 1.5 m*M* $MgCl_2$, each dNTP at a concentration of 200 μ*M*, 10–50 pmol of each primer, and 1–100 ng of template DNA (*see* **Notes 9–11**). Also prepare a mixture without the template DNA to check nonspecific amplification from contaminating DNA.
2. Put the PCR tubes in a thermal cycler, set the thermal cycle program, and start the amplification (*see* **Note 12**). A typical thermal cycle for the Amplitaq Gold polymerase is as follows: 10-min activation of the polymerase at 94°C, followed by 25–40 cycles consisting of 1 min of denaturation at 94°C, 1-min annealing at a temperature optimized for each primer set, 2-min polymerization at 72°C, and finally one cycle of 10-min extension at 72°C (*see* **Note 13**).
3. Check the amplification by electrophoresis. We use a 1.5% agarose gel in TBE buffer for 200- to 1500-bp DNA fragments.
4. After the electrophoresis, stain the gel in ethidium bromide solution for 30 min. To remove unbound ethidium bromide in the gel, soak the gel briefly in water (approx 10 min). Visualize DNA fragments in the gel by using an UV transilluminator.

4. Notes

1. When a water sample is turbid, the use of membranes with larger pore sizes is recommended to remove larger particles.
2. Use gloves and handle with care; ethidium bromide is mutagenic.
3. Filtration should be performed soon after sampling. After filtration, the membranes should be stored in freezer.

4. Careful suspension of cells is essential to increase the yield of extracted DNA.
5. We do not use lysozyme in this procedure, because the effect of this enzyme is selective for a certain group of microorganisms (e.g., Gram-negative bacteria).
6. Alternatively, cells can be disrupted by bead beating ***(5)***. A bead-beating treatment has been considered to be more disruptive than a freezing/thawing treatment, although, in our experience, the DNA recovery by the freezing and thawing treatment was better when DNA was extracted from a small quantity of cells.
7. By using the procedure mentioned here, we have yielded 5–20 µg of DNA from 2 L of groundwater containing 2 to 6 × 10^6 microbial cells per mL. For PCR amplification of 16S rDNA, we have added 0.5 µL of this DNA preparation (100 µL) to a PCR mixture (dilute if necessary).
8. PCR facilitates amplification of DNA fragments from only several copies to electrophoresis-detectable quantities, which may result in recovery of contaminating DNA fragments. To avoid (or minimize) contamination, the authors conduct this step in a clean lab. Bake all glasswares, autoclave all plasticwares, and pass all reagents through 0.22-µm membranes.
9. The magnesium concentration required for optimal PCR amplification is dependent on the specific set of primers and template DNA. When the PCR is inefficient, examine PCR at different $MgCl_2$ concentrations (the typical concentration ranges between 1.0 m*M* and 4.0 m*M*).
10. Excess amounts of primer and template DNA result in the increase of nonspecific amplification and smearing of the electrophoretic pattern.
11. Inefficient amplification caused by contamination with inhibitory substances (e.g., humics) may be circumvented by supplementing the PCR mixture with bovine serum albumin at a final concentration of 0.001–0.005% (w/v).
12. If the thermal cycler is not equipped with a heating lid, overlay the reaction mixture with mineral oil (10 µL). After the PCR, remove the oil from the sample by extraction with a phenol/chloroform/isoamyl alcohol solution (*see* **Subheading 2.1.9.**) and chloroform.
13. We generally set the annealing temperature 1–5°C lower than the theoretical melting temperature (Tm); it should be optimized empirically. For calculation of a theoretical Tm, we have been using a method of Breslauer et al. ***(19)***.
14. Amplification bias may occur owing to primer mismatches or to variations in the secondary structure or in the GC content of 16S rDNA fragments. To reduce amplification bias, the cycle number should be as small as possible.

References

1. Muyzer, G., Hottentrager, S., Teske, A., and Wawer, C. (1996) Denaturing gradient gel electrophoresis of PCR-amplified 16S rDNA. A new molecular approach to analyze the genetic diversity of mixed microbial communities, in *Molecular Microbial Ecology Manual* (Akkermans, A. D. L., van Elsas, J. D., de Bruijn, F. J., eds.), Kluwer Academic Publishing, Dordrecht, pp. 3.4.4.1–3.4.4.22.

2. Marsh, T. L. (1999) Terminal restriction fragment length polymorphism (T-RFLP): an emerging method for characterizing diversity among homologous populations of amplification products. *Cur. Opin. Microbiol.* **2,** 323–327.
3. Wintzingerode, F., Göbel, U. B., and Stackebrandt, E. (1997) Determination of microbial diversity in environmental samples: pitfalls of PCR-based rRNA analysis. *FEMS Microbiol. Rev.* **21,** 213–229.
4. Watanabe, K., Kodama, Y., Syutsubo, K., and Harayama, S. (2000) Molecular characterization of bacterial populations in petroleum-contaminated groundwater discharged from underground crude-oil-storage cavities. *Appl. Environ. Microbiol.* **66,** 4803–4809.
5. Akkermans, A. D. L., van Elsas, J. D., and de Bruijn, F. J. (1999) *Molecular Microbial Ecology Manual.* Kluwer Academic Publishing, Dordrecht.
6. Snaidr, J., Amann, R., Huber, I., Ludwig, W., and Schleifer, K. H. (1997) Phylogenetic analysis and *in situ* identification of bacteria in activated sludge. *Appl. Environ. Microbiol.* **63,** 2884–2896.
7. Dojka, M. A., Hugenholtz, P., Haack, S. K., and Pace, N. R. (1998) Microbial diversity in a hydrocarbon- and chlorinated-solvent-contaminated aquifer undergoing intrinsic bioremediation. *Appl. Environ. Microbiol.* **64,** 3869–3877.
8. Großkopf, R., Janssen, P. H., and Liesack, W. (1998) Diversity and structure of the methanogenic community in anoxic rice paddy soil microcosms as examined by cultivation and direct 16S rRNA gene sequence retrieval. *Appl. Environ. Microbiol.* **64,** 960–969.
9. Achenbach, L. and Woese, C. (1995) 16S and 23S rRNA-like primers, in *Archaea—A Laboratory Manual* (Robb, F. T., Sowers, K. R., DasSarma, S., Place, A. R., Schreier, H. J., Fleischmann, E. M. eds.), Cold Spring Harbor Laboratory Press, Cold Spring Harbor, NY, pp. 521–523.
10. Watanabe, K., Kodama, Y., and Harayama, S. (2001) Design and evaluation of RCR primers to amplify bacterial 16S ribosomal DNA fragments used for community fingerprinting. *J. Microbiol. Methods* **44,** 253–262.
11. Takai, K. and Sako, Y. (1999) A molecular view of archaeal diversity in marine and terrestrial hot water environments. *FEMS Microbiol. Ecol.* **28,** 177–188.
12. Lane, D. J. (1991) 16S/23S rRNA sequencing, in *Nucleic Acid Techniques in Bacterial Systematics* (Stackebrandt, E. and Goodfellow, M. eds.). Wiley, New York, pp. 115–175.
13. Stahl, D. A. and Amann, R. (1991) Development and application of nucleic acid probes, in *Nucleic Acid Techniques in Bacterial Systematics* (Stackebrandt, E. and Goodfellow, M., eds.), Wiley, New York, pp. 115–175.
14. DeLong, E. F. (1992) Archaea in coastal marine environments. *Proc. Natl. Acad. Sci. USA* **89,** 5685–5689.
15. Nübel, U., Engelen, B., Felske, A., Snaidr, J., Wieshuber, A., Amann, R. I., et al. (1996) Sequence heterogeneities of genes encoding 16S rRNAs in *Paenibacillus polymyxa* detected by temperature gradient gel electrophoresis. *J. Bacteriol.* **178,** 5636–5643.

16. Embley, T. M., Finlay, B. J., Thomas, R. H., and Dyal, P. L. (1992) The use of rRNA sequences and fluorescent probes to investigate the phylogenetic positions of the anaerobic ciliate *Metopus palaeformis* and its archaeobacterial endosymbiont. *J. Gen. Microbiol.* **138,** 1479–1487.
17. Barns, S. M., Fundyga, R. E., Jeffries, M. W., and Pace, N. R. (1994) Remarkable archaeal diversity detected in Yellowstone National Park hot spring environment. *Proc. Natl. Acad. Sci. USA* **91,** 1609–1613.
18. Olsen, G. J., Lane, D. J., Giovannoni, S. J., Pace, N. R., and Stahl, D. A. (1986) Microbial ecology and evolution: a ribosomal rRNA approach. *Annu. Rev. Microbiol.* **40,** 337–365.
19. Breslauer, K. J., Frank, R., Blocker, H., and Marky, L. A. (1986) Predicting DNA duplex stability from the base sequence. *Proc. Natl. Acad. Sci. USA* **83,** 3746–3750.

23

Identification of Copper-Resistant Microorganisms by PCR

Virginia H. Albarracín, Juliana M. Benito, Manuel Siñeriz Louis, María J. Amoroso, and Carlos M. Abate

1. Introduction

Copper is an essential trace element required as an enzyme cofactor, and it is involved in diverse biological processes, including respiration, destruction of free radicals, and iron homeostasis *(1)*. It is also used in industrial applications such as production of steel and other alloys, galvanization of iron, electroplating, and manufacture of batteries, TV tubes, and pigments. However, mining and leaching from natural deposits contribute to environmental contamination *(2)*.

This kind of pollution posses a great risk to the survival of natural microorganism populations in soil and sediments where industries discharge their effluents. Exposure to this metal leads microorganisms to develop resistance mechanisms such as reduced influx, facilitated efflux, sequestration, and modification of copper, among others *(3)*.

Investigation assays are performed in many unrelated microorganisms, such as *Synechococcus* ***(4)***, *Saccharomyces cerevisiae* ***(5)***, *Candida albicans* ***(3)***, *Enterococcus hirae* ***(6)***, *Helicobacter* spp. ***(7)***, *Escherichia coli* ***(8)***, *Pseudomonas* spp. ***(9)***, *Methanobacterium bryantii* ***(10)***, *Vibrio alginolyticus* ***(11)***, *Salmonella typhimurium* ***(12)***, *Xanthomonas campestris* ***(13)***, *Alcaligenes eutrophus* ***(14)***, and Gram positives such as actinomycetes ***(15)***.

Trajanovska et al. *(2)* investigated the nature of the genetic systems encoding copper resistance, using primers constructed on the base of the nucleotide sequence of the plasmid-borne copper resistant (*pco*) determinant, which is

From: *Methods in Biotechnology: Environmental Microbiology: Methods and Protocols*
Edited by: J. F. T. Spencer and A. L. Ragout de Spencer

carried by the plasmid pRJ1004, as a growth promoter *(16)*. Nevertheless, DNA isolation and purification of both genomic and plasmid DNA is the crucial step for the successful development of this molecular method. Among the protocols used in the isolation of genomic and plasmid DNA in *Streptomyces* strains, the Kirby mix, the lysozyme treatment, salting out, and CTAB procedures are used for *Streptomyces* genomic DNA isolation *(17)*. For *Streptomyces* plasmid isolation, an adaptation of the alkaline lysis procedure is most commonly used. The hot alkaline lyses–acid phenol method produces CCC *Streptomyces* and *E. coli* plasmids that become suitable for restriction digests *(17)*.

The aim of this chapter is to describe a genomic DNA isolation technique for different actinomycete strains and a PCR amplification assay using pcoA and pcoR primers, useful for identification of copper-resistant actinomycetes genes.

2. Materials

2.1. DNA Isolation, Purification, and Visualization (see *Note 1*)

1. 50 mg (wet weight) of mycelium (*see* **Note 2**).
2. Lysozyme solution: 8 mg/mL lysozyme and 50 µg/mL RNAase in 0.3 *M* sucrose, 25 m*M* Tris, pH 8.0, 25 m*M* EDTA (pH 8.0) (*see* **Note 3**).
3. 2% SDS (w/v).
4. 3 *M* Sodium acetate, pH 4.8.
5. Phenol-chloroform-isoamyl alcohol (25:24:1) *(18)*.
6. Chloroform–isoamyl alcohol 24:1 (v/v).
7. Isopropanol and 70% ethanol.
8. 1X TAE buffer: 0.04 *M* Tris-acetate, 0.001 *M* EDTA, pH 8.0 *(18)*.
9. 0.8% Agarose in 1X TAE buffer.
10. Ethidium bromide staining solution: 1 µg/mL in 1X TAE buffer *(18)*.
11. Molecular-weight markers: K562 High Molecular Weight (Promega).
12. Loading buffer.

2.2. PCR Assays

1. STR buffer (Promega).
2. *Taq* polymerase (Promega).
3. Primers pcoA1-pcoA2 and pcoR1-pcoR2 *(2)*.
4. 1X TAE buffer: 0.04 *M* Tris-acetate, 0.001 *M* EDTA, pH 8.0 *(18)*.
5. 0.8% Agarose in 1X TAE buffer.
6. Ethidium bromide staining solution: 1 µg/mL in 1X TAE buffer *(18)*.
7. Loading buffer (Promega).
8. Molecular-weight marker: 1 kb DNA marker (Promega).

3. Methods

3.1. DNA Isolation, Purification, and Visualization

1. Resuspend 50 mg of mycelium in 500 µL of lysozyme solution and incubate at 37°C for approx 30 min or until the cells become translucent (*see* **Note 4**).
2. Add 250 µL of 2% SDS and vortex mix until the viscosity of the solution has decreased noticeably.
3. Add 1 vol of phenol-chloroform-isoamyl alcohol and mix by inverting the tube several times.
4. Centrifuge (10,000*g* for 15 min). Transfer the aqueous phase (upper) to a new tube, leaving the white interface behind. Repeat phenol extraction once or until no (or very little) interface is seen.
5. Transfer the upper aqueous phase to a clean tube and add an equal volume of chloroform–isoamyl alcohol. Again mix well and centrifuge (10,000*g* for 5 min). Repeat this extraction two times (*see* **Note 5**).
6. Transfer the aqueous phase to a new tube and precipitate the DNA by adding 1/10 vol of 3 *M* sodium acetate and 0.6 to 1 vol of 2-propanol. Mix gently and incubate at –20°C from 1 h to overnight.
7. Centrifuge (10,000*g* for 15 min). Discard the supernatant and wash DNA with 500 µL of 70% ethanol to remove residual salts and isopropanol. Centrifuge (10,000*g* for 5 min), carefully discard the ethanol, and dry until ethanol has been removed.
8. Resuspend DNA in 20–30 µL of double-distilled sterile water and 0.1–0.2 µL RNase A. Allow to dissolve at 37°C at least 3 h (*see* **Note 6**).
9. Visualize the extracted DNA by running 5 µL of the sample in a 0.8% agarose gel electrophoresis using 1X TAE electrophoresis buffer and K562 DNA marker. Electrophoresis is carried out at 10 V/cm for 1 h at room temperature; gels should be stained with ethidium bromide solution and observe under UV light.

3.2. PCR Amplification

1. Prepare the reaction mixture by adding 2.5 µL of STR buffer, 0.2 µL of *Taq* polymerase, and 0.1 µL of each primer, and complete the volume with double-distilled sterile water.
2. Amplifications should be performed in 25 µL reaction volumes for 35 cycles, using a thermal cycler (e.g., Gene Amp PCR 9700 System, Applied Biosystems, CA).
3. The following temperature profile is used: initial denaturation at 95°C for 5 min; followed by 30 cycles of 95°C for 30 s, 57°C for 90 s, and 72°C for 90 s; and a final extension at 72°C for 12 min.
4. As a positive control, it is recommended to use plasmid DNA of pRJ1004 *(16)*.
5. A negative control (sterile water) has to be included in the assay.
6. Visualize the PCR products by running 10 µL of the sample in a 0.8% agarose gel electrophoresis using 1X TAE electrophoresis buffer and 1 Kb DNA marker. Electrophoresis is carried out at 10 V/cm for 35 min at room temperature; gels should be stained with ethidium bromide solution and observed under UV light.

4. Notes

1. When manipulating DNA, it is always recommended to wear gloves in order to avoid contamination of the sample with DNAases present in hands.
2. The mycelium might be either a fresh liquid culture or a frozen stock. When working with liquid cultures, the suspension has to be centrifuged at 8000*g* for 20 min. Subsequently, the pellet is resuspended in saline solution by vortexing, and centrifuged at 8000*g* for 20 min. The supernatant is discarded and the pellet is washed once more.
3. The lysis solution without lysozyme and RNAase can be aseptically prepared and stored at 4°C for long periods of time. The lysozyme is added just before using.
4. The incubation time in this step is variable, but in general we have obtained better results when it is longer than 30 min.
5. The chloroform–isoamyl alcohol step is used to remove any phenol remaining in the aqueous solution. This step can be repeated as many times as desired, depending on what the DNA is to be used for.
6. DNA samples obtained by the procedure described above are usually suitable for PCR amplification. Concentration and quality should be tested by gel electrophoresis and absorbance measurements. After electrophoretic evaluation, the concentrations of the samples should be adjusted to be similar.
7. It is mandatory to use gloves that have no residue of talc in this step, because talc inhibits *Taq* polymerase.

References

1. Linder, M. C. and Hazegh-Azam, M. (1996) Copper biochemistry and molecular biology. *Am. J. Clin. Nutr.* **63,** 797S–811S.
2. Trajanovska, S., Britz, M. L., and Bhave, M. (1997) Detection of heavy metal ion resistance genes in Gram-positive and Gram-negative bacteria isolated from a lead contaminated soil. *Biodegradation* **8,** 113–124.
3. Riggle, P. J. and Kumamoto, C. A. (2000) Role of a *Candida albicans* P1-type ATPase in resistance to copper and silver ion toxicity. *J. Bacteriol.* **182,** 4899–4905.
4. Phung, L. T., Ajlani, G., and Haselkorn, R. (1994) P-type ATPase from the cyanobacterium *Synechococcus* 7942 related to the human Menkes and Wilson disease products. *Proc. Natl. Acad. Sci. USA* **91,** 9651–9654.
5. Okuyama, M., Kobayashi, Y., Inouhe, M., Tohoyama, H., and Joho, M. (1999) Effect of some heavy metal ions on copper-induced metallothionein synthesis in the yeast *Saccharomyces cerevisiae. Biometals.* **12,** 307–314.
6. Bissig, K., Wunderli-Ye, H., Duda, P. W., and Solioz, M. (2001) Structure-function analysis of purified *Enterococcus hirae* CopB copper ATPase: effect of Menkes/Wilson disease mutation homologues. *Biochem. J.* **357,** 217–223.
7. Bayle, D., Wängler, S., Weitzenegger, T., Steinhilber, W., Volz, J. Przybylski, M., et al. (1998) Properties of the P-type ATPases encoded by the *copAP* operons of *Helicobacter pylori* and *Helicobacter felis. J Bacteriol.* **180,** 317–329.

8. Rensing, C., Fan, B., Sharma, R., Mitra, B., and Rosen, B. P. (2000) CopA: an *Escherichia coli* Cu(I)-translocating P-type ATPase. *Cell Biology* **97,** 652–656.
9. Lim, C. K. and Cooksey, D. A. (1993) Characterization of chromosomal homologs of the plasmid-borne copper resistance operon of *Pseudomonas syringae. J. Bacteriol.* **175,** 4492–4498.
10. Kim, B. K., Pihl, T. D., Reeve, J. N., and Daniels, L. (1995) Purification of the copper response extracellular proteins secreted by the copper-resistant methanogen *Methanobacterium bryantii* BKYH and cloning, sequencing, and transcription of the gene encoding these proteins. *J. Bacteriol.* **177,** 7178–7185.
11. Harwood-Sears, V. and Gordon, A. (1990) Copper-induced production of copper-binding supernatant proteins by the marine bacterium *Vibrio alginolyticus. Appl. and Environmental Microbiol.* **56,** 1327–1332.
12. Gupta, S. D., Wu, H. C., and Rick, P. D. (1997) A *Salmonella typhimurium* genetic locus which confers copper tolerance on copper-sensitive mutants of *Escherichia coli. J. Bacteriol.* **179,** 4977–4984.
13. Lee, Y., Hendson, M., Panopoulos, N. J., and Schrotch, M. N. (1994) Molecular cloning, chromosomal mapping, and sequence analysis of copper resistance genes from *Xanthomonas campestris pv. juglandis:* homology with small blue copper proteins and multicopper oxidase. *J. Bacteriol.* **176,** 173–188.
14. Diels, L. and Mergeay, M. (1990) DNA probe-mediated detection of resistant bacteria from soils highly polluted by heavy metals. *Appl. and Environmental Microbiol.* **56,** 1485–1491.
15. Erardi, F. X., Failla, M. L., and Falkinham, J. O. (1987) Plasmid-encoded copper resistance and precipitation by *Mycobacterium scrofulaceum. Appl. Env. Microbiol.* **53,** 1951–1954.
16. Tetaz, J. T. and Luke, R. K. (1983) Plasmid-controlled resistance to copper in *Escherichia coli. J. Bacteriol.* **154,** 1263–1268.
17. Keiser, T., Bibb, M. J., Buttner, M. J., Chater, K. F., Hopwood, D. A. (2000) Preparation and analysis of genomic and plasmid DNA, in *Practical Streptomyces Genetics* (Kieser, T., Bibb, M. J., Buttner, M. J., Chater, K. F., Hopwood, D. A., eds.), Norwich, England, pp. 161–210.
18. Sambrook, J., Maniatis, T., and Fritsch, E. F. (1989) *Molecular Cloning. A Laboratory Manual* (Sambrook, J. and Maniatis, T., eds.), Cold Spring Harbor Laboratory Press, Cold Spring Harbor, NY.

24

Selection of Tolerant Heavy Metal Yeasts From Different Polluted Sites

Liliana Villegas, María J. Amoroso, and Lucía I. C. de Figueroa

1. Introduction

As a result of many industrial and waste-disposal operations, our environment is becoming more and more contaminated by heavy metals—mercury, chromium, arsenic, cadmium, nickel, copper, and so on. Water and soil pollution has become a major concern in the world, as much of the population relies on groundwater for drinking and irrigation. Heavy-metals contamination can cause metal toxicity in animals and humans. Trace elements such cadmium, copper, mercury, or arsenic are very toxic heavy metals and have been found in the human environment at increased concentrations, because a wide variety of industrial activities have accelerated the release of these metals at higher rates than natural geochemical cycling processes can tolerate *(1,2)*.

Although many heavy-metal cations are essential components of biological systems, all are potentially toxic. They play an important role in sophisticated biochemical reactions such as nitrogen fixation, water cleavage during oxygenic photosynthesis, respiration with oxygen or nitrate, hydrogen assimilation, cleavage of urea, transcription of genes into mRNA, and so on. These reactions are based on catalysis by biochemical heavy-metal complex compounds. Essential or not, most of the heavy-metal ions are toxic at higher concentrations *(3)*. The environmental mercury cycle is mediated by both geological and biological processes. Mercury compounds are leached from natural sources by weathering of rocks and soil. However, the sources of environmental contamination of mercury arise from human activities, such as burning coal and petroleum products, use of mercurial fungicides in papermaking

From: *Methods in Biotechnology: Environmental Microbiology: Methods and Protocols*
Edited by: J. F. T. Spencer and A. L. Ragout de Spencer

and agriculture, and mercury catalysis in industry, which can increase local mercury levels several thousand-fold above background *(4)*. The main industrial uses of chromium are in the manufacture of metallic alloys for structural and protective purposes, in fabric dyes, and in the tanning of leathers. Chromium is usually found in nature in its trivalent state, but hexavalent chromate is found generally as a result of human activities *(5)*. Arsenicals are toxic compounds, which are commonly present in the environment at increasing concentrations as a result of industrial pollution. Arsenicals are well-established carcinogens that are able to induce gene amplification and morphological transformation of eukaryotic cells *(6)*. Arsenic is a highly toxic metalloid that is naturally present in a number of minerals; it is also released into the environment by industrial and agricultural activities. Geochemical contamination of drinking water by arsenic is a serious health problem affecting millions of people worldwide. Long-term exposure to arsenic is known to cause a number of human diseases *(7)*. Cadmium is a heavy metal that causes both acute and chronic toxicity. In cases of severe acute exposure, death can result in a few days, and long-term exposure to lesser amounts can result in damage to kidney, lung, and bone tissue. Cadmium has no known beneficial effects in the human body, but it is used extensively in the production of human consumables like batteries, plastics, various pigments, and metal coatings *(8)*. In many developing countries, nickel-cadmium batteries are still used extensively, resulting in important contamination. They are classified as hazardous waste because nickel and cadmium are heavy metals and suspected carcinogens *(9)*. Copper, an essential trace metal, is utilized as a cofactor in a variety of redox and hydrolytic proteins. Excess copper, however, is highly toxic to most organisms, either eukaryotes or prokaryotes *(10)*.

Microorganisms may be used to remediate wastewaters or soils contaminated with heavy metals. The metal-processing capacity of microorganisms can be used to concentrate, remove, and recover metals from streams and enhance the efficiency of wastewater treatment processes *(2)*. The types of microbial biomass of interest can pragmatically be those that can be easily obtained in larger quantities. One of the most ubiquitous biomass types utilized on a large scale by man for centuries is yeasts. Earlier observations noting the capacity of yeast strains to sequester heavy metals have been largely scattered and inconsistent, focusing mainly on the nutritional requirements or toxicology of the yeast *(11)*.

The study of the interactions between metals and fungi has long been of scientific interest. In an environmental context, accelerating pollution by toxic metals, metalloids, radionuclides, and organometalloids has influenced research towards the biotechnological potential of utilizing microorganisms for metal removal and/or recovery from the biosphere. Yeasts possess an acknowledged capability of accumulating a range of metal cations. Fungal accumulation is essentially a biphasic process consisting of a metabolism-independent and a metabolism-dependent step. The initial biosorption step is rapid, typically only a few minutes in duration, and is independent of temperature, metabolic energy, the presence of a metabolizable energy source, and the presence of metabolic inhibitors. The second step, called *bioaccumulation,* is a slower, metabolism-dependent step, influenced by factors such as temperature and the presence of metabolic inhibitors. Greater amounts of metal may be accumulated by this means in some organisms, especially yeasts ***(12)***. The influence of pH on metal uptake by yeasts, algae, and bacteria is very similar. Extremes of pH generally decrease the rate and extent of metal uptake. A pH between 4.0 and 8.0 is widely accepted as being optimal for metal uptake for almost all types of biomass. Temperature effects are confined to metabolism-dependent metal accumulation. At low temperature (0–5°C), little or no metal is sequestered through metabolic processes by viable biomass. Most laboratory experiments are carried out in the temperature range 20–35°C, which has been reported optimal for metal accumulation ***(13)***. Yeasts have contributed to the understanding of metal uptake and toxicity, and how metal resistance may be achieved in eukaryotic cells. In a biotechnological context, yeasts may be useful in the treatment of metal-containing effluents ***(12)***.

The objective of this chapter is to describe procedures used for yeast isolation from several sites contaminated with industrial effluents containing heavy metals, with the main purpose of isolating yeasts able to live in these unfavorable environments.

2. Materials

2.1. Materials, Habitats

1. Material for isolating mercury-tolerant yeasts: agricultural and papermaking soil contaminated with petroleum products and mercurial fungicides.
2. Material for isolating chromium-tolerant yeasts: wastes from tanneries and dye factories.
3. Material for isolating arsenic-tolerant yeasts: soil contaminated with pesticides and sediment from an arsenic-endemic area.
4. Material for isolating cadmium-tolerant yeasts: soil and water contaminated with spent nickel-cadmium batteries.
5. Material for isolating copper-, mercury-, cadmium-, chromium-, iron-, nickel-, and lead-tolerant yeasts: soil and sediments exposed to effluents from mining and other industries.

2.2. Culture Media and Solutions

1. Enrichment solution: 1 g/L glucose and 0.5 g/L yeast extract.
2. Isolation medium: 3 g/L yeast extract, 3 g/L malt extract, 5 g/L peptone, 10 g/L dextrose, 20 g/L agar (YM, Difco); plus 100 µg/mL of tetracycline, erythromycin, and chloramphenicol.
3. Maintenance culture medium : 10 g/L yeast extract, 20 g/L peptone, 15 g/L dextrose, and 20 g/L agar (YEPD).
4. Base culture medium for determining heavy-metal tolerance: 6.7 g/L yeast nitrogen base (YNB, Difco), 10 g/L dextrose, and 1.5 g/L agar. It is used as a minimal medium.
5. Heavy-metal stock solutions: 1 *M* $CuSO_4$, $NiSO_4$, $CdCl_2$, $K_2Cr_2O_7$, As_2O_3, As_2O_5, and $HgCl_2$ (*see* **Note 1**).

3. Methods

3.1. Yeasts Isolation

Samples obtained from different habitats should be 10–20 g. They are transported in sterile plastic bags, and kept at 4°C until they are used. The samples are suspended in sterile enrichment solution (1:10 w/v) supplemented with the antibiotics (*see* **Note 2**). After shaking (200 rpm) 24–72 h at 30°C, suspensions are poured into petri dishes containing YM agar medium supplemented with antibiotics. The inoculated plates are incubated at 30°C and, after approx 3–5 d, well-isolated colonies are transferred onto YEPD agar until growth is observed. Colonies have to be restreaked and picked on maintenance medium until their ability to tolerate heavy metals can be determined (*see* **Note 3**).

3.2. Qualitative Screening for Heavy-Metals Resistance

Heavy metals tolerance may be tested using an agar diffusion assay on YNB-glucose-agar *(2)* (*see* **Note 4**). A ditch in the center of the plate is made with a sterile needle or sterile spatula (*see* **Note 5**, **Figs. 1** and **2**). The ditch of each plate is filled with 500 µL of one of the heavy-metal solutions, while the isolations are inoculated in lines extending from the ditch at a right angle by means of a platinum loop (*see* **Notes 6** and **7**). Yeast growth is assumed as a qualitative parameter of heavy-metal resistance. Growth up to the ditch is considered resistant to the concentration supplied, while no growth is sensitive. The size of growth inhibition extending from the ditch gives an approximate idea about the level of tolerance in cases where incomplete tolerance is observed. As a control, distilled and sterilized water is used (*see* **Note 8**).

3.3. Semiquantitative Screening for Heavy-Metals Resistance

Sensitivity of the selected yeasts to heavy metals is tested by agar diffusion assay. Disks of approx 6-mm diameter are formed in Petri dishes containing

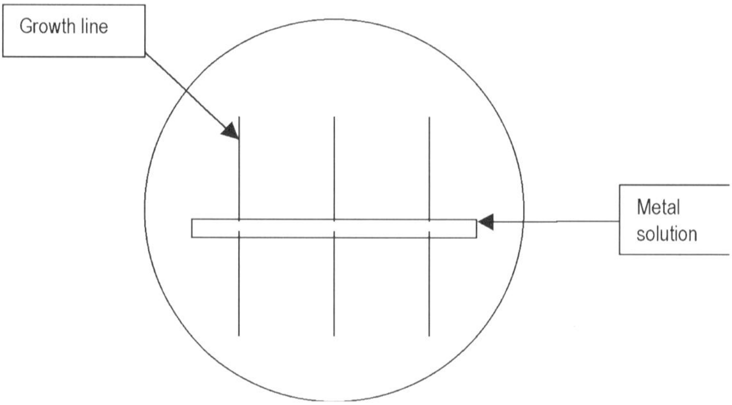

Fig. 1. Scheme of the qualitative screening for heavy-metals tolerance.

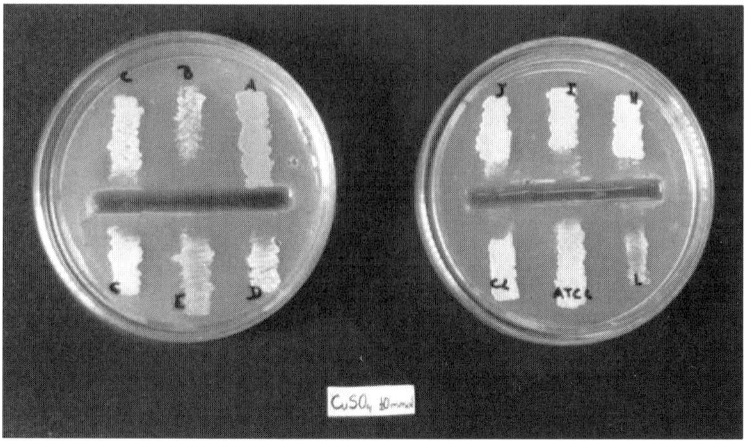

Fig. 2. Yeasts growth in the presence of a $CuSO_4$ solution.

YNB- glucose-agar previously inoculated with 100 µL of cell suspensions. Disks are saturated with heavy-metal solutions up to a final concentration of 100 mmol/L. The diameter of the growth inhibition is measured after incubation at 20, 25 and 30°C for 3–4 d. Disks saturated with water are used as controls.

An inhibition zone longer than 10-mm diameter is arbitrarily considered as a metal-sensitive response, whereas zones of inhibition of tolerant strains are lower than 10 mm (*see* **Note 9**).

3.4. Quantitative Assays in Liquid Medium

Tolerance is evaluated on YNB-glucose supplemented with heavy-metal solutions up to a final concentration of 5 mmol/L.

One aliquot from freshly grown cultures is used to inoculate Erlenmeyer flasks containing 100 mL of YNB-glucose with different metal concentrations (*see* **Note 10**). Growth medium without heavy metals is used as a control. Flasks should be incubated at 30°C in a rotary shaker at 300 rpm.

Growth is monitored every 4 h by optical density at 620 nm, glucose consumed, dry weight, and UFC (*see* **Note 11**).

3.5. Heavy-Metals Tolerance Induction

Heavy-metals tolerance can be induced by inoculating the isolates into 100-mL Erlenmeyer flasks containing 10 mL YNB-glucose supplemented with each of the heavy-metal solutions. Flasks should be incubated at 30°C, 300 rpm, for 48 h. 500 μL from these cultures are then transferred to flasks containing 100 mL of YNB-glucose supplemented with the respective metal solution. Yeast growth is followed using turbidity as the sensibility assay. Growth is monitored at intervals of 4 h for 3 d. A shorter lag phase will be obtained in induced yeasts. Induction allows the cells to tolerate the highest metals concentrations (*see* **Note 12**).

3.6. Heavy-Metals Determination

Heavy-metals analysis is carried out by atomic absorption spectrophotometer. Amounts are determined in the supernatant of the culture medium during the growth phase. The final concentration will indicate whether or not yeasts are able to reduce the heavy metal that was in the medium.

4. Notes

1. Heavy-metal stock solutions can be sterilized by autoclaving and stored in the dark. Mercury solution has to be prepared at concentrations 10 times lower than the other metals solutions because of its toxicity.
2. Antibiotic stock solutions dissolved in ethanol do not need to be sterilized. Solutions are stored in light-tight containers at –20°C. The following concentrations are recommended: 34 mg/mL chloramphenicol, 40 mg/mL erythromycin, and 5 mg/mL tetracycline.
3. Yeast cultures can be kept indefinitely in sterile glycerol 30% at –20°C until they are used. This reduces the risk of possible mutational changes and contaminations.
4. In tolerance assays, it is better to use minimal medium instead of YEPD owing to the presence in this medium of antioxidants such as glutathione. Moreover, YNB also minimizes complex formation of the heavy metal ions.

5. Scheme of the agar diffusion assay using a Petri dish with YNB-glucose-agar (*see* **Fig. 1**). The authors have made an investigation, in the Tucumán region of Northwest Argentina, of yeasts tolerant to heavy metals, isolated from sites contaminated with industrial effluents. **Figure 2** shows the screening of the yeasts using the method described previously. The ditch of the plate was filled with 500 µL of one of the heavy-metal solutions, while the isolations were inoculated in lines extending from the ditch at a right angle (A: control plate; B: with 10 m*M* copper solution).
6. Plates should be set up using 20 mL of YNB-glucose-agar, so the ditch will have the right depth. A control plate is made by filling the ditch with sterile water.
7. Yeast cell suspensions are prepared with distilled water. The strains should be inoculated as a line no more than 0.5 cm wide to avoid growing over to the position of an adjacent sample.
8. Growth determination with this methodology will indicate a high or low degree of heavy-metals tolerance.
9. Growth determination with this test is useful for comparing the heavy-metal tolerance among the isolates and for selecting the best strains in order to study their heavy-metal susceptibility in liquid medium.
10. A culture in exponential phase is recommended. The inoculum is prepared in the same physical conditions to avoid cellular stress; otherwise, the lag phase will be very long because the isolate needs adaptation time.
11. Turbidity is a good measure of the growth of yeasts in the presence of different heavy-metal solutions because changes are observed in the presence of the metals. When yeasts produce pseudomycelium, the turbidity is heterogeneous and it is not a good way to measure growth.
12. Heavy metals produce reactive oxygen species that cause damage to all of the major cellular constituents, including lipids peroxidation. The oxidative stress affects the progression of cells through the cell division cycle. Reactions of cells to sudden changes in their environment, and the mechanisms protecting them against damage caused by stress conditions, have received enormous attention. Studies on stress are not only interesting on a scientific basis, but also have medical, economic, and bioremediation relevance *(14)*.

References

1. Volesky, B. and Holan, Z. R. (1995) Biosorption of heavy metals. *Biotechnol. Prog.* **11,** 235–250.
2. Amoroso, M. J., Castro, R. G., Carlino, F. J., Romero, N. C., Hill, R. T., and Oliver, G. (1998) Screening of heavy metal-tolerant actinomycetes isolated from the Salí River. *J. Gen. Appl. Microbiol.* **44,** 129–132.
3. Villegas, L. B. and de Figueroa, L. I. C. (2002) Communities of yeasts tolerant to heavy metals from different polluted sites of Argentina. 102nd General Meeting of the American Society for Microbiology. Salt Palace Convention Center, Salt Lake City, Utah, May 19–23, 2002.

4. Revel, J., Amoroso M. J., Colwell R. R., and Hill, R. T. (1998) Mercury-resistant actinomycetes from the Chesapeake Bay. *FEMS Microbiol. Lett.* **162,** 177–184.
5. Amoroso, M. J., Castro, R. G., Duran, A., Peraud, O., Oliver G., and Hill, R. T. (2001) Chromium accumulation by two *Streptomyces* spp. isolated from riverine sediments. *J. Ind. Microbiol. Biotechnol.* **26,** 210–215.
6. Wysocki, R., Bobrowicz, P., and Ulszewski, S. (1997) The *Saccharomyces cerevisiae ACR3* gene encodes a putative membrane protein involved in arsenite transport. *J. Biol. Chem.* **272,** 30,061–30,066.
7. Tamás, M. J. and Wysoki, R. (2001) Mechanisms involved in metalloid transport and tolerance acquisition. *Curr. Genet.* **40,** 2–12.
8. Chrestensen, C. A., Starke, D. W., and Mieyal, J. J. (2000) Acute cadmium exposure inactivates thioltransferase, inhibits intracellular reduction of protein-glutathionyl-mixed disulfides, and initiates apoptosis. *J. Biol. Chem.* **275,** 26,556–26,565.
9. Cerruti, C., Gurutchet, G., and Donati, E. (1998) Bio-dissolution of spent nickel-cadmium batteries using *Thiobacillus ferrooxidans. J. Biotechnol.* **62,** 209–219.
10. Arnesano, F., Banci, L., and Bertini, E. (2001) Solution structure of the Cu(I) and apo forms of the yeast metallochaperone. AtxI. *Biochem.* **40,** 1528–1539.
11. Volesky, B. and May-Phillips, H. A. (1995) Biosorption of heavy metals by *Saccharomyces cerevisiae. Appl. Microbiol. Biotechnol.* **42,** 797–806.
12. Blackwell, K. J., Singlenton, I., and Tobin, J. M. (1995) Metal cation uptake by yeast: a review. *Appl. Microbiol. Biotechnol.* **43,** 579–584.
13. Brady, D. and Duncan, J. R. (1994) Bioaccumulation of metal cations by *Saccharomyces cerevisiae. Appl. Microbiol. Biotechnol.* **41,** 149–154.
14. Ruis, H. and Schüller, C. (1995) Stress signaling in yeast. *Bioassays* **17,** 959–965.

25

Medium for Differential Enumeration of *Lactobacillus casei* and *Lactobacillus acidophilus* From Lyophilized Mixed Cultures

Mónica M. Locascio, Rosalía Alesso, Vilma I. Morata, and Silvia N. González

1. Introduction

It is very important for the dairy industry to have methods to determine the microbial content of different products *(1–3)*. Probiotics, prebiotics, and eubiotics are foods that improve health status through selected bacteria *(4)*. The nutritinal benefit of probiotics has been investigated in relation to fermentation, quantity, availability, and nutrient absorption. These kinds of foods are fermented products, but they are also augmented with beneficial microorganisms *(5–10)*. The addition of probiotic microorganisms to milk products produces not only organoleptic but also biological and therapeutic characteristics. Differential cell counts in a product with mixed cultures are indispensable to control quality during its shelf life. To be suitable for industrial applications, a method for viable microbe counting and identification should be reliable, easy to prepare, quick, and of low cost; this is the selective medium. In some cases, there are no differences between lactobacillus sugar fermentation profiles that permit the differential growth of one without development of the other one. Generally, fluid fermented products use concentrated frozen cells as starter cultures.

Lyophilization is a method that permits large bacterial concentrations in reduced volumes, without the need for cold storage. The basic principle of lyophilization is cellular drying in a glass container, removing frozen water by sublimation. That means that the water is eliminated as vapor, using a system of high vacuum as a desiccator absorbs water vapor. Finally, the container of

From: *Methods in Biotechnology: Environmental Microbiology: Methods and Protocols*
Edited by: J. F. T. Spencer and A. L. Ragout de Spencer

dry cells is closed under vacuum. When the cover of the vial is removed or the bladder is broken under aseptic conditions, cultivable-viable microorganisms can be recovered after resuspension in appropriate solutions and seeding in adequate media.

In this chapter we describe a culture medium designed to differentiate *Lactobacillus casei* and *Lactobacillus acidophilus* at a macroscopic level when these strains are used as lyophilized starters in a dry milk product.

2. Materials

2.1. Growth Media

1. M.R.S. agar ***(11)***: 10 g/L peptone, 8 g/L meat extract, 0 4 g/L yeast extract, 20 g/L D(+) glucose, 2 g/L dipotassium hydrogen phosphate, 1 mL/L Tween 80, 2 g/L triamonic citrate, 5 g/L sodium acetate•$3H_2O$, 0.1 g/L magnesium sulphate•$7H_2O$, 0.05/L manganese sulphate•$4H_2O$, 15 g/L agar; pH 6.8. Sterilize by autoclaving at 121°C for 15 min.
2. Trehalose-MRS: 10 g/L peptone, 8 g/L meat extract, 0.4 g/L yeast extract, 20 g/L trehalose, 2 g/L dipotassium hydrogen phosphate, 1 mL/L Tween 80, 2 g/L triamonic citrate, 5 g/L sodium acetate•$3H_2O$, 0.1 g/L magnesium sulphate•$7H_2O$, 0.05/L manganese sulphate $4H_2O$, 15 g/L agar; pH 6.8. Sterilize by autoclaving at 121°C for 15 min.
3. BCG-Trehalose-MRS: 10 g/L peptone, 8 g/L meat extract, 0.4 g/L yeast extract, 20 g/L trehalose, 2 g/L dipotassium hydrogen phosphate, 1 mL/L Tween 80, 2 g/L triamonic citrate, 5 g/L sodium acetate•$3H_2O$, 0.1 g/L magnesium sulphate•$7H_2O$, 0.05/L manganese sulphate•$4H_2O$, 15 g/L agar; pH 6.8. Sterilize by autoclaving at 121°C for 15 min. Before use, sterile brome cresol green solution was added to obtain 0.08 g/L final concentration.
4. The threalose solution (40% w/v) was sterilized by filtration. Each sterile basal medium was added of threalose solution at 2% final concentration.
5. The coloring solution: brome cresol green (0.06 g) was dissolved in 15 mL NaOH to obtain a 0.01 *M* solution (final concentration), which was sterilized by filtration. In all cases 2 mL of stain solution (0.01 *M*) were added to 100 mL of basal MRS medium.
6. Microorganisms: lyophilized and frozen cells of *Lactobacillus casei* CRL 431 and *Lactobacillus acidophilus* CRL 730 (CERELA culture collection) were used as individual strains and also as mixed cultures. Other lyophilized and frozen cells were provided by SanCor (dairy industry). The cells in all cases were resuspended in both sterile physiological solution and 20% glutamate solution before carrying out the tests. The glutamate content of the solution allowed us to decrease the cellular osmotic stress when the samples are reconstituted from the frozen or lyophilized cultures.

3. Methods

The differential count of frozen mixed cultures containing *L. casei* and *L. acidophilus* was obtained by seeding of appropriate dilutions on three Petri

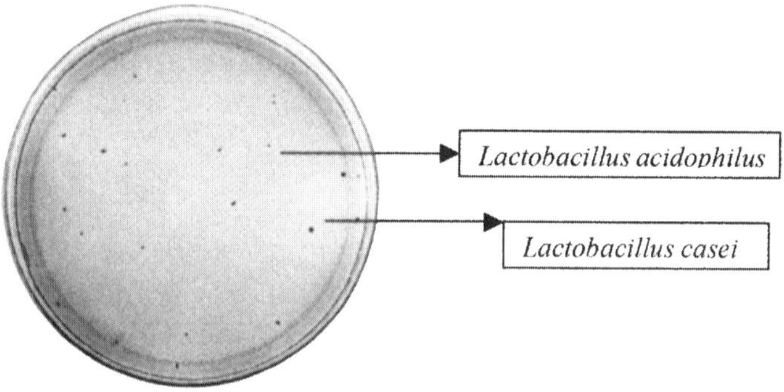

Fig. 1. Growth of *L. casei* on BCG-trehalose MRS characterized by middling colonies surrounded by a halo. Growth of *L. acidophilus* on BCG-trehalose MRS characterized by small colonies without uncolored area.

dishes with MRS-agar. For the *L. casei* count, one of them was incubated at 22°C for 72 h, and for the *L. acidophilus* count the other one was incubated at 43°C for 48 h. The third Petri dish was incubated at 37°C for 48 h for a total lactobacillus count. In all cases, frozen pure cultures were used as controls. The freezing process did not affect cellular viability, and the temperature of incubation can be used as a differential parameter (*see* **Note 1**).

The problem was to develop a method able to differentiate *L. casei* and *L. acidophilus* from mixed lyophilized cultures. Because trehalose (-D-glucopyranosyl-(-D-glucopyranose) is a sugar without a free hydroxyl glycosyl group, it is unable to reduce Fehling's solution. For this reason, this sugar is considered the nonreducing disaccharide standard. When trehalose is fermented, the resulting glucose comes from a specific trehalase activity but not from a glucosidase activity. Then, the fermentation of glucose produces lactic acid with a consequent pH decrease. This process can be observed in a culture medium containing a pH indicator such as brome cresol green. The use of MRS-trehalose as a culture medium for detection and counting of *Lactobacillus acidophilus* is permitted by the International Dairy Federation (IDF). This culture medium is suggested for obtaining a differential count of *Lactobacillus acidophilus* when the dairy product also contains *Streptococcus thermophilus* and bifidobacteria *(12)* (*see* **Note 2**).

On trehalose-MRS-agar, *Lactobacillus casei* forms lenticular middling colonies after incubation for 48 h at 37°C. Under similar conditions *Lactobacillus acidophilus* develops small colonies.

On trehalose-MRS-agar with brome cresol green, after incubation for 48 h at 37°C, *Lactobacillus casei* forms middling colonies surrounded by a halo, the uncolored area indicating a pH decrease. Under similar conditions *Lactobacillus acidophilus* develops small colonies without an uncolored area (**Fig. 1**).

BCG-trehalose-MRS was an optimal culture medium to differentiate macroscopically *Lactobacillus casei* and *Lactobacillus acidophilus* from lyophilized mixed cells, which can be found in dry dairy products (*see* **Note 3**).

4. Notes

1. The freezing of bacteria to temperatures less than 0°C can produce cellular damage due to ice-crystal formation, which exerts a high pressure on cellular membranes. The technique generally used includes a quick freezing to 50°C to –70°C because under these conditions cellular damage is reduced *(13)*. The reconstitution of frozen cultures allows recovery of viable-cultivable cells without changes in the inherent properties of each species.

   The lyophilization of cells produces a more important damage than a freezing process. The reconstitution of some lyophilized cultures permits recovery of cells with modifications of their physiological characteristics. In the case of *Lactobacillus acidophilus* we can observe that reconstituted cells from lyophilized cultures were unable to grow at temperatures above 41°C.

   On the other hand, the use of an osmotic solution, such as 20% glutamate instead of a physiological medium to carried out the cellular reconstitution from lyophilized cultures, can improve the recovery of cultivable cells.
2. From trehalose-MRS the macroscopic differentiation between *Lactobacillus acidophilus* and *Lactobacillus casei* required an experienced microbiologist because the colonies were uncolored.
3. The use of brome cresol green as a pH indicator is accepted by the IDF. This stain allows differentiation between homo- and heterofermentative metabolism. *Lactobacillus acidophilus* possesses a homofermentative metabolism, while *Lactobacillus casei* is a microorganism with heterofermentative metabolism.

References

1. Dave, R. I., and Shah, N. P. (1996) Evaluation of media for selective enumeration of *Streptococcus thermophilus, Lactobacillus delbruekii* spp *bulgaricus, Lactobacillus acidophilus* and bifidobacteria. *J. Dairy Sci.* **79,** 1529–1536.
2. Francés, M. C., Villanua, P., and Miranda, A. (1997) Alimentos preparados para lactantes sanos y preparados de continuación. *Información Tecnológica* **8,** 187–193.
3. Pérez Chaia, A., Zárate, G., and Oliver, G. (1999) Probiotic properties of *Propionibacteria. Lait* **79,** 175–185.
4. Morata de Ambrosini, V., Perdigón, G., González, S., de Ruiz Holgado, A. P., and Oliver, G. (1998) Immunostimulating activity of cell walls from lactic acid bacteria and related species. *Food Agric. Imm.* **10,** 183–191.
5. International Dairy Federation (IDF) (1999) Cultured and culture-containing dairy products in health. *Bulletin of the IDF* 294. Int. Dairy Fed. Annual Sess. Athens, Greece.
6. Oliver, G., González, S., and Moreno, R. (1999) Fermented milk for biotherapy. *Interciencia* **24,** 177–181.

7. Reid, G. (1999) The scientific basis for probiotic strains of *Lactobacillus. Applied and Environmental Microbiology* **65,** 3763–3766.
8. DeMan, J. C., Rogosa, M., and Sharpe, E. (1960) A medium for the cultivation of lactobacilli. *J. Appl. Bacteriol.* **23,** 130–155.
9. International Dairy Federation (IDF) (1995) Detection and enumeration of *Lactobacillus acidophilus. Bulletin of the IDF* 306. Int. Dairy Fed. General Secretariat, Brussels, Belgium.
10. Farkas, J. (1997) Physical methods of food preservation in food microbiology, in *Fundamentals and Frontiers* (Doyle, M. P., Beuchat, L. R., and Montville, T. J., eds.), ASM Press, Washington, DC, p. 500.
11. Ingham, S. C. (1999) Use of modified *Lactobacillus* selective medium and *Bifidobacterium* iodoacetate medium for differential enumeration of *Lactobacillus acidophilus* and *Bifidobacterium* spp. *J. Food Prot.* **62,** 77–80.
12. McCann, T., Egan, T., and Weber, G. H. (1996) Assay procedures for commercial probiotic cultures. *J. Food Prot.* **59,** 41–45.
13. Lankaputhra, W. E. V. and Shah, N. P. (1996) A simple method for selective enumeration of *Lactobacillus acidophilus* in yogurt supplemented with *L. acidophilus* and *Bifidobacterium* spp. *Milchwissenschaft* **51,** 446–451.

26

Silk as a Means of Recovering Bacteriocin From a Culture Medium

M. Carina Audisio and María C. Apella

1. Introduction

Bacteriocins are thermostable molecules of a proteinaceous nature produced by different microorganisms, such as lactic acid bacteria (LAB), with an inhibitory action spectrum directed toward related strains *(1–3)*. However, different *Enterococcus faecium* strains are bacteriocin producers, and their activity spectra are not only directed toward related microorganisms *(4)*. Within the last decade, biopreservation has received increased attention as a new method to control pathogenic and spoilage bacteria in foods, and this trend includes the use of bacteriocin-producing LAB or addition of their bacteriocins. For this reason, techniques to purify and concentrate these molecules have become very important, but few assays have been published *(5,6)*. A simple procedure is proposed here that can easily be carried out in the laboratory. Silk, produced by *Bombix mori* L., is made up of two different proteins: sericin and fibroin; the first is heat labile, while the second is very resistant to heat and acid or alkaline media. Silk has the property of catching and interacting with proteins, and it has been employed to immobilize enzymes *(7)*. Because of this characteristic, silk could be considered as a new alternative to recover, concentrate, and purify bacteriocin out of a culture medium, since the interaction between silk and bacteriocin can be reversible *(8)*.

2. Materials

2.1. Bacteriocin Preparation

1. LAB bacteriocin producer strain.
2. Sterile 2 *N* NaOH solution.

From: *Methods in Biotechnology: Environmental Microbiology: Methods and Protocols*
Edited by: J. F. T. Spencer and A. L. Ragout de Spencer

3. Membrane filter of 0.22-µm pore size.
4. Centrifuge.

2.2. Bacteriocin Activity Detection

1. Distilled and sterile water.
2. Agarized selective media (1.5% w/v agar) (*see* **Note 1**).
3. Sterile hollow punch (*see* **Note 2**).
4. Indicator strain(s).

2.3. Silk Preparation

1. Crude silk cocoons (*see* **Note 3**).
2. Low-suds washing powder (*see* **Note 4**).

2.4. Bacteriocin Recovery From Silk (see Note 5)

1. 0.5 *M* Sodium phosphate buffer, pH 7.0.
2. 0.5 *M* Sodium phosphate buffer, pH 6.0.
3. Methanol.
4. Isopropanol.
5. 0.5 *M* Sodium acetate buffer, pH 5.5.
6. 0.5 *M* Sodium acetate buffer pH 4.5.

3. Methods

3.1. Bacteriocin Extraction From the Producer Strain

1. Centrifuge (1500*g*, 20 min) an overnight culture of the LAB bacteriocin producer strain (*see* **Note 6**).
2. Remove cells and neutralize supernatant to pH 6.5–7.0 with 2 *N* NaOH solution.
3. Sterilize by filtration through a 0.22-µm membrane (*see* **Note 7**).
4. Store bacteriocin suspension at 4°C (*see* **Note 8**).

3.2. Bacteriocin Activity Determination (9–11)

1. Prepare serial twofold dilutions of bacteriocin suspension in sterile distilled water (1/2, 1/4, 1/8, 1/*n*) obtained according to **Subheading 3.1.**
2. Gently mix 100 µL of an overnight culture of the indicator strain with 10 mL of melted agar medium culture (45°C) and pour onto a sterile Petri dish. Let solidify and dry at room temperature (*see* **Note 9**).
3. Cut wells of known diameter with sterile hollow punch in the agar.
4. Fill one well with 20 µL of undiluted bacteriocin. Fill the others with the bacteriocin suspension dilutions described in **step 1**.
5. Allow sample to diffuse at room temperature before incubating dishes at the growth temperature of the sensitive strain (*see* **Note 10**).
6. Bacteriocin action is detected by the presence of growth inhibition zones of the indicator strain around the wells.

7. Calculate bacteriocin activity, expressed in arbitrary units (AU) per milliliter, as follows: AU/mL = (1/0.02 mL) × maximal dilution that inhibits growth of the indicator strain.

3.3. Silk Sample Preparation

1. Take 1 g of crude silk cocoon (*see* **Note 11**).
2. Add 10 mL of distilled water and heat to boiling for 10 min (*see* **Note 12**).
3. Add 7 g of low-suds soap, more distilled water (final volume 20 mL), and heat to boiling (96–100°C) for 1 h (*see* **Note 13**).
4. Rinse several times with distilled water to eliminate soap.
5. Heat-sterilize (121°C for 15 min) prior to use (*see* **Note 14**).

3.4. Silk Treatment

1. Take 1 mL of bacteriocin suspension and put it in contact with 0.01 g of treated silk.
2. Incubate at 4°C without agitating for 18–24 h (*see* **Note 15**).
3. Take 50 μL sample.
4. Determine total proteins.
5. Calculate bacteriocin titer according to **Subheading 3.2.**

3.5. Bacteriocin Recovery From Silk

1. Place the silk with the attached bacteriocin molecules in a clean and sterile container (*see* **Note 16**).
2. Add 1 mL of the selected solvent or buffer according to the physicochemical properties of the bacteriocin (*see* **Note 5**).
3. Shake at 4°C for at least 2 h and recover the supernatant by centrifugation at 1000*g* for 15 min (*see* **Note 17**).
4. Calculate bacteriocin titer according to **Subheading 3.2.**

4. Notes

1. The chemical composition of the culture media depends on the indicator strain selected.
2. The diameter of the hollow punch must be well determined in order to quantify the bacteriocin titer.
3. Silk cocoons may be manually or mechanically combed.
4. Ordinary low-suds soap can be employed.
5. The selected solvent or buffer used to recover the bacteriocin molecule from silk depends on the physicochemical features of each bacteriocin. Here, many alternatives are proposed.
6. Incubation time and optimal temperature for maximal bacteriocin synthesis by the selected bacteria must be determined prior to this assay.
7. Bacteriocin may be sterilized by autoclaving (121°C for 15 min) if it is sufficiently thermostable to withstand temperatures that kill the producer cells.

8. The bacteriocin suspension may be stored at 4°C for 1 mo, although other bacteriocins have been known to remain active even after 6 mo at 4°C *(4)*. Lyophilization must be used for longer periods.
9. Concentration of the sensitive strain has to be previously optimized. The cell lawn in the agar layer must show a marked inhibition zone around the seeded well.
10. This procedure helps bacteriocin diffusion before the indicator strain develops *(12)*.
11. This amount is variable, depending on the volume of the bacteriocin tested and the number of assays.
12. This step is taken to remove impurities naturally present in silk.
13. Silk treatment with soap and heat contributes to expose silk fibroin, which has a high affinity for proteinaceous molecules.
14. This step is taken to avoid contamination.
15. Shaking must be avoided to prevent bacteriocin from leaving its support (silk). Contact time should be optimized for each bacteriocin molecule.
16. Carefully press silk against the container wall with a sterile glass rod. For full bacteriocin recovery, repeat procedure in the remaining supernatant with a new silk sample.
17. Contact time should be optimized for each bacteriocin molecule.

References

1. Jack, R., Tagg, H., and Ray, B. (1995) Bacteriocins of Gram-positive bacteria. *Microbiol. Rev.* **59,** 171–200.
2. Klaenhammer, T. R. (1988) Bacteriocins of lactic acid bacteria. *Biochemie* **70,** 337–349.
3. Tagg, H. A., Dajani, A. S., and Wannamaker, L. W. (1976) Bacteriocins of Gram (+) bacteria. *Bacteriol. Rev.* **40,** 722–756.
4. Audisio, M. C., Oliver, G., and Apella, M. C. (1999) Antagonistic effect of *Ent. faecium* J96 against human and poultry pathogenic *Salmonella* spp. *J. Food Prot.* **62,** 751–755.
5. Yang, R., Johnson, M. C., and Ray, B. (1992) Novel method to extract large amounts of bacteriocins from lactic acid bacteria. *Appl. Environ. Microbiol.* **58,** 3355–3359.
6. Janes, M. E., Nannapaneni, R., Proctor, A., and Johnson, M. G. (1998) Rice hull ash and silicic acid as adsorbents for concentration of bacteriocins. *Appl. Environ. Microbiol.* **64,** 4403–4409.
7. Ellenrieder, G. and Daz, M. (1996) Thermostabilization of naringinase from *Penicillium decumbens* by proteins in solution and immobilization on insoluble proteins. *Biocatal. & Biotransf.* **14,** 113–123.
8. Audisio, M. C., Pailhé, L. A., Oliver, G., and Apella, M. C. (1999) Silk as a means of capturing bacteriocins. *Biocell* **23,** 55.
9. Daba, H., Lacroix, C., Huang, J., and Simard, R. E. (1993) Influence of growth conditions on production and activity of mesenterocin 5 by a strain of *Leuconostoc mesenteroides*. *Appl. Microbiol. Biotechnol.* **39,** 166–173.

10. Lewus, C. B. and Montville, T. J. (1991) Detection of bacteriocins produced by lactic acid bacteria. *J. Microbiol. Methods* **13,** 145–150.
11. Tagg, J. R. and McGiven, A. R. (1971) Assay system for bacteriocins. *Appl. Microbiol.* **21,** 943.
12. Abdel-Bar, N., Harris, N. D., and Rill, R. (1987) Purification and properties of an antimicrobial substance produced by *Lactobacillus bulgaricus*. *J. Food Sci.* **52,** 411–415.

27

Optimization of a Method for Isolating Plasmid DNA From Lactic Acid Bacteria From Wine

Fabiana M. Saguir and María C. Manca de Nadra

1. Introduction

Lactic acid bacteria play an important role in the manufacture of wines. They produce the deacidification of wine, contributing to the flavor and the microbiological stability of the final product *(1,2)*. Many of the phenotypes that are paramount from a technological point of view could depend on plasmid-encoded determinants. These include exopolysaccharide or bacteriocin production *(3,4)*, proteinase activity *(5)*, biogenic amines formation *(6)*, malate or citrate fermentation *(7)*, and so on. This fact justifies the research on the presence and characterization of wine lactic acid bacteria. Physical evidence for plasmids in bacteria was first presented in the early 1960s from the results of buoyant-density centrifugation in the presence of intercalating dyes, and electron microscope techniques *(8)*. The discovery of plasmid DNA in the lactic acid bacteria was generally attributed to Cords et al. *(9)*. Procedures for the detection and analysis of plasmids in bacteria were greatly simplified by the development of agarose gel electrophoresis methods for plasmid DNA *(10)*, applied to the lactic acid bacteria by Klaenhammer et al. *(11)*.

We investigated the presence of plasmids in strains of wine lactic acid bacteria belonging to the genera *Lactobacillus, Oenococcus,* and *Pediococcus,* and only in *Lactobacillus hilgardii* 5w did we find plasmid DNA. We associated the low yield with a deficient purification procedure by an inefficient cellular lysis or loss of them together with chromosomal DNA.

This study was conducted to optimize the method applied for the isolation and purification of plasmids in wine lactic acid bacteria.

From: *Methods in Biotechnology: Environmental Microbiology: Methods and Protocols*
Edited by: J. F. T. Spencer and A. L. Ragout de Spencer

2. Materials

1. Lactic acid bacteria were isolated from Argentinian wines ***(12,13)***.
2. Strains from which plasmid DNA was to be extracted were grown in the appropriate MRS broth ***(14)***, supplemented with 15% tomato juice at 30°C for 24 h, and were harvested by centrifugation. Pelleted cells were used immediately for plasmid DNA preparation or frozen at –20°C for use the following days.
3. TES buffer: 25 m*M* Tris, pH 8.0, 10 m*M* disodium EDTA, pH 8.0, 50 m*M* sucrose. TE buffer: 10 m*M* Tris, pH 8.0, disodium EDTA, pH 8.0. 1X Buffer TAE: 40 m*M* Tris, 20 m*M* acetic acid, and 2 m*M* EDTA, pH 8.0.
4. The reagent phenol-chloroform-isoamyl alcohol must be stored at –20°C in small aliquots, or may be stored at 4°C for up to 1 mo.

3. Methods

3.1. Plasmid Purification Protocol

The optimized protocol for purification of plasmid DNA is presented. It was based on the method of Birnboim and Doly ***(15)***. The first modification was the addition of lysozyme (15 mg/mL) and incubation for 30 min at 37°C (*see* **Note 1**). This procedure was again changed to a method optimization (*see* **Note 2**). The lysozyme concentration was increased from 15 to 25 mg/mL and the incubation time from 30 to 60 min; then, to complete the cellular lysis, 25 m*M* Tris-HCl–125 m*M* EDTA and a solution of 1% SDS–0.2 *N* NaOH were added to the reaction mixture (*see* **Note 3**). The screening protocol was designed to be performed in a 1.5 mL Eppendorf centrifuge tube. Centrifugations were performed at room temperature in an Eppendorf centrifuge. *Lacto-bacillus hilgardii* 5w was used as a positive control.

3.2. Agarose Gel Electrophoresis

Agarose gel electrophoresis was performed in Tris acetate buffer containing 40 m*M* Tris, 20 m*M* acetic acid, and 2 m*M* EDTA, pH 8.1. Gels contained 0.8% agarose, and electrophoresis was performed at 60 V for 4 h. Gels were stained with 0.5 µg of ethidium bromide per milliliter and photographed (*see* **Note 4**).

3.3. Details of Optimized Purification Protocol for Lactic Acid Bacteria From Wine

1. Resuspend pelleted cells in 400 µL TES buffer, pH 8.0. Transfer to Eppendorf tube and centrifuge.
2. Add 150 µL of lysozyme: 25 mg/mL in TES buffer, pH 8.0. Incubate 60 min at 37°C.
3. Add 125 µL of 0.25 *M* EDTA, 50 m*M* Tris-HCl, pH 8.0. Mix immediately by inversion.

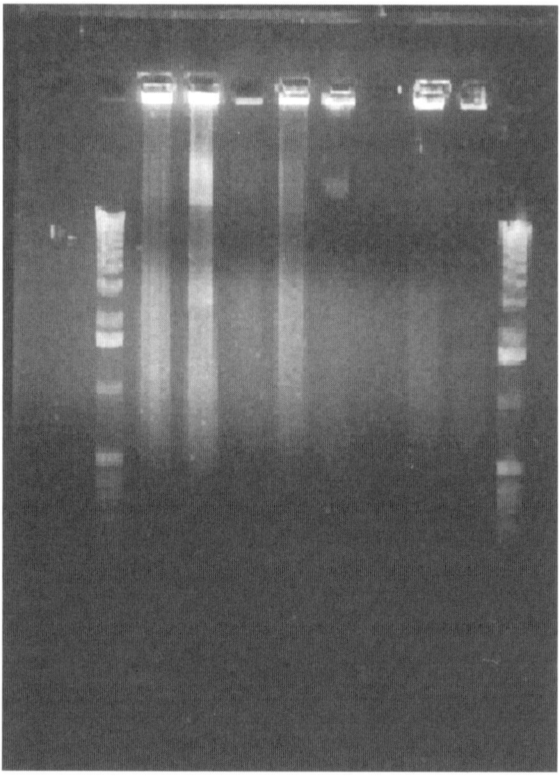

Fig. 1. Plasmid content of lactic acid bacteria from wine. Line 1,1 kb DNA ladder; line 2, *Lactobacillus casei* (positive control); line 3, *Lactobacillus hilgardii* 5w; line 4, *Lactobacillus hilgardii* X_1B; line 5, *Pediococcus pentosaceus* N_5P; line 6, *Pediococcus pentosaceus* 12 p; line 7, *Oenococcus oeni* m; line 8, *Oenococcus oeni* ST; line 9, *Oenococcus oeni* X_2L; line 10, 1 kb DNA ladder.

4. Add 200 µL of alkaline SDS solution (1% wt/v sodium dodecyl sulfate, 0.2 *N* NaOH). Gently mix immediately by intermittent inversion.
5. Add 150 µL of 5 *M* potassium acetate, 11.5% v/v. Continue gently mixing for 3 min. Incubate 5 min at 4°C. Centrifuge.
6. Transfer supernatant to new Eppendorf tube and add 650 µL of phenol-saturated chloroform-isoamyl alcohol (25:24:1). Mix thoroughly. Centrifuge.
7. Remove upper phase and extract with chloroform. Mix thoroughly. Centrifuge.
8. Remove upper phase, precipitate with 1 vol of ethanol or isopropanol. Incubate at –20°C overnight. Centrifuge.
9. Remove excess of ethanol and add 200 µL of 70% ethanol. Centrifuge.
10. Remove excess of 70% ethanol and resuspend in 30 µL of TE buffer.

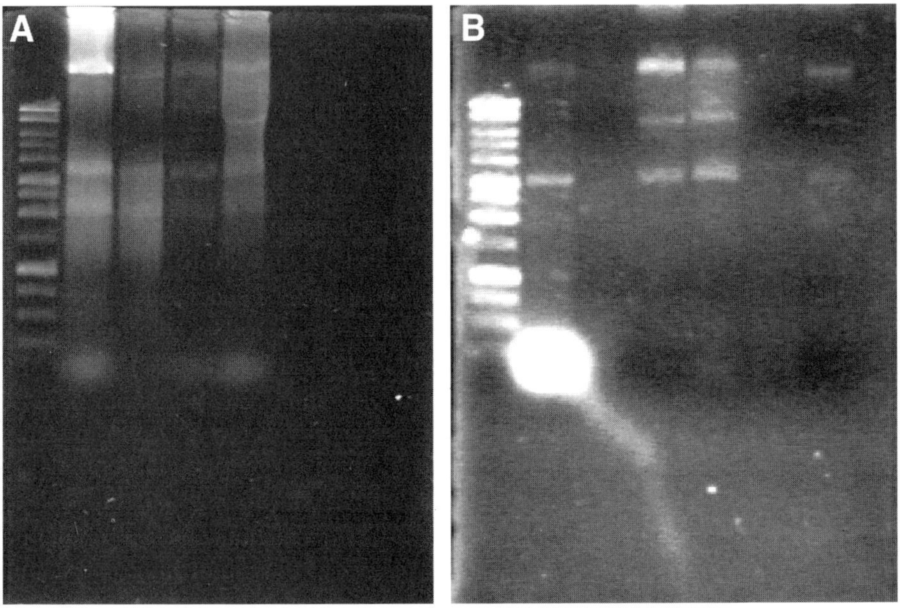

Fig. 2. Plasmid content of lactic acid bacteria from wine obtained by optimized protocol. A: line 1, 1 kb DNA ladder; line 2 and 3, *Lactobacillus hilgardii* 5w (positive control); line 4, *Lactobacillus hilgardii* X_1B; line 5, *Pediococcus pentosaceus* 12 p. B: line 1, 1 kb DNA ladder; line 2, *Lactobacillus hilgardii* 5w (positive control); line 3, *Oenococcus oeni* X_2L; line 4, *Oenococcus oeni* m; line 5, *Oenococcus oeni* ST; line 6, *Pediococcus pentosaceus* N_5P; line 7, *Pediococcus pentosaceus* 12 p.

4. Notes

1. Before the optimization of the method, from eight different lactic acid bacteria from wine, a plasmid DNA profile was observed only in *Lactobacillus hilgardii* 5w. Bands corresponding to 2500, 5000, and 10,000 bp were identified (*see* **Fig. 1**).
2. To demonstrate the effectiveness of optimized protocol detailed previously (**Subheading 3.3.**), we isolated plasmid DNA from *Lactobacillus hilgardii* 5w (positive control) and from *Lactobabillus hilgardii* X_1B, *Pediococcus pentosaceus* 12p and N5p, and *Oenococcus oeni* X_2L, m, and ST.
3. Key developments in optimizing this protocol were (1) determining an increase of 10 mg/mL in the quantity of lysozyme (25 mg/mL) as very efficient to carry out the cellular lysis, (2) increasing the time of incubation from 30 to 60 min, and (3) adding the Tris-HCl-EDTA mixture previously to the alkaline denaturation-renaturation steps.
4. The results obtained after agarose gel electrophoresis demonstrated the DNA plasmid bands in the control strain, and we visualized plasmid bands in *Lactobacillus hilgardii* X1B, *Pediococcus pentosaceus* 12p, and *Oenococcus oeni* m and ST (*see* **Fig. 2**).

References

1. Kunkee, R. E. (1967) Malolactic fermentation. *Adv. Appl. Microbiol.* **9,** 235–279.
2. Rodriguez, S. B., Amberg, E., and Thornton, R. J. (1990) Malolactic fermentation in Chardonnay: growth and sensory effects of commercial strains of *Leuconostoc oenos*. *J. Appl. Bacteriol.* **68,** 139–144.
3. Manca de Nadra, M. C. and Strasser de Saad, A. M. (1995) Polysaccharide production by *Pediococcus pentosaceus* from wine. *Int. J. Food Microbiol.* **27,** 101–106.
4. Strasser de Saad, A. M., Manca de Nadra, M. C., and Pasteris, S. E. (1995) Production and stability of pediocin N5p in grape juice medium. *J. Appl. Microbiol.* **78,** 473–476.
5. Manca de Nadra, M. C., Farías, M. E., Moreno-Arribas, M. V., Pueyo, E., and Polo, M. C. (1999) Proteolytic effect of *Leuconostoc oenos* on nitrogenous macromolecular fraction of red wine. *FEMS Microbiol. Lett.* **174,** 41–47.
6. Farías, M. E, Strasser de Saad, A. M., and Manca de Nadra, M. C. (1996) Viability and histidine decarboxylase activity of *Lactobacillus hilgardii* 5w during storage. *Microbiologie-Aliments-Nutrition* **14,** 351–355.
7. Saguir, F. M. and Manca de Nadra, M. C. (1996) Organic acid metabolism under different glucose concentrations of *Leuconostoc oenos* from wine. *J. Appl. Bacteriol.* **81,** 393–397.
8. Helinski, D. R. and Clewell, D. B. (1971) Circular DNA. *Annu. Rev. Biochem.* **40,** 899–942.
9. Cords, B. R., McKay, L. L., and Guerry, P. (1974) Extrachromosomal elements in group N streptococci. *J. Bacteriol.* **117,** 1149–1152.
10. Meyers, J. A., Sanchez, D., Elwell, L. P. and Falkow, S. (1976) Simple agarose gel electropheretic method for identification and characterization of plasmid deoxyribonucleic acid. *J. Bacteriol.* **127,** 1529–1537.
11. Klaenhammer, T. R., McKay, L. L., and Baldwin, K. A. (1978) Improved lysis of group N streptococci for isolation and rapid characterization of plasmid deoxyribonucleic acid. *Appl. Environ. Microbiol.* **35,** 592–600.
12. Manca de Nadra, M. C. and Strasser de Saad, A. M. (1987) Evolution of lactic acid bacteria during the different stages of vinification of Cafayate (Argentina) wine. *Microbiologie-Aliments-Nutrition* **5,** 235–240.
13. Strasser de Saad, A. M. and Manca de Nadra, M. C. (1987) Isolation and identification of the lactic acid bacteria from Cafayate (Argentina) wines. *Microbiologie-Aliments-Nutrition* **5,** 45–49.
14. De Man, J. C., Rogosa, M., and Sharpe, M. E. (1960) A medium for the cultivation of Lactobacilli. *Journal of Applied Bacteriology* **23,** 130–135.
15. Sambrook, J., Fritsch, E. F., and Maniatis, T. (1989) *Molecular Cloning: A Laboratory Manual* (2nd ed), Cold Spring Harbor Laboratory, Cold Spring Harbor, NY.

28

Simultaneous and Sequential Methods to Study Interactions Between Yeast and Lactic Acid Bacteria From Wine

Marta E. Farías and María C. Manca de Nadra

1. Introduction

Wine is the result of a complex interaction between yeast, lactic acid bacteria (LAB), grape must, and physical conditions. Understanding the roles of the yeasts in wine will help in the quest to develop cultured wines that have similar characteristics to traditional fermented grape juice. Alcoholic fermentation of natural grape juice is carried out by the sequential action of different yeast genera and species; initially, apiculate yeasts *(Kloeckera/Hanseniaspora)* are present, but after 3–4 d, these are replaced by *Saccharomyces cerevisiae* ***(1)***. *Hanseniaspora uvarum (Kloeckera apiculata)* and *Candida stellata* survived during the fermentation for longer periods than previously thought, and grew to maximum populations of 10^6–10^7 cfu/mL ***(2,3)***.

The principal LAB species associated with wines are *Oenococcus oeni, Pediococcus pentosaceus,* and various species of *Lactobacillus.* There are many reports about these lactic acid bacteria interactions ***(4–6)***.

This study was conducted to study yeast–LAB interaction in simultaneous (both microorganisms in the same medium) and in sequential (growth of LAB in the supernatant modified by yeast growth) growth methods using reproducible and simple techniques.

2. Materials

1. *Lactobacillus hilgardii* 5w ***(7)*** and *Hanseniaspora uvarum* ca12 *(Kloeckera apiculata)* ***(8)*** were isolated from Argentinean wines.
2. Basal medium: 10 g/L yeast extract, 5.0 g/L glucose, 1 mL/L Tween-80, and 57 mL/L grape juice ***(9)***, pH 4.8, sterilized by autoclaving for 15 min at 121°C (*see* **Note 1**).

From: *Methods in Biotechnology: Environmental Microbiology: Methods and Protocols*
Edited by: J. F. T. Spencer and A. L. Ragout de Spencer

3. MRS agar plus tomato juice: 10 g/L polypeptone, 10 g/L meat extract, 5 g/L yeast extract, 20 g/L glucose, 5 g/L sodium acetate, 1 mL/L Tween 80, 2 g/L ammonium citrate, 2 g/L K_2HPO_4, 0.2 g/L $MgSO_4•7H_2O$, 0.05 g/L $MnSO_4•4H_2O$, 150 mL/L tomato juice, and 20 g/L agar, pH 5.0, sterilized by autoclaving for 15 min at 121°C.
4. MYPG agar supplemented with antibiotics: 5 g/L malt extract, 10 g/L yeast extract, 20 g/L glucose, 20 g/L peptone, 20 g/L agar, 0.05 g/L gentamicin, 0.1 g/L oxytetracycline, pH 5.5, sterilized by autoclaving for 15 min at 121°C.
5. MRS agar supplemented with 0.05 g/L cycloheximide and 0.025 g/L pimaricin, pH 4.8, sterilized by autoclaving for 15 min at 121°C.

3. Methods

3.1. Simultaneous Cultures in Liquid Media

1. *Lactobacillus hilgardii* and *Hanseniaspora uvarum* were grown separately in 250 mL stoppered flasks containing 50 mL of basal medium.
2. 4-mL portions of late-log-phase cultures were transferred to a 1-L stoppered flask containing basal medium and incubated at 30°C.
3. Bacterial and yeast cells (1000 mL) were aseptically harvested by centrifugation at 10,000*g* for 10 min at 4°C.
4. The pellets were washed twice with 0.1% (w/v) peptone water and resuspended in 50 mL of 0.1% (w/v) peptone water to avoid dilution of nutrients (*see* **Note 2**).
5. Appropriate quantities of 20-fold-concentrated cell suspensions were added to 250 mL stoppered flasks with 200 mL of basal medium to give a viable yeast cell concentration of 10^8 cells/mL and final viable bacterial cell numbers of 10^2, 10^4, and 10^8 cells/mL, and then incubated at 30°C.
6. 5-mL aliquots of culture were taken at different time intervals, and the viable cell counts were determined. The culture supernatants were utilized for analytical measurements (*see* **Note 3**).

3.2. Sequential Cultures in Liquid Media

1. A stoppered 250-mL flask containing 200 mL of basal medium was inoculated with a yeast strain and incubated at 30°C.
2. At different times, 20 mL of fermented broth was collected and the number of viable cells counted (*see* **Note 4**).
3. The broth was filtered aseptically through a 0.2-μm pore-size membrane (Millipore) or heated for a few seconds at 115°C (3/4 atm) in autoclave (*see* **Note 5**).
4. The broth was dispensed into a 75-mL flask to which 10% of basal medium culture of *Lactobacillus hilgardii* 5w was added, and incubated for 48 h at 30°C.
5. 5-mL aliquots of culture were taken at different time intervals and the viable cell counts were determined. The yeast and bacteria culture supernatants were utilized for analytical measurements (*see* **Note 3**).

3.3. Counting of Bacterial and Yeast Cells

Populations of yeast and bacteria were followed by counting viable cells.

3.3.1. Counts of Viable Bacteria and Yeast Cells in Simultaneous Cultures

1. Serial dilutions were done in saline solution, plating duplicates of statistically significant dilutions on selective media.
2. Yeasts were counted by surface spreading 0.1-mL samples on MYPG agar, supplemented with appropriate concentrations of antibiotics (gentamicin and oxytetracycline) to inhibit bacterial growth. The plates were incubated at 30°C for 48 h.
3. Lactic acid bacteria were counted by surface spreading 0.1 mL samples on MRS agar medium, supplemented with 15% tomato juice. Before pouring, appropriate concentrations of cycloheximide and pimaricin were added to the medium to inhibit the growth of yeast. The plates were incubated at 30°C for 5 d (*see* **Note 6**).

3.3.2. Counts of Viable Bacteria and Yeast Cells in Sequential Cultures

1. Serial dilutions were done in saline solution, plating duplicates of statistically significant dilutions on selective media.
2. Yeasts were counted by surface spreading 0.1-mL samples on MYPG agar. The plates were incubated at 30°C for 48 h.
3. Lactic acid bacteria were counted by surface spreading 0.1-mL samples on MRS agar medium, supplemented with 15% tomato juice. The plates were incubated at 30°C for 5 d.

3.4. Proteins Determination

The proteins of the samples were determined by the reaction with Coomassie® Brilliant Blue G-250. Calibration was carried out using bovine serum albumin (BSA) ***(10)*** (*see* **Note 7**).

4. Notes

1. The individual growth of yeast and lactic acid bacteria in the same basal medium with added grape juice is possible because they share the same natural ecological niche.
2. In simultaneous yeast–lactic acid bacteria cultures, the proportions of both inoculums were 1:1, 1:0.5, or 1:0.25.
3. As a control of the results of interaction between yeast and lactic acid bacteria in simultaneous and sequential inoculation methods, we measured organic nitrogen modification in the media.
4. In sequential cultures, samples were taken at different yeast growth phases (0, 3, 8, 12, 24, and 34 h), and each one was utilized to bacteria inoculation.

5. During sequential inoculation, after yeast growth, the eucaryotic cells were removed from the broth by filtration through a 0.2-µm-pore membrane and by heating at 3/4 atm for a few seconds. No significant differences in the bacterial counts and analytical determinations were obtained between the two methods of yeast elimination. We prefer the heat procedure because it is a simpler and more rapid method.
6. In mixed cultures, for lactic acid bacteria determinations, a combination of two antibiotics (cycloheximide and pimaricin) was used to inhibit the yeast.
7. The modification of nitrogen in the media was monitored by protein measurements. The initial protein concentrations correspond to yeast extract plus grape juice.

References

1. Fleet, G. H. (1997) Wine, in *Food Microbiology, Fundamentals and Frontiers* (Doyle, M. P., Beuchat, L. R., and Montville, T. J., eds.), American Society for Microbiology, Washington DC, pp. 671–694.
2. Fleet, G. H. and Heard, G. M. (1993) Yeast growth during fermentation, in *Wine Microbiology and Biotechnology* (Fleet, G. H. ed.), Chur, Harwood Academic Publishers, Chuv, Switzerland, pp. 27–54.
3. Magasanik B. (1992) Regulation of nitrogen utilization, in *The Molecular and Cellular Biology of the Yeast Saccharomyces,* vol. 2 (Jones, E. W., Pringle, J. R., and Broach, J. B. eds.), Cold Spring Harbor Laboratory Press, Cold Spring Harbor, NY, pp. 283–317.
4. Rodriguez, A. V. and Manca de Nadra, M. C. (1994) Sugar and organic acid metabolism in mixed cultures of *Pediococcus pentosaceus* and *Leuconostoc oenos* isolated from wine. *J. Appl. Bacteriol.* **77,** 61–66.
5. Rodriguez, A. V. and Manca de Nadra, M. C. (1995) Interactions between *Lactobacillus hilgardii* and *Pediococcus pentosaceus* and their metabolism of sugars and organic acids. *World J. Microbiol. Biotechnol.* **11,** 349–350.
6. Rodriguez, A. V. and Manca de Nadra, M. C. (1995) Mixed culture of *Lactobacillus hilgardii* and *Leuconostoc oenos* isolated from wine. *J. Appl. Bacteriol.* **78,** 521–525.
7. Manca de Nadra, M. C. and Strasser de Saad, A. M. (1987) Evolution of lactic acid bacteria during the different stages of vinification of Cafayate (Argentina) wine. *Microbiol.-Alim.-Nutr.* **5,** 235–240.
8. Farías, M. E., Sosa, O. A., and Manca de Nadra, M. C. (2001) Flocculation of *Kloeckera apiculata* mc1 from wine, in *100th General Meeting American Society for Microbiology* (ASM), Florida, pp. 438–439.
9. Manca de Nadra, M. C., Farías, M. E., Moreno-Arribas, M. V., Pueyo, E., and Polo, M. C. (1999) Proteolytic effect of *Leuconostoc oenos* on nitrogenous macromolecular fraction of red wine. *FEMS Microbiol. Lett.* **174,** 41–47.
10. Bradford, M. M. (1976) A rapid and sensitive method for quantification of microgram quantities of protein using the principle of protein dye-binding. *Anal. Biochem.* **72,** 248–255.

29

Method for Determining Lindane Concentration in Water and Solid Samples

Claudia S. Benimeli, Adriana P. Chaile, and María J. Amoroso

1. Introduction

Pesticides are largely used in agriculture, as well as in public health, for the prevention of infectious and transmissible diseases. The use of pesticides in agriculture may represent a risk to human health and the environment. The general population is mainly exposed to pesticide residues through the ingestion of contaminated foods. Foods like fruits, vegetables, and cereals may be contaminated because they are directly treated with pesticides or are grown on contaminated fields *(1)*. Pesticide contamination of surface waters and groundwaters from agricultural use has been well documented around the world. The persistence of pesticide residues in the soil and their movement in the water-soil system are key aspects of their environmental behavior. Pesticides are primarily moved from agricultural fields to surface waters in surface run-off *(2)*. This stimulates, on the one hand, the establishment of legal directives to control their levels through the establishment of maximum residue levels, and on the other, a continuous look for pesticides less persistent and toxic for human beings. This fact has increased extraordinarily the number of pesticides registered and/or recommended, and the analytical difficulties for their control *(3)*.

The insecticide lindane (1α,2α,3β,4α,5α,6β-hexachlorocyclohexane γ-HCH) has been used extensively worldwide, despite its persistence in the environment, tendency to bioaccumulation, and toxicity for non-target organisms, including human beings, as well as possible toxicological and environmental problems of its residues *(4,5)*.

From: *Methods in Biotechnology: Environmental Microbiology: Methods and Protocols*
Edited by: J. F. T. Spencer and A. L. Ragout de Spencer

In this chapter, we describe a very sensitive method for determining lindane concentration in liquid and solid samples like water and soils.

2. Materials

1. Lindane calibration standards (ULTRAScientific, North Kingstown, RI) (*see* **Notes 1** and **2**).
2. Distilled water.
3. Methanol (*see* **Notes 3** and **4**).
4. Hexane (*see* **Notes 3** and **4**).
5. Ethyl acetate (*see* **Notes 3** and **4**).
6. 25 m*M* Tris-EDTA buffer, pH 8.0.
7. Bond Elut C18 cartridge 6 mL-500 mg.
8. Gas chromatograph (Hewlett-Packard 6890, Wilmington, DE) equipped with electron capture detector (^{63}Ni ECD), split/splitless injector HP 7694, HP 5 capillary column (30 m × 0.53 mm × 0.35 μm), and ChemStation Vectra XM software.

3. Methods

In recent years, capillary columns have been almost completely replaced by packed columns owing to their high resolving power, which allows the separation of a large number of pesticides with similar physical-chemical characteristics.

In the past 30 yr, the ECD has been the detector most used in pesticide residue analysis. It presents a very high sensitivity to polychlorinated hydrocarbons and other halogenated pesticides. GC detectors using ECD afford detection limits typically between 1 and 5 ng/L.

3.1. Water

For lindane determination in liquid substances, it is necessary to extract this pesticide from the sample; a solid phase extraction is described in the following steps (*see* **Note 5**).

1. Solid phase extraction (SPE): extract lindane from liquid samples using C18 cartridges. Prior to the extraction, the C18 bonded phase must be washed with hexane, distilled water, and methanol, respectively, under vacuum. The columns are not allowed to dry, as recommended. Allow the sample to percolate through the columns at a flow-rate of 5 mL/min under vacuum.
2. After sample extraction, collect the pesticide trapped in the cartridge by using hexane as eluting solvent.
3. Evaporate fractions to 1 mL in a gentle stream of nitrogen.
4. Transfer the residue into a vial and inject directly in the GC-ECD system.

3.2. Solid Samples

The compound of interest (lindane) must first be removed from the solid matrix and isolated from any interference using ethyl acetate as described in the following steps:

1. Wash the sample once with 25 m*M* Tris-EDTA buffer, pH 8.0.
2. Wash the sample with ethyl acetate to remove extracellular adsorbed lindane.
3. Extract intracellular lindane adding ethyl acetate (20%), leaving this suspension for 24 h at room temperature.
4. Separate solvent phase by centrifuging or with a pipet.
5. Evaporate completely the ethyl acetate of the isolated samples of **steps 2** and **4** with a gentle stream of nitrogen.
6. Dissolve the extracts in 1 mL hexane.
7. Transfer the extract into a vial and inject directly in the GC-ECD system.

3.3. Gas Chromatography Analysis

1. Inject the extracts from liquid and solid samples in the gas chromatograph with the following program: 90°C (1 min) to 180°C (3 min) at 30°C/min and to 290°C (2 min) at 20°C/min (*see* **Note 6**).
2. Helium was used as the carrier at 25 cm/s and nitrogen was used as make-up gas at 25 mL/min.
3. Perform quantification by external and internal calibration using dilutions of authentic lindane standard.

4. Notes

1. The lindane calibration standard must be stored in a freezer.
2. Lindane is a moderately dangerous pesticide, so it should be handled with latex gloves.
3. All the solvents must be HPLC grade.
4. Exercise care when handling organic solvents.
5. The liquid samples with high organic matter concentrations have to be filtered with 0.4 μm acetate cellulose membranes.
6. Certain areas of the GC system are heated. Avoid bodily contact with these areas and use care in handling flammable solvents in and around the GC system.
7. The liquid samples with high organic matter concentrations have to be filtered with 0.4 μm acetate cellulose membranes.

References

1. Hura, C., Leanca, M., Rusu, L., and Hura, B. A. (1999) Risk assessment of pollution with pesticides in food in the Eastern Romania area (1996–1997). *Toxicol. Lett.* **107,** 103–107.
2. Albanis, T. A., Hela, D. G., Sakellarides T. M., and Konstantinou, I. K. (1998) Monitoring of pesticide residues and their metabolites in surface and underground

waters of Imanthia (N. Greece) by means of solid-phase extraction disks and gas chromatography. *J. Chromatogr. A* **823,** 59–71.

3. Torres, C. M., Picó, Y., and Mañes, J. (1996) Determination of pesticide residues in fruit and vegetables. *J. Chromatogr. A* **754,** 301–331.
4. Singh, B. K. and Kuhad, R. C. (1999) Biodegradation of lindane (γ hexachlorocyclohexane) by the white-rot fungus *Trametes hirsutus. Lett. Appl. Microbiol.* **28,** 238–241.
5. Johri, A. K., Dua, M., Tuteja, D., Saxena, R., Saxena D. M., and Lal, R. (1996) Genetic manipulations of microorganisms for the degradation of hexachlorocyclohexane. *FEMS Microbiol. Rev.* **19,** 69–84.

30

Extracellular Hydrolytic Enzymes Produced by Yeasts

Fabio Vazquez, Martha Dina Vallejo Herrera, Lucía I. C. de Figueroa, and María Eugenia Toro

1. Introduction

Different yeasts are able to utilize different carbon sources, and nutritional selectivity determines yeast species diversity in particular niches. In other words, yeasts exhibit great specialization for habitat *(1)*.

Most yeasts have a saprophytic lifestyle in association with other organisms producing death or decaying plant material, but few plant parasitic (endophytic) yeasts are known. Both saprophytic and parasitic yeasts living on vegetation depend on the host that supplies essential nutrients for yeast growth. Several yeasts, like *Pichia minuta, Candida ernobii,* or *Cryptococcus skinneri,* appear to be truly wood inhabiting, whereas other yeast-wood associations involve tree exudates or decomposing wood *(2,3)*.

As far as it is known, enzymes such as xylanases, pectinases, proteases, amylases, and lipases could be involved in substrate colonization (e.g., fruits and other plant tissues), causing the breakdown of plant tissues with a concomitant release of sugars from plant cells, which, in turn, can be utilized for yeast growth *(4,5)*. Ecologically, one of the most important substrates is xylan, which is, after cellulose, the most abundant renewable polysaccharide on earth. Xylanase is produced by some genera of yeasts: *Cryptococcus* ***(6,7)***, *Candida* ***(8)***, *Sporothrix* ***(8)***, *Pichia* ***(9,10)***, *Aureobasidium* ***(11,12)***, and *Debaryomyces* ***(8)***. In the same way, pectinolytic activity is shown in several yeasts, e.g., *Saccharomyces cerevisiae* ***(13,14)***, *Candida boidinii* ***(15)***, *Trichosporon penicillatum* ***(16)***, *Kluyveromyces marxianus* ***(17)***, and *Saccharomycopsis fibuligera* ***(18)***.

From: *Methods in Biotechnology: Environmental Microbiology: Methods and Protocols*
Edited by: J. F. T. Spencer and A. L. Ragout de Spencer

Recent studies have demonstrated that in food ecosystems, lipolytic, proteolytic, and amylolytic activities are some of the yeast characteristics considered important for colonization of habitats and for development of different strains, as well as for some interactions with other microorganisms ***(19–22)***. In the winemaking process, it is important to ascertain the potential of wild wine yeasts for producing extracellular enzymes, in order to alter certain components of the musts, and thus enhance the sensory attributes of the wines. Pectinases, proteases, and glycosidases are enzymes secreted by yeasts that are of interest in winemaking because of their technological effects and their contribution to aroma formation ***(20,23)***. The smear microorganisms that develop on a cheese surface play an important role in the ripening process through the action of both proteolytic and lipolytic enzymes. It is known that yeasts are involved directly and indirectly in the ripening process ***(20,21,24)***.

Lipases secreted by *Candida albicans* may play a role in adhesion, interactions with the immune system, and nutrition, providing a carbon and energy source for this pathogenic yeast ***(25)***.

1.1. Enzymes and Their Substrates

Plants synthesize starch as part of photosynthesis. It is synthesized in plastids found in leaves as a storage compound for use in respiration during dark periods, and also in amyloplasts found in tubers, seeds, and roots as a long-term storage compound. Starch is a polymer of glucose linked one to another through the C1 oxygen, known as a glycosidic bond. This bond is stable at high pH but hydrolyzes at low pH. At the end of the polymeric chain there is a latent aldehyde group known as the reducing end. Two types of glucose polymers are present in starch: amylose and amylopectin. Amylose is a linear polymer consisting of up to 6000 glucose units with α,1-4 glycosidic bonds. The number of glucose residues, also called the degree of polymerization (DP), varies with the botanical origin. Amylopectin consists of short α,1-4-linked linear chains of 10–60 glucose units and α,1-6-linked side chains with 15–45 glucose units. The average number of branching points in amylopectin is 5%, but it varies according to the botanical origin. The complete amylopectin molecule contains about 2 million glucose units, and is one of the largest molecules in nature ***(26)***.

There are basically four groups of starch-converting enzymes: endoamylases, exoamylases, debranching enzymes, and transferases.

Endoamylases are able to cleave the α,1-4 glycosidic bonds present in the inner part (endo-) of the amylose or amylopectin chain. α-Amylase (EC 3.2.1.1) is a well-known endoamylase. It is produced by many microorganisms belonging to Archaea, Bacteria, as well as Fungi ***(27)***. End products of α-amylase activity are small oligosaccharides with an α configuration and α-limit dextrins, which constitute branched oligosaccharides.

Enzymes belonging to the second group, the exoamylases, either exclusively cleave α,1-4 glycosidic bonds such as β-amylase (EC 3.2.1.2), or cleave both α,1-4 and α,1-6 glycosidic bonds like amyloglucosidase or glucoamylase (EC 3.2.1.3) and α-glucosidase (EC 3.2.1.20). Exoamylases act on the external glucose residues of amylose or amylopectin, and thus produce only glucose (glucoamylase and α-glucosidase) or maltose and β-limit dextrin (β-amylase).

β-amylase and glucoamylase also convert the anomeric configuration of the liberated maltose from α to β. Glucoamylase and α-glucosidase differ in their substrate preference: α-glucosidase acts better on short maltooligosaccharides and liberates glucose with α-configuration while glucoamylase better hydrolyzes long-chain polysaccharides ***(27)***.

Lipases (EC 3.1.1.3) catalyze the hydrolysis of triacylglycerols and the synthesis of esters from glycerol and long-chain fatty acids. Lipases bind at the interface between aqueous and organic phases, and catalyze hydrolysis at this interface. This binding not only places the lipase close to the substrate but also increases the catalytic power of the lipase, a phenomenon called interfacial activation—e.g., the enhancement of catalytic activity on lipid aggregates (micelles) rather than on lipid monomers in aqueous solution ***(28–30)***.

Purification and characterization of multiple intracellular lipases of yeast *(Candida rugosa, Candida antarctica, Geotrichum candidum)* have become rampant in microbial molecular biology ***(31,32)***. The asporic yeast *C. rugosa* secretes several closely related exolipase isoforms. There have been varying reports on the number of lipase isoforms (two, three, or five) and their molecular weight ***(33,34)***.

Lipases of *C. rugosa* and *G. candidum* have been studied together, since they show great similarities in many aspects. These yeast species possess different lipase-encoding genes, which might account for proteins differing in their amino-acid sequences and possibly in their biochemical and enzymatic properties ***(31)***. It is possible to speculate that the availability of related but nonidentical enzymes (lipase isoenzymes and others) would improve the adaptation of the producing organisms to different substrate sources, due to the physiological relevance of those enzymes for those organisms ***(35)***.

Proteases are the single class of enzymes that occupy a pivotal position with respect to their applications in both physiological and commercial fields. Proteolytic enzymes catalyze the cleavage of peptide bonds in other proteins. Proteases are degradative enzymes, which catalyze the total hydrolysis of proteins ***(36)***. As enzymes are proteins, they can be degraded by proteases.

Proteases are classified on the basis of three major criteria: type of reaction catalyzed, chemical nature of catalytic site, and evolutionary relationship with reference to structure ***(37)***.

Proteases are grossly subdivided into two major groups: exopeptidases and endopeptidases, depending on their action site. Exopeptidases cleave the peptide bond proximal to the amino or carboxy termini of the substrate, whereas endopeptidases cleave peptide bonds distant from the termini of the substrate. Based on the functional group present at the active site, proteases are further classified into four prominent groups: serine proteases, aspartic proteases, cysteine proteases, and metalloproteases. There are a few miscellaneous proteases, which do not precisely fit into the standard classification, e.g., ATP-dependent proteases, which require ATP for their activity *(38)*.

1.2. Industrial Applications of Enzymes

During the last three decades, amylases have been exploited in the starch-processing industry, replacing acid hydrolysis in starch hydrolysates production. These enzymes are also used for removing starch from beer, fruit juices, clothes, or porcelain. Thermostable pullulanase is another starch-hydrolyzing enzyme used widely for the debranching of amylopectin. A new and recent application is to use maltogenic amylase as an antistaling agent, to prevent starch retrogradation in bakery products. Only one type of starch-modifying enzyme has found its way into the commercial market: cyclodextrin glycosyltransferase, either for cyclodextrins production in non-food applications, or for starch hydrolysis during the saccharification process. Other starch-modifying enzymes, e.g., amylomaltase and branching enzyme, are not yet used industrially, although potentially interesting applications have been described in patents and scientific literature *(26,27)*.

Lipases are not only capable of hydrolyzing the ester bonds of water-insoluble substrates at the interface between substrate and water, but also catalyze the reverse reaction (ester synthesis) in nonaqueous systems. Since lipases can evidently catalyze many different reactions, they have been widely used in industrial applications, such as foods, chemicals, pharmaceuticals, oil de-esterification, digestive aids, industrial detergents, and so on *(39)*. When used as ingredients in laundry detergents, lipases that are stable at pH 10.0–11.0 from 30 to 60°C in presence of surfactants are preferred *(30)*. Recently, lipase-catalyzed esterification reactions have growing importance due to the numerous products that can be obtained, e.g., monoacylglycerol, flavor esters, or surfactants esters *(40)*.

Advances in analytical techniques have demonstrated that proteases conduct highly specific and selective modifications of proteins, such as activation of zymogenic forms of enzymes by limited proteolysis, blood clotting and lysis of fibrin clots, and processing and transport of secretory proteins across membranes. The current estimated worldwide yearly sales of industrial enzymes is

US $1 billion; hydrolytic enzymes account for 75% of this. Proteases represent one of the three largest groups of industrial enzymes, accounting for nearly 60% of the worldwide enzymes market ***(36)***. Yeast extracellular proteases are of particular interest because of their direct commercial applications, and their potential use in expression systems for heterologous proteins production. *Yarrowia lipolytica XPR2* promoter and some parts of the *XPR2* gene that provide secretory and processing signals have been used in constructs to direct the synthesis and secretion of prochymosin, human anaphylatoxin C5a, porcine *a*1-interferon, invertase, tissue plasminogen activator, and hepatitis B virus middle surface antigen ***(41–44)***. Extracellular proteases produced by *Candida albicans* and by some species of *Aspergillus* have received special attention for their role in pathogenesis ***(45)***.

The objective of this chapter is to describe techniques to qualify and quantify enzymatic activities (amylases, lipases, and proteases) involved in yeast's colonization of different habitats.

2. Materials

2.1. Strains

Yeasts to be tested may be isolated and characterized from different niches: nature (e.g., plant surfaces, plant-root environment, decaying wood), manufactured (e.g., cheese surfaces, fermenting grape must, ripening fruits), or from specific cultures collections (*see* **Note 1**).

2.2. Media

1. General growth media: MYGP: 3.0 g/L malt extract, 3.0 g/L yeast extract, 5.0 g/L peptone, and 20 g/L glucose. YPD: 10 g/L yeast extract, 20 g/L peptone, and 20 g/L glucose. pH is adjusted with 1 *M* HCl to 4.5.
2. Medium for α-amylase production: 10 g/L yeast extract, 0.6 g/L $MgCl_2$, 0.1 g/L $CaCl_2$, and 10 g/L soluble starch. All components should be dissolved in 0.2 *M* acetate buffer, pH 5.5, before autoclaving.
3. Medium for amylase activity determination: 10 g/L RBB-starch (prepared as in **Subheading 3.1., item 2**), commercial nutrient agar. pH adjusted to 4.5 with 0.1 *M* H_2SO_4.
4. Media for lipase secretion. YPDH: 10 g/L yeast extract, 10 g/L bacto-peptone, 10 g/L glucose, and 10 g/L olive oil. YTDH: 10 g/L yeast extract, 10 g/L bacto-tryptone, 10 g/L glucose, and 50 g/L olive oil ***(46,47)***. Media are buffered with 50 m*M* phosphate buffer, pH 4.5.
5. Medium for protease activity detection: 10 g/L glucose, 7 g/L yeast nitrogen base (YNB), and 20 g/L agar.
6. Casein growth medium: 10 g/L casein; 10 g/L glucose; 1 g/L yeast nitrogen base (YNB) without amino acids, and 0.66 g/L $(NH_4)_2SO_4$.

2.3. Solutions and Devices

2.3.1. Amylase Activity Determination in Liquid Extract

1. Buffer solutions: 100 m*M* sodium-acetate (NaOH-CH_3COOH) buffer (AcB), pH 5.0, or citric acid–Na_2HPO_4 buffer (CPB), pH 3.0 or 4.5, maintained at 4°C (*see* **Note 2**).
2. Substrate for α-amylase activity (AAase) determination: solution of 10 g/L soluble starch in AcB, pH 5.0, or CPB, pH 3.0 or 4.5 (*see* **item 1** of this list).
3. Somogyi reagent ***(48)***.
4. Nelson reagent ***(49)***.
5. Glucose stock solution: 2.0 g/L anhydrous glucose, in the buffer used.

2.3.2. Amylase Activity Detection in Solid Medium

1. Remazol brilliant blue (RBB): 10 g/L RBB in water.
2. Corn starch.
3. Na_3PO_4.
4. $(NH_4)_2SO_4$.
5. Methanol.

2.3.3. Lipase Esterification Activity

1. Butyric acid.
2. Butanol.
3. *n*-Heptane.
4. Stock solution: 1.35 mL butyric acid (0.16 *M*) and 2.7 mL butanol (0.33 *M*) in 85.95 mL *n*-heptane (butyric acid, butanol, and *n*-heptane are distilled once before use).
5. Methanol.
6. 0.02 *M* NaOH in water.
7. Phenolphthalein.

2.3.4. Hydrolytic Lipase Activity

1. 0.1 *M* McIlvaine's buffer solution (citric acid–Na_2HPO_4, pH 7).
2. SB solution: 20 g/L sodium benzoate in buffer.
3. 5 g/L Gum acacia in ethanol.
4. SDS solution: 100 g/L sodium dodecyl sulphate (SDS) solution in water.
5. 0.33 *M* Tributyrin emulsion is prepared in phosphate buffer. SB solution is used as a preservative, and acacia gum is used as an emulsifying agent. SDS solution 50 μL is also added.

2.3.5. Proteinase Detection in Liquid Medium

1. Buffer: 0.1 *M* Na_2HPO_4–citric acid, pH 3.2.
2. Acid-denatured hemoglobin: It is prepared by adjusting a solution of bovine hemoglobin (1 g/20 mL) to pH 3.2 with 0.3 *M* HCl. The solution is made up to 100 mL with 0.066 *M* sodium citrate buffer (pH 3.2).

3. 0.5 g/L trichloroacetic acid (TCA) in water.
4. Lowry reagent ***(50)***.
5. Folin reagent ***(50)***.
6. Ciocalteu's phenol reagent ***(50)***.
7. Tyrosine standards: they are prepared in the buffer used.

2.3.6. Proteolytic Activity in Solid Medium

1. Buffer A: 24 g/L KH_2PO_4 and 35 g/L $Na_2HPO_4 \cdot 7H_2O$.
2. Citrate-phosphate buffer 0.05 *M*, pH 3.5.
3. Skim milk solution: 100 g/L skim milk powder in citrate-phosphate buffer (0.05 *M*, pH 3.5).
4. 1 *N* HCl.

3. Methods

All crude extracts employed for enzymatic determinations must be stored at 4°C for short periods or at –22°C (in freezer) for long periods.

3.1. Amylase Activity Assays

3.1.1. α-Amylase Activity Determination in Liquid Extract

1. α-Amylase (AAase) activity is determined in culture extracts of solid-state fermentations, culture filtrates, or supernatants of liquid fermentations. If enzymes are obtained from liquid fermentations, yeasts are grown overnight at 25°–30°C in 20 mL YPD (*see* **Subheading 2.2., item 1**), then inoculated into 250-mL Erlenmeyer flask containing 50 mL of culture medium for α-amylase production (*see* **Subheading 2.2., item 2**) and incubated at 30°C in a rotary shaker (250 rpm). Cells are harvested by centrifugation after an overnight. Crude supernatant may be concentrated or partially purified by ultrafiltration.
2. The enzymatic reaction is run in test tubes containing 450 µL of starch solution (*see* **Subheading 2.3.1., item 2**)—keeping the test tubes in crushed ice bath—and 50 µL of enzyme sample.
3. Each sample needs a proper blank with starch solution and without the enzyme.
4. Test tubes are incubated 30 min at 37°C in water bath.
5. The reaction is stopped by cooling the tubes in an ice-water bath (10 min).
6. 50 µL of enzyme sample is added to the blank tubes.

Reducing groups released from the starch can be quantified by the Somogyi method ***(48)*** as follows:

1. Keeping the tubes in ice-water bath, add 500 µL of Somogyi reagent.
2. Heat the tubes for 10 min in a boiling-water bath and cool in ice-water bath.
3. Mix with 500 µL of Nelson reagent at room temperature.
4. Allow to stand 30 min at room temperature, then add to each tube 4500 µL of water.

5. Measure absorbance at 660 nm against a colorimetric reaction blank, replacing the sample and the starch solution by 500 µL of the buffer used in the enzymatic reaction (AcB or CPB) (*see* **Subheading 2.3.1., item 1**).
6. The equivalent reducing amount of the samples is determined from a standard curve of glucose (*see* **Subheading 2.3.1., item 5**) done up to 250 mg/L, subtracting from the total absorbance the respective blank sample absorbance.

One unit of AAase activity is the activity that liberates reducing groups equivalent to 1 µmol of glucose per minute, under the conditions of the assay.

3.1.2. Amylase Activity Detection in Solid Medium (Plate Method)

Remazol brilliant blue (RBB)–starch is prepared as follows:

1. Corn starch (5 g) is suspended in 50 mL water (60°C) with vigorous stirring. To this, 100 mL RBB solution **(Subheading 2.3.2., item 1)** is added.
2. Add 8 g $(NH_4)_2SO_4$, and stir at 60°C for 30 min, until complete dissolution.
3. The dye is fixed to the corn starch by the addition of 1 g Na_3PO_4.
4. The mixture is maintained at 60°C for 2 h with continuous stirring.
5. Insoluble dyed RBB-starch is collected by centrifugation and washed with water (55°C) until the supernatant is clear, and then washed with methanol twice.
6. The powder is dried and stored under desiccation at room temperature.
7. Petri dishes with the described medium (*see* **Subheading 2.2., item 3**) are seeded and incubated at 30°C for 48–72 h.
8. Amylase production is detected by the disappearance of the color of the blue starch around microbial colonies ***(51)***.

3.2. Lipase Activity Assays

3.2.1. Esterification Activity Measurements

1. Yeast are grown overnight at 25°C in 10 mL MYGP (YPD may be used) (*see* **Subheading 2.2., item 1**). Then the cells are transferred into 250-mL Erlenmeyer flask containing 50 mL YPDH or YTDH medium (*see* **Subheading 2.2., item 4**). Cultures are incubated at 30°C in a rotary shaker (250 rpm) and cells are harvested by centrifuging after 24 h. Crude supernatant may be concentrated or partially purified by ultrafiltration.
2. Lipase detection is set up in 50 mL stoppered conical flasks containing 3 mL stock solution (*see* **Subheading 2.3.3., item 4**) with the appropriate quantity of the lipase extract.
3. Flasks are incubated at 30–60°C in a shaking water bath for 120 min.
4. 1 mL Methanol is added and the flask content is immediately titrated with NaOH (**Subheading 2.3.3., item 6)** using phenolphthalein as an indicator.
4. A blank flask with 3 mL of stock solution without enzyme is titrated as mentioned above to determine the total acid content of the reaction mixture.

Butanol does not react with butyric acid in the absence of lipase within the 2 h employed for measuring the activity. The esterification activity is determined by the following relationship:

$$\text{Esterification activity} = \frac{V \times M \times 100}{E \times T} \text{ units} \qquad [1]$$

V = difference in volume (mL) of NaOH between the blank and the samples, which is a measure of the butyric acid consumed due to esterification

T = period of incubation

M = molarity of NaOH

E = amount of the enzyme employed (mg) (or mL if samples are liquid)

One unit of esterification activity is defined as 1 μmol butyric acid consumed in the esterification reaction per min per mg (or mL) lipase ***(52)***.

3.2.2. Hydrolytic Lipase Activity Measurements

1. 4 mL Tributyrin (*see* **Subheading 2.3.4., item 5**) and 8 mL McIlvaine's buffer (*see* **Subheading 2.3.4., item 1**) mixture is incubated with the respective lipase extract (8 mL) at 50°C and agitated at 175 rpm.
2. A blank is also tested in the absence of enzyme. Samples are withdrawn at intervals of time. The hydrolytic activity is determined from **Eq. 1** (*see* **Subheading 3.2.1.**), where V is now the measure of the butyric acid released due to hydrolysis.

One unit of hydrolytic activity is defined as 1 μmol butyric acid releases per min per milligram or milliliter lipase (*see* **Note 3**).

3.3. Protease Activity Assays

3.3.1. Protease Activity Determination in Liquid Extract **(19,53)**

3.3.1.1. Preparation of Crude Enzyme Solution

Yeasts are grown overnight at 25°C in 10 mL MYGP (YPD may be used) (*see* **Subheading 2.2., item 1**). Then the cells are inoculated into a 250-mL Erlenmeyer flask containing 50 mL casein growth medium (*see* **Subheading 2.2., item 6**). Cultures are incubated at 30°C in a rotary shaker (120 rpm) and yeasts are harvested by centrifuging after 3 d. Crude supernatant may be concentrated or partially purified by ultrafiltration.

3.3.1.2. Assay Procedure

Proteolytic activity must be measured by the hydrolysis of acid-denatured hemoglobin (*see* **Note 4**).

1. One mL of enzyme solution is incubated with 2 mL acid-denatured hemoglobin (*see* **Subheading 2.3.5., item 2**) for 60 min at 37°C.
2. The reaction is stopped by adding 5 mL of TCA (*see* **Subheading 2.3.5., item 3**).
3. After 30 min at room temperature the mixture is centrifuged at 75*g*, 20 min, to remove any precipitates.
4. One mL of supernatant is then mixed with 5 mL Lowry reagent.
5. After 10 min of reaction, 0.5 mL of Folin and Ciocalteu's phenol reagent (1:1) is added to the solution and allowed to react 30 min.
6. Intensity of blue color is read at 700 nm in a spectrophotometer.
7. Blanks are prepared by adding TCA (*see* **Subheading 2.3.5., item 3**) before the addition of the enzyme.
8. A calibration curve is prepared using different concentrations of tyrosine (*see* **Subheading 2.3.5., item 7**).

Protease activity is expressed as μmol of tyrosine produced per milliliter crude enzyme per minute of incubation.

3.3.2. Plate Method for Proteolytic Activity Detection

Proteolytic activity is tested on two different substrates: casein and gelatin. A modification of the method proposed by Bilinsky et al. ***(54)*** is employed.

1. Filter-sterilized buffer A (60 mL) (*see* **Subheading 2.3.6., item 1**) is added to 70 mL skim milk powder in citrate-phosphate buffer (*see* **Subheading 2.3.6., items 2** and **3**) and autoclaved at 100°C for 10 min.
2. The resulting mixture is combined with 480 mL of an autoclaved agarized protease-detection medium (*see* **Subheading 2.2., item 5**) (*see* **Note 5**). Plates seeded with a loopful of yeasts are incubated at 28°C for 2 d and the presence of clear zones is observed.
3. In the same way, gelatin agar plates may be prepared, replacing skim milk with 15 g/L gelatin. Plates are incubated at room temperature for 10 d, and gelatin breakdown is observed as the emergence of a clear zone around the colonies.
4. Addition of HCl (*see* **Subheading 2.3.6., item 4**) (acetic acid, 50 g/L, also may be used) highlights the hydrolysis zone ***(19,54)***.

4. Notes

1. Examples of yeast strains having amylolytic activity are *Saccharomycopsis fibuligera* ***(55)***, *Schwanniomyces alluvius, Candida tsukubaensis* ***(56)***, and *Lipomyces starkeyi* ***(57)***. Some yeast species producing proteolytic activity are *Yarrowia lipolytica* ***(44)***, *Saccharomycopsis fibuligera* ***(36)***, *Phaffia rhodozyma* ***(58)***, *Candida tropicalis,* and *C. albicans* ***(59–61)***. Distinct yeasts can secrete

lipolytic enzymes, e.g. *Candida rugosa* ***(62)***, *C. antarctica* ***(63)***, *Yarrowia lipolytica* ***(46)***, and so on.

2. Most of the polymeric substrate-degrading enzymes act at acidic pH, but it is possible to find enzymatic activities at other pH values. Buffers and temperatures suggested here are only a general orientation in searching for optimum enzymatic reaction conditions.
3. As an alternative, a plate method may be used for qualitative lipase detection. Tributyrin agar is used. The prepared nutrient agar is liquefied by heating, tributyrin is added to give a final concentration of 10 g/L, and pH is adjusted to 3.5 with 0.1 *M* HCl. A loopful of each yeast is inoculated onto the agar plate, which is incubated for 2–4 d at 25°C. Hydrolysis of tributyrin is observed as a clear zone surrounding the yeast growth on an opaque background ***(46,64)***.
4. Determinations are focused on acid proteases because many yeasts having proteolytic activity are normally isolated from fruits at diverse degrees of ripeness.
5. For detection of proteases that can be repressed by readily utilizable nitrogen sources, YNB without amino acids may be used instead of the standard YNB.

References

1. Walker, G. M. (1999) Introduction to yeast, in *Yeast Physiology and Biotechnology,* John Wiley and Sons Ltd., England, pp. 1–10.
2. Lachance, M.-A., Phaff, H. J., and Starmer, W. T. (1993) *Kluyveromyces basillisporus sp. nov.* a yeast from Enony Oak exudate. *Int. J. Systematic Bacteriol.* **43,** 115–119.
3. Hatchison, L. J. and Miratsuka, Y. (1994) Some wood inhabiting yeasts of trembling aspen *(Populus tremuloides)* from Alberta and North Eastern British Columbia. *Mycologia* **86,** 386–391.
4. Morais, P. B., Martins, M. B., Klaczko, L. B., Mendonça-Hagler, L. C., and Hagler, A. N. (1995) Yeast succession in the Amazon fruit *Parahancornia amapa* as resource partitioning among *Drosophila* spp. *App. Environ. Microbiol.* **61,** 4251–4257.
5. Blanco, P., Sieiro, C., and Villa, T. G. (1999) Production of pectic enzymes in yeasts. *FEMS Microbiol. Lett.* **175,** 1–9.
6. Shubakov, A. A., Mikhaleva, N. I., Boev, A. V., and Okunev, O. N. (1994) Formation of xylanase by the yeast *Cryptococcus podzolicus. Appl. Biochem. Microbiol.* **30,** 658–664.
7. Scorzetti, G., Petrescu, I., Yarrow, D., and Fell, J. W. (2000) *Cryptococcus adeliensis sp. nov.,* a xylanase producing basidiomycetous yeast from Antarctica. *Antonie van Leeuwenhoek* **77,** 153–157.
8. Schäfer, A., Konrad, R., Kuhnig, T., Kämfer, P., Hertel, H., and König, H. (1996) Hemicellulose-degrading bacteria and yeasts from the termite gut. *J. Appl. Bacteriol.* **80,** 471–478.
9. Lee H., Biely, P., Latta, R. K., Barbosa, M. F., and Schneider, H. (1986) Utilization of xylan by yeasts and its conversion to ethanol by *Pichia stipitis* strains. *Appl. Environ. Microbiol.* **52,** 320–324.

10. Basaran, P., Basaran, N., and Hang, Y. D. (2000) Isolation and characterization of *Pichia stipitis* mutants with enhanced xylanase activity. *World J. Microbiol. Biotechnol.* **16,** 545–550.
11. Dobberstein, J. and Emeis, C. C. (1989) β-Xylanase produced by *Aureobasidium pullulans* CBS 58475. *Appl. Microbiol. Biotechnol.* **32,** 262–268.
12. Bastawde, K. B., Puntambekar, U. S., and Gokhale, D. V. (1994) Optimization of cellulase-free xylanase production by a novel yeast strain. *J. Ind. Microbiol.* **13,** 220–224.
13. Blanco, P., Sieiro, C., Díaz, A., and Villa, T. G. (1994) Production and partial characterization of an endopolygalacturonase from *Saccharomyces cerevisiae. Can. J. Microbiol.* **40,** 974–977.
14. Karam, G. A., Lequart, C., and Belarbi, A. (1994) Use of *Saccharomyces cerevisiae* for the clarification of fruit juices. *Biotechnol. Lett.* **16,** 1329–1334.
15. Stratilová, E., Breierová, E., Vadkertiová, R., Machová, E., Malavíková, A., and Slaviková, E. (1998) The adaptability of the methylotrophic yeast *Candida boidinii* on media containing pectic substances. *Can. J. Microbiol.* **44,** 116–120.
16. Iguchi, K., Hirano, H., Kishida, M., Kawasaki, H., and Sakai, T. (1997) Cloning of a protopectinase gene of *Trichosporon penicillatum* and its expression in *Saccharomyces cerevisiae. Microbiology* **143,** 1657–1664.
17. Schwan, R. F. and Rose, A. H. (1994) Polygalacturonase production by *Kluyveromyces marxianus:* effect of medium composition. *J. Appl. Bacteriol.* **76,** 62–67.
18. Fellows, P. J. and Worgan, J. T. (1984). An investigation into the pectolytic activity of the yeast *Saccharomycopsis fibuligera. Enzyme Microb. Technol.* **6,** 405–410.
19. Charoenchai, C., Fleet, G. H., Henschke, P. A., and Todd, B. E. N. (1997) Screening of non-*Saccharomyces* wine yeasts for the presence of extracellular hydrolytic enzymes. *Australian J. Grape and Wine Research* **3,** 2–8.
20. Fernández, M., Úbeda, J. F., and Briones, A. I. (2000) Typing of non-*Saccharomyces* yeasts with enzymatic activities of interest in wine-making. *Int. J. Food Microbiol.* **59,** 29–36.
21. Corsetti, A., Rossi, J., and Gobbetti, M. (2001) Interactions between yeasts and bacteria in the smear surface–ripened cheeses. *Int. J. Food Microbiol.* **69,** 1–10.
22. Petersen, K. M., Møller, P. L., and Jespersen, L. (2001) DNA typing methods for differentiation of *Debaryomyces hansenii* strains and other yeasts related to surface ripened cheeses. *Int. J. Food Microbiol.* **69,** 11–24.
23. Esteve-Zarzoso, B., Manzanares, P., Ramon, D., and Querol, A. (1998) The role of non-*Saccharomyces* yeasts in industrial wine-making. *Int. Microbiol.* **1,** 143–148.
24. Smacchi, E., Fox, P. F., and Gobbetti, M. (1999) Purification and characterization of two extracellular proteinases from *Arthrobacter nicotianae* 9458. *FEMS Microbiol. Lett.* **170,** 327–333.
25. Hube, B., Stehr, F., Bossenz, M., Mazur, A., Kretschmar, M., and Schäfer, W. (2000) Secreted lipases of *Candida albicans:* cloning, characterisation and expres-

sion analysis of a new gene family with at least ten members. *Arch. Microbiol.* **174,** 362–374.

26. van der Maarel, M. J. E. C., van der Veen, B., Uitdehaag, J. C. M., Leemhuis, H., and Dijkhuizen, L. (2002) Properties and applications of starch-converting enzymes of the α-amylase family. *J. Biotechnol.* **94,** 137–155.
27. Pandey, A., Nigam, P., Soccol, C. R., Soccol, V. Y., Singh, D., and Mohan, R. (2000) Advances in microbial amylases. *Biotechnol. Appl. Biochem.* **31,** 135–152.
28. Kazlauskas, R. J. (1994) Elucidating structure mechanism relationships in lipases. Prospects for predicting and engineering catalytic properties. *Trends in Biotechnol.* **12,** 464–472.
29. Mingarro, I., Gonzalez-Navarro, H., and Braco, L. (1996) Trapping of different lipase conformers in water-restricted environments. *Biochemistry* **35,** 9935–9944.
30. Maheshwari, R., Bharadwaj, G., and Bhat, M. K. (2000) Thermophilic fungi: their physiology and enzymes. *Microbiol. Molec. Biol. Rev.* **64,** 461–488.
31. Benjamin, S. and Pandey, A. (1998) *Candida rugosa* lipase: molecular biology and versatility in biotechnology. *Yeast* **14,** 1069–1087.
32. Wu, J. C., Zhang, G. F., and He, Z. M. (2001) Enhanced activity of *Candida rugosa* lipase modified by polyethylene glycol derivatives. *Biotechnol. Lett.* **23,** 211–214.
33. Shaw, J. F. and Chang, C. H. (1989) Characterization of three distinct forms of lipolytic enzymes in a commercial *Candida* lipase preparation. *Biotechnol. Lett.* **11,** 779–784.
34. Rua, M. L. and Ballesteros, A. (1994) Rapid purification of two lipase isoenzymes from *Candida rugosa. Biotechnol. Tech.* **8,** 21–26.
35. Bertolini, M. C., Schrag, J. D., Cyller, M., Thomas, D. Y., and Vernet, T. (1995) Expression in yeast, characterization and comparison of lipases from *Geotrichum candidum. Yeast* **11,** Spec. Iss. S523.
36. Rao, M. B., Tanksale, A. M., Ghatge, M. S., and Deshpande, V. V. (1998) Molecular and biotechnological aspects of microbial proteases. *Microbiol. Molec. Biol. Rev.* **62,** 597–635.
37. Barett, A. J. (1994) Proteolytic enzymes: serine and cysteine peptidases. *Methods Enzymol.* **244,** 1–15.
38. Menon, A. S. and Goldberg, A. L. (1987) Protein substrates activate the ATP-dependent protease La by promoting nucleotide binding and release of bound ADP. *J. Biol. Chem.* **262,** 14,929–14,934.
39. Takahashi, S., Ueda, M., Atomi, H., Beer, H. D., Bornscheuer, U. T., Schmid, R. D., et al. (1998) Extracellular production of active *Rhizopus oryzae* lipase by *Saccharomyces cerevisiae. J. Ferm. Bioeng.* **86,** 164–168.
40. Hari Krishna, S., Manohar, B., Divakar, S., Prapulla, S. G., and Karanth, N. G. (1999) Optimization of isoamylacetate production using immobilized lipase from *Rhyzomucor miehei* by response surface methodology. *Enzyme Microbiol. Tech.* **26,** 132–137.
41. Tharaud, C., Ribet, A.-M., Costes, C., and Gaillardin, C. (1992) Secretion of human blood coagulation factor XIIIa by the yeast *Yarrowia lipolytica. Gene* **121,** 111–119.

42. Ogrydziak, D. M. (1993) Yeast extracellular proteases. *Crit. Rev. Biotechnol.* **13,** 1–55.
43. Hamsa, P. V. and Chattoo, B. B. (1994) Cloning and growth regulated expression of the gene encoding the hepatitis B virus middle surface antigen in *Yarrowia lipolytica. Gene* **143,** 165–170.
44. Glover, D. J., McEwen, R. K., Thomas, C. R., and Young, T. W. (1997) pH-regulated expression of the acid and alkaline extracellular proteases of *Yarrowia lipolytica. Microbiology* **143,** 3045–3054.
45. Hensel, M., Tang, C. M., Arst, H. N., Jr., and Holden, D. W. (1995) Regulation of fungal extracellular proteases and their role in mammalian pathogenesis. *Can. J. Bot.* **73,** 1065–1070.
46. Pignède, G., Wang, H., Fudalej, F., Gaillardin, C., Seman, M., and Nicaud, J.-M. (2000) Characterization of an extracellular lipase encoded by *LIP2* in *Yarrowia lipolytica. J. Bacteriol.* **182,** 2802–2810.
47. Novotny, C., Dolezalova L., Musil P., and Novak M. (1988) The production of lipases by some *Candida* and *Yarrowia* yeasts. *J. Basic Microbiol.* **4,** 221–227.
48. Somogyi, M. (1952) Notes on sugar determination. *J. Biol. Chem.* **195,** 19–23.
49. Nelson, N. (1944) A photometric adaptation of the Somogyi method for determination of glucose. *J. Biol. Chem.* **153,** 376–380.
50. Lowry, O. H., Rosebrough, N. J., Farr, A. L., and Randall, R. J. (1951) Protein measurement with folin phenol reagent. *J. Biol. Chem.* **193,** 265–275.
51. Akpan, I., Bankole, M. O., and Adesemowo, A. M. (1999) A rapid plate culture method for screening of amylase producing micro-organisms. *Biotechnol. Tech.* **13,** 411–413.
52. Kiran, K. R., Hari Krishna, S., Suresh Babu, C. V., Karanth, N. G., and Divakar, S. (2000) An esterification method for determination of lipase activity. *Biotechnol. Lett.* **22,** 1511–1514.
53. Lagace, L. S. and Bisson, L. F. (1990) Survey of yeast acid proteases for effectiveness of wine haze reduction. *Am. J. Enol. & Vitic.* **41,** 147–155.
54. Bilinsky, C. A., Russell, I., and Stewart, G. G. (1987) Applicability of yeast extracellular proteinases in brewing: physiological and biochemical aspects. *Appl. Environ. Microbiol.* **53,** 495–499.
55. González, C. F., Fariña, J. I., and Figueroa, L. I. C. (2002) A critical assessment of a viscometric assay for measuring *Saccharomycopsis fibuligera* α-amylase activity on gelatinised cassava starch. *Enzyme Microb. Technol.* **30,** 169–175.
56. Nigam, P. and Singh, D. (1995) Enzyme and microbial systems involved in starch processing. *Enzyme Microb. Technol.* **17,** 770–778.
57. Bignell, G. R., Bruce, I. J., and Evans, I. H. (2000) Amylolytic enzymes of *Lipomyces starkeyi:* purification and size-determination. *Biotechnol. Lett.* **22,** 1713–1718.
58. Bang, M. L., Villiadsen, I., and Sandall, T. (1999) Cloning and characterization of an endo-β-1,3 (4) glucanase and an aspartic protease from *Phaffia rhodozyma* CBS 6938. *Appl. Microbiol. Biotechnol.* **51,** 215–222.

59. Hube, B., Turver, C. J., Odds, F. C., Eiffert, H., Boulnois, G. J., Kochel, H., and Ruchel, R. (1991) Sequence of the *Candida albicans* gene encoding the secretory aspartate protease. *J. Med. Vet. Mycol.* **29,** 129.
60. Togni, G., Sanglard, D., Falchetto, R., and Monod, M. (1991) Isolation and nucleotide sequence of the extracellular acid protease gene (ACP) from yeast *Candida tropicalis. FEBS Lett.* **286,** 181–185.
61. Muruyama, S. Y., Takeda, O., Mukai, H., Takesako, K., Sano, E., Kato, I., et al. (1995). *Candida albicans* aspartic proteinase: cDNA cloning and comparison among strains. *Adv. Exp. Med. Biol.* **362,** 573–576.
62. de la Casa, R. M., Guisán, J. M., Sánchez-Montero, J. M., and Sinisterra, J. V. (2002) Modification of the activities of two different lipases from *Candida rugosa* with dextrans. *Enzyme Microb. Technol.* **30,** 30–40.
63. Rubio, C., Latxague, L., Déléris, G., and Coulon, D. (2001) Enzymatic hydrolysis of (chloromethyldimethylsilyl)-2-propenyl acetate isomers: atypic specificity of *Candida antarctica* lipase. *J. Biotechnol.* **92,** 61–66.
64. Shelley, A. W., Deeth, H. C., and MacRey, I. C. (1987) Review of methods of enumeration, detection and isolation of lipolytic microorganisms with special reference to dairy applications. *J. Microbiol. Meth.* **6,** 123 – 137.

31

Extracellular Hydrolytic Enzymes Produced by Phytopathogenic Fungi

Martha Dina Vallejo Herrera, María Eugenia Toro, Lucía I. C. de Figueroa, and Fabio Vazquez

1. Introduction

Plant cell walls give plants shape and support, help to regulate physiological processes including defense responses, and act as physical barriers to pathogen invasion. Most plant pathogens produce an array of enzymes capable of degrading plant cell-wall components *(1,2)*.

Phytopathogenous fungi use different mechanisms to breach the surface of the host plant. Fungi penetration requires contact and adherence of spores and/or the first hypha resulting from germination (germination tube) to the vegetal surface. To penetrate the host, fungi must excrete enzymes like cutinases and esterases that alter the vegetal surface and facilitate fungi adherence.

Plant diseases produced by phytopathogenic fungi are due to the individual or combined action of four mechanisms *(2)*:

1. Synthesis and liberation of enzymes that degrade the cell wall, such as polygalacturonases, pectin lyases, hemicellulases, and cellulases. Sometimes they are liberated sequentially. These enzymes degrade the pectic substances that bind together the cell walls in parenchymatic tissues; as a result, fibers are bound more loosely, the cell wall is degraded, and the cells die.
2. Toxin production by fungi. These toxins are non-enzymatic products. They have low molecular weight and interfere with plant metabolism or affect the normal structure of protoplasm. Toxins are characterized in two categories: specific host toxins, which show the same specificity as the pathogen; and nonspecific host toxins, where the pathogen is not specifically related with the host plant, so symptoms are not specific.

From: *Methods in Biotechnology: Environmental Microbiology: Methods and Protocols*
Edited by: J. F. T. Spencer and A. L. Ragout de Spencer

3. Synthesis and liberation of products that produce interference with the normal control of growth and development. Many pathogenic conditions produced by fungi (dwarfism, elongation, cellular proliferation, and so on) are similar to those produced by hormonal disorders.
4. Growth of the fungus in the normal movement of water, nutrients, and metabolites.

Despite the importance of these four mechanisms, the scope of this chapter is limited to the first one.

The plant cell wall is a complex insoluble matrix highly recalcitrant to biological degradation. Thus, microbial enzymes that attack it need to be in intimate and prolonged contact with this composite structure. To facilitate this interaction, plant cell-wall hydrolases expressed by aerobic microorganisms are generally modular in structure, comprising a catalytic module appended to one or more noncatalytic carbohydrate-binding modules (CBMs). Based on primary structure similarities, CBMs have been grouped into 27 families. Most of the CBMs bind to cellulose, xylan, chitin, or starch *(3)*. On the other hand, microbial plant cell-wall hydrolases recycle photosynthetically fixed carbon. This is a pivotal biological process, an integral part of one of the major geochemical cycles in the biosphere. This process is also of considerable industrial importance, as it releases annually 10–11 t/yr of monosaccharides that could be exploited in the chemicals and fuel industries.

Lignocellulose, the most abundant renewable resource in nature, is composed of three major structural polymers: cellulose (a homopolymer built of D-glucosyl residues), hemicellulose (a group of heteropolymers that includes xylans and mannans), and lignin (a complex polyphenolic polymer). The main carbohydrate constituents of lignocellulosic material—i.e., cellulose, mannan, and xylan—consist of principal chains of β-1,4-linked pyranosyl units that can be variously substituted. These β-1,4-glycosidic bonds within the polysaccharide backbones are respectively hydrolyzed by cellulases, mannanases, and xylanases, the synthesis of which is usually subjected to induction and/or catabolic repression.

Polysaccharides are far too large to pass through cell membranes and trigger the response in a microbial cell leading to enhanced synthesis of endoglycanases. It is generally accepted that low-molecular-mass soluble catabolites can easily enter the cell, signal the presence of an extracellular substrate, and provide the stimulus for the accelerated synthesis of the respective enzymes *(4)*. Such catabolites are usually released from polymeric compounds by the action of low, constitutive amounts of these hydrolases.

It is known that many fungi produce extracellular hydrolytic enzymes with different kinetic and physicochemical properties, depending on growth method (submerged or solid-state culture). Presumably, the regulatory mechanisms of extracellular enzyme production are different ***(5,6)***. Solid-state culture is close to the natural mode of growth of most fungi on vegetal substrates. Therefore, some characteristics of phytopathogenesis may be inferred from this technique.

The concept that extracellular enzymes of fungi can degrade physical barriers during the invasion of a host has been controversial. There is some evidence for their role in pathogenesis, including the correlation between the level of such enzymes and virulence; immunocytochemical evidence for secretion of enzymes during infection, protection of the host by inhibition of enzymes with selective inhibitors or specific antibodies, and enhancement of virulence by gene transfer. These facts have been challenged on the basis of the results of disruption experiments on genes that encode cutinase and cell-wall-degrading enzymes. However, highly pathogenic fungi usually have multiple genes encoding enzymes that degrade the polymer barriers of the host, and some of them may be expressed only in the host ***(7,8)***. For example, *Nectria hematococca* (a plant pathogen) infects pea plants and produces several pectin-degrading enzymes, presumably to help penetration of cell-wall carbohydrate barriers in the host during infection. Such enzymes include endopectate lyases that are encoded by at least four pectate lyase (PL) genes *(pel)* ***(7)***. Another example is the tomato vascular wilt pathogen *Fusarium oxysporum* f. sp. *lycopersici,* which produces an array of pectinolytic enzymes that may contribute to penetration and colonization of the host plant ***(8)***. The soilborne plant pathogen *F. oxysporum* causes vascular wilt disease in a wide variety of crops ***(8)***. It produces diverse extracellular cell-wall-degrading enzymes, including xylanases, cellulases, proteases, pectate lyases, and exo- and endopolygalacturonases, which contribute to the degradation of structural barriers constituted by plant cell walls ***(9)***.

Fungal plant pathogens also produce a film involved in their invasion strategy, and present different kinds of enzyme activities. *Botrytis cinerea,* an important fungal plant pathogen, produces an extracellular matrix (ECM), or unsheathing film, that serves partially in the attachment of its germlings. Isolated ECM exhibited polygalacturonase and laccase activity and was able to catalyze the hydrolysis of *p*-nitrophenyl butyrate, a model substrate for assessing cutinase activity. Cellulase, pectin lyase, and pectin methyl esterase activities were noted with both heated and unheated ECM preparations ***(10)***. Conidia of *B. cinerea,* the primary inoculum source, first attach the substrate by a hydrophobic interaction that is easily disrupted, and then, upon germination, attach strongly through secretion of ECM ***(11)***. ECM, also referred to as an unsheathing film or fungal sheath, is secreted by germ tubes and appressoria but not by conidia themselves ***(11)***. ECM is very resistant to removal from various substrata, brief exposure to a solution of strong base being the only known chemical treatment that can cause detachment of germlings ***(11)***.

Extracellular enzymes that degrade physical barriers of the host may be suitable targets for antifungal therapy *(7)*.

Active plant defense responses include the rapid death of the plant cells that first come into contact with the pathogen (hypersensitive response), a rapid oxidative burst, cross-linking and strengthening of plant cell wall, the induction of the phenylpropanoid pathway and synthesis of lignin, synthesis and accumulation of antimicrobial compounds called phytoalexins, and synthesis of hydroxyproline-rich glycoproteins and fungal-wall-degrading enzymes (chitinases, glucanases). Other molecules, called elicitors, have been shown to induce defense responses when applied to plant tissues ***(12–14)***.

Hydrolytic enzymes secreted by fungi are also involved in fungus–fungus interaction. *Mycoparasitism* is a general term to describe the multistep degradation and final assimilation of phytopathogenic fungi. Since this, process requires the degradation of the cell wall of the fungal host, hydrolases secreted by fungi such as *Trichoderma* (including chitinases and β-glucanases) have been considered major determinants of biocontrol activity. While it has now become increasingly evident that biocontrol is the result of various cellular activities and not only enzymatic hydrolysis, recent findings clearly indicate that particularly chitinases play a major role ***(15)***.

1.1. Cell-Wall Components and Related Enzymes

Cellulose is a linear, essentially insoluble β-1,4-glucosidically linked homopolymer of approx 8000–12,000 glucose units. It is used as an energy source by numerous and diverse microorganisms, including fungal plant pathogens, which produce functionally complete cellulase enzyme systems. Inducible cellulases of the saprophytic fungus *Trichoderma reesei* are among the best characterized systems. They consist of at least two 1,4-β-D-glucan cellobiohydrolases, four endo-1,4-β-D-glucanases, and two 1,4-β-D-glucosidases. They are formed adaptively in the presence of cellulose to synergistically cooperate in its degradation. This fungus is able to carry out efficient hydrolysis of crystalline cellulose to glucose through the action of a number of secreted cellulase enzymes, which can be categorized as endo- and exoglucanases based on their mode of action. Cellobiohydrolases (CBH) are exoglucanases degrading cellulose sequentially from the ends of the glucose chains, thus producing cellobiose as the major product. CBH have a central role in the degradation of crystalline cellulose. Endoglucanases, on the other hand, are capable of hydrolyzing only amorphous cellulose in a random manner, and as a result cellooligosaccharides are produced ***(16)***.

The fungal mechanism by which an insoluble substrate triggers the formation of degrading enzymes has been a matter of speculation for three decades. The presence of cellulases bound to the conidial surface of *T. reesei* has been demonstrated. They are essential for growing on cellulose as the sole carbon source. There is evidence that the major role of one of these cellulases (CBH II) is to enable the fungus to start growing on cellulose. CBH II releases small amounts of cellobiose. This compound, directly or after further conversion, may induce cellulase formation ***(17,18)***.

The naturally occurring lignocellulosic biomass consists of 20–30% hemicellulosic materials that are heterogeneous polysaccharides found in association with cellulose and constitute the second most abundant renewable resource. It has high potential for degradation to useful end products ***(19)***. Xylan is the major constituent of hemicellulose. β-1,4-xylans are heteropolysaccharides with a homopolymeric backbone chain of 1,4-linked β-D-xylopyranose units. *O*-acetyl, α-L-arabinofuranosyl, α-1,2-linked glucuronic, or 4-*O*-methylglucuronic acid are the most frequent substituents in the backbone ***(20)***.

It has frequently been suggested that the catalytic mechanism of glycosidases resembles that of lysozyme. The hydrolysis reaction catalyzed by xylanases as well as cellulases proceeds through an acid-base mechanism involving two residues. The first residue acts as general catalyst and protonates the oxygen of the osidic bond. The second residue acts as a nucleophile that, in the case of retaining enzymes, interacts with the oxocarbonium intermediate or promotes the formation of a hydroxyl ion from a water molecule, as observed for inverting enzymes. Reaction with retention of configuration involves a two-step mechanism in which proton transfer occurs to and from an oxygen atom in an equatorial position at the anomeric center ***(21)***.

Lignin is a three-dimensional biopolymer composed of oxygenated phenylpropanoid units. The most common linkage between the units is an arylglycerol-β-aryl ether or β-*O*-4 bond. Degradation of recalcitrant lignin is an oxidative and nonspecific process carried out by white-rot basidiomycetes, including the best-known ligninolytic organism, *Phanerochaete chrysosporium* ***(22)***. The most prominent enzymes associated with lignin degradation are lignin peroxidase (LiP) and manganese-dependent peroxidase (MnP), as well as laccases and H_2O_2-producing oxidases ***(22,23)***. Some fungi lack one or more of these enzymes, suggesting that there is more than one mechanism for fungal degradation of lignin. The fact that different fungal species produce the same type of ligninolytic enzyme, such as laccase, with different properties further complicates the degradation mechanism.

In structural terms, laccases are either monomeric or multimeric copper-containing glycoproteins, which may exhibit additional heterogeneity owing to variable carbohydrate content, differences in copper content, or because laccases are expressed as the products of multiple genes. The highest redox potential of laccases reported so far is approx 0.8 V, too low to oxidize nonphenolic lignin substructures. However, laccase can oxidize some compounds with redox potentials above 0.8 V, such as 1-hydroxybenzotriazole (HBT) ***(24)***. Furthermore, laccases represent an interesting enzyme family that exhibits rather broad (or "relaxed") substrate specificities, which may be expanded by inclusion of redox mediators into their reaction mixtures ***(25–27)***. These enzymes are also involved in different aspects of fungal physiology (breakdown of plant polymers, detoxification of phenolic compounds, conidial pigmentation, morphogenesis, and pathogenesis) ***(28)***.

Pectic substances occur as structural polysaccharides in the primary cell walls and middle lamella of higher plants. Pectin is a complex heteropolysaccharide (molecular weight 30,000–300,000) composed mainly of D-galacturonic acid residues joined by α-1,4 linkages forming homogalacturonan chains. This backbone structure ("smooth regions") alternates with branched regions ("hairy regions") composed of rhamnose, arabinanes, and arabinogalactanes as side chains. Carboxyl groups of pectin are partially methyl-esterified, and hydroxyl groups are sometimes partially acetylated ***(29)***. Enzymatic pectin degradation involves a group of enzymes called *pectinases:* pectin methyl esterase (PME) (pectinesterase, pectase), which removes methoxyl groups; pectin lyase (PL), which cuts the internal glycosidic bond of highly esterified pectic polymers by a β-elimination reaction; and polygalacturonases (PGs), which hydrolyze polygalacturonic acid. Endo-PG catalyzes the hydrolysis of 1,4-α-D-galacturosiduronic linkages between two nonmethylated galacturonic acid residues, while exo-PG removes terminal residues. Lyases also present as endo-acting (cleave randomly) or exo-acting (cleave at the extremities) on the appropriate polymers. Rhamnogalacturonase specifically cuts rhamnose side chains. Rombouts and Pilnik ***(30)*** and Sakai et al. (1993) ***(29)*** reported on different aspects of pectic substances and pectic enzymes.

There have been preliminary reports of the involvement of fungal PG in phytopathogenesis (ability to cause disease) or virulence (level of disease induced), but only indirect evidence could be provided. Even in the nonpathogenic *Aspergillus nidulans,* a phytopathogenic potential could be observed when PG synthesis is induced. In several model systems, PG activity *in planta* is correlated with disease severity in the host *(Fusarium solani, F. oxysporum). Aspergillus flavus* harbors two glucose-repressible and one constitutive endo-PG gene. The expression of one of the glucose-repressible genes *(pecA)* could be required for the invasiveness and thus pathogenicity of the fungus. Furthermore, an elevated level of PG (among other enzymes) was found in aggressive versus nonaggressive strains of the rapeseed pathogen *Phoma lingam.* An inverse correlation has, however, been observed in *Leptosphaeria maculans* (teleomorph of *P. lingam*): less virulent isolated strains exhibited a higher activity in cell-wall-degrading enzymes than highly virulent strains.

Although genetic approaches available today could help to demonstrate more directly that PG serves as a virulence factor, there is still significant controversy over the role of PG in fungal virulence and the infection process. Direct demonstration of the involvement of PG—as one of several virulence factors—requires the application of recombinant genetic techniques such as targeted gene deletion and deregulated overexpression of a single gene *(31)*.

When fungi are grown on plant cell-wall material in vitro, pectic enzymes are invariably the first enzymes to be secreted, followed by hemicellulases and cellulases. The action of pectic enzymes, and in particular of endopolygalacturonase on cell walls, appears to be a prerequisite for wall degradation by other enzymes. Only after pectic enzymes have acted on their substrates does the cellulose-xyloglucan framework—which is normally embedded in the pectin matrix—become accessible, and inducers for cellulase and hemicellulase can be released. Endopolygalacturonases exhibit a great variety of isoenzymatic forms. The molecular basis of this polymorphism has been elucidated for only a few enzymes *(13)*. For example, endopolygalacturonase secreted by the phytopathogenic fungus *Fusarium moniliforme* consists of four molecular mass glycoforms derived from a single endopolygalacturonase gene product. Purified pectic enzymes capable of cleaving the α-1,4-glycosidic bonds of homogalacturonan in an "endo" manner cause plant tissue maceration (cell separation) *(13)*.

Sakai and Okoshima *(32)* and Sakai and Sakamoto *(33)* isolated different microorganisms able to produce enzymes that attack protopectin (the water-insoluble precursor of pectin found in plant tissues), releasing (or solubilizing) water-soluble pectin, with the resultant separation of the plant cells from each other. A long time ago, the term *protopectinase* (PPase) was tentatively applied to these enzymes *(34)*. Nowadays, PPase is considered a synonym for a pectin-releasing or pectin-solubilizing enzyme.

Studies of lignin biodegradation are also of great importance for possible biotechnological applications, because lignin polymers are a major obstacle to the efficient utilization of lignocellulosic materials in a wide range of industrial processes *(35)*. Delignification of lignocellulosic materials by white-rot fungi is, for example, of great interest and has been investigated to improve the digestibility of wood or straw for animal feed *(36)* and to reduce costs for the pulp and paper industry *(37,38)*. Annual plants, such as cereal straw, could serve as alternative types of raw materials for papermaking, and if they were pretreated with selective lignin degraders or enzymes derived from them, the use of such plants could save both forest resources and energy in pulp refining *(38)*.

Xylanases receive considerable attention because of their application in biobleaching of pulp in the paper industry, wherein enzymatic removal of xylan from lignin–carbohydrate complexes facilitates the leaching of lignin from the fibrous cell wall, obviating the need for chlorine for pulp bleaching in the brightening process ***(39,40)***. They also have applications in the pretreatment of animal feed to improve its digestibility ***(19)***.

As cellulose is the most abundant renewable resource, enzymatic conversion of cellulosic materials to glucose is a very promising process. Enzymatic hydrolysis has the advantage of being energy sparing and avoids the use of toxic substances or corrosive acids because of the relatively mild reaction conditions. Therefore, cellulose hydrolysis catalyzed by cellulase has been widely investigated ***(40)***. Today several commercial preparations are available for biostoning and biofinishing of cotton fabrics. Many of them are based on *T. reesei* cellulases ***(16)***.

Commercial pectinases (containing diverse mixtures of pectin-degrading activities) are mostly used in food industries, like fruit-juice processing (in reduction of turbidity, prevention of cloud formation, cell maceration, and so on) or wine production, sometimes combined with cellulases. Another application is in the cosmetics industry. Only a limited number of microorganisms (with GRAS status) are available for application in foods and cosmetics. This concept must be applied to other enzymes described in this chapter. Pectinases also play an important role in the retting process of some textile fibers production. Protopectinases are employed for cell maceration and in pectin production, because they are pectin-liberating enzymes. Detailed information on commercial production of pectinases is confidential and not released by the producers ***(29,30)***.

The objective of this chapter is to introduce readers to studies of relevant enzymes involved in plant infections by filamentous fungi. It describes the methodology to determine enzymatic activities, but enzyme purification and characterization is not included.

2. Materials

2.1. Strains

Pathogenic filamentous fungi can be isolated from infected plants or obtained from fungal collections. Reference strains must be previously assayed and generally are from culture collections suppliers (*see* **Note 1**).

2.2. Media and Culture Conditions

Media and culture conditions to be used must correspond to the specific microorganism and products to be obtained. Liquid cultures are used most often because they are easily manipulated and controlled. Solid-state fermentations are also increasingly used. Solid-state media need nutrients and support (sometimes the solid support is also the carbon and nitrogen source).

The following media are examples:

1. Medium for spores production: commercial PDA (potato-glucose agar) or Czapek can be used. Cultures are incubated at 25–30°C; incubation period depends on the strain.
2. Propagation medium for cellulolytic fungi: 3.0 g/L $(NH_4)_2SO_4$, 2.0 g/L KH_2PO_4, 1.0 g/L yeast extract. Carbon sources for cellulase production are: carboxymethyl cellulose (CMC) sodium salt and pretreated or untreated lignocellulosic substrate (wheat straw, wheat bran, waste newspaper, and so on). Endoglucanase formation is induced by 0.5 g/L sophorose. Conidia for preparation of the inoculum are obtained by growing the fungus (e.g., *T. reesei*) on potato-dextrose agar at 30°C for 7 d. Cultivation is carried out in 1-L Erlenmeyer flasks containing 250 mL of the respective medium, inoculated with 1.6×10^5 spores/mL and incubated at 30°C on a rotary shaker operating at 250 rpm ***(18,42)***.
3. Propagation medium for xylanolytic fungi: 0.5 g/L $MgSO_4 \cdot 7H_2O$, 0.5 g/L $Ca(NO_3)_2$, 0.5 g/L KH_2PO_4, 1.0 g/L yeast extract, and traces of $FeSO_4 \cdot 7H_2O$. Adjust pH to 6.5. Autoclave 15 min at 121°C. The medium must be supplemented with either xylan (10.0 g/L), extracted hemicellulose (10.0 g/L), wheat straw (50.0 g/L), or rice husk (50.0 g/L) ***(43)***. Flasks are incubated at 30°C.
4. Medium for xylanase activity determination: 20.0 g/L agar, 1.5 g/L birchwood 4-*O*-methyl glucuronoxylan, covalently linked to Remazol Brilliant Blue R dye, 6.7 g/L yeast nitrogen base (YNB) without amino acids, 2.0 g/L D-L-asparagine, and 5.0 g/L KH_2PO_4.
5. Medium for laccase production: 40 g/L glucose, 7 g/L glycerol, 0.5 g/L L-histidine, 0.1 g/L $CuSO_4$, 1.8 g/L $NaNO_3$, 1.8 g/L NaCl, 0.5 g/L KCl, 0.5 g/L $CaCl_2 \cdot H_2O$, 0.05 g/L $FeSO_4 \cdot 7H_2O$, 1.0 g/L KH_2PO_4, and 0.5 g/L $MgSO_4 \cdot 7H_2O$. Adjust pH to 5.0. For routine enzyme production in shake flasks, 100 mL of this medium in 500-mL wide-mouthed Erlenmeyer flask is inoculated with 10% (v/v) fungal spore suspension and incubated for up to 19 d at 24°C on a shaker at 250 rpm. Xylidine (2.5–10 μM), a putative laccase inducer, is added to culture medium to study its inductive effect on laccase production.
6. Medium for ligninolytic enzyme production by *Trichophyton rubrum* ***(44)***: 30.0 g/L glucose, 10.0 g/L peptone, 1.5 g/L KH_2PO_4, 0.5 g/L $MgSO_4 \cdot 7H_2O$, 0.020 g/L $CuSO_4 \cdot 7H_2O$, and 0.002 g/L thiamine-HCl. Adjust pH to 5.0.
7. Basal mineral medium for pectinase production: 1.3 g/L KH_2PO_4, 0.60 g/L Na_2HPO_4, 0.75 g/L $(NH_4)_2SO_4$, 0.25 g/L urea, 0.30 g/L $MgSO_4$, and 0.30 g/L $CaCl_2$. As carbon source, 20.0 g/L pectin is added. The initial pH of the medium is adjusted to 5.0–5.5 by addition of 0.1 M H_3PO_4 or 0.1 M NaOH. If necessary, 15.0 g/L agar is added. The medium is sterilized in autoclave at 121°C for 30 min. Culture flasks are grown in a shaking incubator at 25–30°C (*see* **Note 2**).

8. Medium for solid-state cultivation: wheat bran, humidified to approx 50–60% (w/w) with distilled water or with 0.15 *N* HCl. Instead of water or 15 *N* HCl, wheat bran may be humidified with a similar mineral medium to that previously detailed (*see* **Subheading 2.2.7.**). Initial pH and moisture content also depend on the requirements of the strain employed. The medium is sterilized at 121°C for 15 min, cooled, and inoculated with a spore suspension. Cultures are carried out in Petri dishes, flasks, packed-bed columns, or rotary-drum fermentors (*see* **Note 3**).

2.3. Solutions, Substrates, and Equipment

Enzyme solution is the aqueous (or buffered) extract of solid-state fermentations or culture filtrates of liquid fermentations, obtained according to **Subheadings 3.1.** or **3.2.**

2.3.1. Cellulase Activity (Endo-β-1,4-Glucanase) Determination

1. Na-citrate buffer: 0.05 *M* NaOH-citric acid, buffered to pH 4.8.
2. Cellulosic substrate: 20.0 g/L carboxymethyl cellulose (CMC) sodium salt (ultralow viscosity) in Na-citrate buffer (*see* **item 1** of this protocol).
3. Glucose stock solution: 2.0 g/L anhydrous glucose.
4. DNS reagent ***(45)***: 1.416 mL distilled water, 10.6 g 3-5 dinitrosalicylic acid, and 19.8 g NaOH. These components are dissolved completely. Add 306.0 g Na-K-tartrate, 7.6 g phenol (water saturated), and 8.3 g Na-metabisulfite. Titrate 3 mL with 0.1 *N* HCl (they must consume 5–6 mL). Phenolphthalein is used as an indicator.

2.3.2. Paper Assay for Saccharifying Cellulase (FPU Assay)

1. Substrate for FPU assay: Whatman No. 1 filter paper strip, 1.0 × 6.0 cm (approx 50 mg).
2. Buffer for FPU assay: Na-citrate (*see* **Subheading 2.3.1., item 1**).
3. DNS reagent, prepared as outlined in **Subheading 2.3.1., item 4.**
4. Deionized or distilled water.
5. Glucose stock solution: 2.0 g/L anhydrous glucose.

2.3.3. Xylanase Activity Determination

1. Buffer: 0.025 *M* Na_2HPO_4–0.05 *M* citric acid, buffered to pH 5.3.
2. Substrate solution: 10.0 g/L birchwood 4-*O*-methyl glucuronoxylan in 0.05 *M* Na-citrate buffer, pH 5.3, prepared as in **Subheading 3.4., item 1.**
3. Stock solution for standard curve: D-xylose in 0.05 *M* Na-citrate buffer as in item 1 of this protocol.
4. DNS reagent (*see* **Subheading 2.3.1., item 4**).

2.3.4. Laccase Activity–Dimerization of 2,6-Dimethoxyphenol (2,6-DMOP) Determination

1. Buffer: 100 m*M* citrate-NaOH, pH 3.5.
2. Substrate: 5 m*M* solution of 2,6-DMOP in pH 3.5 buffer.

2.3.5. Laccase Activity–Syringaldazine Oxidation Determination

1. Buffer: 0.1*M* citrate-phosphate buffer, pH 5.0.
2. Syringaldazine solution: 4-hydroxy-3,5-dimethoxybenzaldehyde azine, 2.5 *M*.

2.3.6. PGase Activity Determination

1. Buffer AcB: 20 m*M* acetate buffer, pH 5.0, maintained at 4°C.
2. Buffer CPB: 50 m*M* citric acid–25 m*M* Na_2HPO_4, pH 3.0, maintained at 4°C (*see* **Note 4**).
3. Substrate: 2.0 g/L polygalacturonic acid (PGA) solution in AcB, pH 5.0, or CPB, pH 3.0, maintained at 4°C.
4. Somogyi reagent.
5. Nelson reagent.
6. Standard curve: 0.5 g/L D-galacturonic acid monohydrate in the buffer used, maintained in –20°C freezer, and diluted in the same buffer to set the standards.

2.3.7. Pectin Lyase Determination in Liquid Extract

1. Substrate: 10.0 g/L pectin with the pH of interest (may be 3.0, 4.5, 7.0, and so on) in citrate-phosphate buffer.
2. Enzyme sample, dialyzed overnight against CPB or distilled water.

2.3.8. Pectinase Activity Determination

1. Buffer: *see* **Subheading 2.3.6., item 1** and **Subheading 2.3.6., item 2**.
2. Substrate: 2.0 g/L polymethylgalacturonic acid (PMGA) solution in AcB, pH 5.0, or CPB, pH 3.0, maintained at 4°C.
3. Somogyi reagent.
4. Nelson reagent.
5. Standard curve: D-galacturonic acid monohydrate, prepared as described in **Subheading 2.3.6., item 6**.

2.3.9. Pectinase Activity Detection in Solid Medium

1. Microbial colonies ground in Petri dishes.
2. Hexadecyltrimethylammonium bromide solution: 10 g/L hexadecyltrimethylammonium bromide in water.

2.3.10. Pectin Esterase Determination

1. NaOH solution: 0.1 *N* NaOH.
2. Substrate: 10.0 g/L citrus pectin solution in 0.1 *N* NaCl.
3. Phenol Red solution: 28 mL of 3.54 g/L Phenol Red in 0.01 *N* NaOH. Add 225 mL of distilled water.

2.3.11. Protopectin Preparation

1. Commercial lemons.
2. Ethanol 96% (v/v), ice-cold.
3. Distilled water, cold.
4. Stainless steel sieves, meshes 10, 20, 50, and 60.
5. Acetone.
6. 32 *N* H_2SO_4, maintained at 4°C.
7. Carbazole solution: 2 g/L carbazole solution in absolute ethanol, maintained at –20°C.

2.3.12. PPase Activity Determination

1. Buffer: AcB or CPB (*see* **Subheading 2.3.6., item 1** and **Subheading 2.3.6., item 2**).
2. Dried protopectin, prepared as described in **Subheading 3.10., Step 1**.
3. Filter paper: Whatman no. 1, circles folded in a conical shape, appropriate size for the top of the test tube.
4. Sulfuric-sodium-tetraborate solution: 0.0125 *M* solution of sodium tetraborate in concentrated H_2SO_4, kept at 4°C.
5. *m*-Hydroxydiphenyl solution: 1.5 g/L solution of meta-hydroxydiphenyl in 5.0 g/L NaOH, kept covered with aluminum foil at 4°C.
6. Standard curve: D-galacturonic acid monohydrate, prepared as described in **Subheading 2.3.6., item 6.**

3. Methods

3.1. Enzyme Extraction for Enzymatic Assays From Submerged Cultures

The culture is filtered through Whatman no. 4 filter paper (if desired enzymes do not absorb) or membrane filter (0.45 μm). The filtrate is collected and stored at freezing temperature for enzyme assays. If necessary, filtrate is dialyzed, further purified, and concentrated (*see* **Note 4**).

3.2. Enzyme Extraction for Enzymatic Assays From Solid-State Cultures

A fixed amount of culture (10 g) is mixed with 10–20 mL distilled water (or buffer solution), and pressed in a hydraulic press. The liquid extract is then filtered and treated as described in **Subheading 3.1.** Another possibility is to mix 5–10 g water per g of fermented solid and stir the flask for 30 min. The suspension is filtered through a plastic mesh. The filtrate is centrifuged and processed as a submerged culture filtrate **Subheading 3.1** (*see* **Note 5**).

3.3. Cellulase Activity Assays (for Endo-β-1,4-Glucanase)

3.3.1. Liquid Assay for Cellulase (Endo-β-1,4-Glucanase) Activity Determination (46)

1. Add 0.5 mL of enzyme sample diluted in citrate buffer (*see* **Subheading 2.3.1., item 1**) to a 25-mL or larger test tube, keeping it in an ice-water bath.
2. Raise temperature to 50°C.
3. Add 0.5 mL substrate solution (*see* **Subheading 2.3.1., item 2**), mix, and incubate at 50°C for 30 min.
4. Add 3.0 mL DNS (*see* **Subheading 2.3.1., item 4**), mix.
5. Boil for exactly 5.0 min in a vigorously boiling water bath containing enough water. Samples, enzyme blanks, glucose standards, and the reagent blank should be boiled together. After boiling, transfer immediately to a cold-water bath.
6. Add 20 mL of deionized or distilled water.
7. Mix by completely inverting the tubes several times so that solution separates from the bottom of the tube at each inversion.
8. Measure the color formed against the reagent blank at 540 nm. When small dilutions are used, the color formed in the enzyme blank is subtracted from that of the sample tube.
9. Reagent blank follows the same procedure, replacing enzyme sample with 0.2 mL of buffer. For enzyme blank, DNS must be added before the enzyme dilution.
10. Calibration curve is made with glucose solutions (*see* **Subheading 2.3.1., item 3**).
11. One unit of enzyme activity is defined as the amount of enzyme producing 1 μmol of glucose equivalents per minute, under the given conditions, and corresponds to 16.67 nanokatal (nkat).

3.3.2. Filter Paper Assay for Saccharifying Cellulase (FPU Assay) Determination (46)

1. Add 1.0 mL 0.05 *M* Na citrate, pH 4.8 (*see* **Subheading 2.3.1., item 1**) to a 25-mL or larger test tube.
2. Add 0.5 mL enzyme, diluted in citrate buffer. Two dilutions, at a minimum, of each enzyme sample must be made. One dilution should release slightly more and one slightly less than 2.0 mg of glucose under the reaction conditions.
3. Adjust temperature to 50°C.
4. Add one filter paper strip (*see* **Subheading 2.3.2., item 1**) and mix (if the paper "winds" up the tube it must be pushed down again).
5. Incubate at 50°C for 60 min.
6. Add 3.0 mL DNS (*see* **Subheading 2.3.1., item 4**) and mix.
7. Boil for 5 min in a water bath, samples, enzyme blanks, glucose standard, and reagent blank. After that, cool them in a cold-water bath.
8. Add 20 mL of deionized or distilled water. Mix by completely inverting the tubes several times so that the solution separates from the bottom of the tube at each inversion.

9. When the “pulp” has settled well, after 20 min at least, the color formed is measured against the reagent blank at 540 nm. If the paper pulp does not settle, it will do so after stirring with a glass rod.

From the calibration curve, translate the absorbance values of the samples into glucose, considering the dilution used.

Plot glucose liberated against enzyme concentration on semilogarithmic scale and estimate the concentration of the enzyme which would have released exactly 2.0 mg glucose.

$$FPU = \frac{0.37}{\text{enzyme concentration to release 2.0 mg glucose}} \text{units/mL}$$

3.4. Xylanase Activity Assays

3.4.1. Substrate Preparation for Xylanase Activity Determination (47)

1. Homogenize 1.0 g birchwood xylan in 80 mL buffer at 60°C (*see* **Note 6**).
2. Heat to boiling point on a heating magnetic stirrer.
3. Cool with continuous stirring (overnight, keeping the device covered).
4. Make up to 100 mL with buffer. Substrate may be stored at 4°C for 1 wk or aliquots may be frozen.

3.4.2. Xylanase Activity Determination (47,48)

1. Add 1.8 mL substrate solution (*see* **Subheading 3.1., step 4**) to a 15-mL or larger test tube. Heat to 50°C.
2. Add 200 µL enzyme diluted in citrate buffer and mix.
3. Incubate for 5 min at 50°C.
4. Add 3 mL of DNS (*see* **Subheading 2.3.1., item 4**), mix, and boil for 5 min; cool in cold water.
5. Measure the color produced at 540 nm against the reagent blank.
6. Reagent blank and enzyme blank are prepared as in **Subheading 3.3.1., step 9**.
7. Standard curve is made with pure D-xylose solutions (*see* **Note 7**).

One unit of enzyme activity is defined as the amount of enzyme producing 1 µmol of D-xylose equivalents per minute, under the given conditions, and corresponds to 16.67 nkat.

To measure xylanase activity, *see* **Note 8.**

3.5. Laccases Activity Assays

Measure of enzyme activity is based either on the oxidation or on the oxidative dimerization of such compounds as syringaldazine, 2,2'-azino-bis (3-ethylbenzothiazoline-6-sulfonic) acid (ABTS), or 2,6-dimethoxyphenol ***(10,27,44,49,50)***. The reaction is monitored spectrophotometrically. Xylidine is usually used as an inducer.

3.5.1. Enzymatic Dimerization of 2,6-Dimethoxyphenol (2,6-DMOP) Determination

Measurement of enzyme activity is based on the oxidative dimerization of 2,6-dimethoxyphenol (2,6-DMOP).

1. To 500 mL of 5 m*M* 2,6-DMOP, add 500 mL of 100 m*M* citrate-NaOH (pH 3.5).
2. Add 10-mL aliquot of enzyme dilution.
3. Reaction mixture is monitored at 477 nm with a spectrophotometer.
4. To evaluate laccase activity, the absorbance change caused by dimerization of 2,6-dimethoxyphenol is measured. Therefore, specific laccase activities are recorded as microkatals per milligram of protein.

3.5.2. Laccase Activity Determination on Syringaldazin

Laccase activity is measured with syringaldazine at 25°C.

1. Reaction mixture (total volume, 1 mL) contains 0.1 mL of culture filtrate in 0.1 *M* citrate-phosphate buffer, pH 5.0.
2. Start the reaction by adding of 10 µL 2.5 *M* of syringaldazine solution.
3. The increase in A_{525} is recorded, and the activity is expressed as nanokatals per liter.

3.6. Polygalacturonase Activity (Endopoligacturonase) Determination

Polygacturonase (PGase) activity is determined in culture extracts of solid-state fermentations or culture filtrates of liquid fermentations, using as substrate PGA solution in AcB or in CPB, maintained at 4°C (*see* **Subheading 2.3.6., items 1** and **2** and **Subheading 2.3.6., item 3**).

1. Enzymatic reaction courses in test tubes containing 450 µL of PGA solution (keeping the test tubes in crushed ice bath) and 50 µL of enzyme sample.
2. Each sample needs a proper blank with PGA solution and without the enzyme.
3. Test tubes are incubated for 10 min at 37°C in water bath.
4. Stop the reaction by cooling the tubes in a ice-water bath (10 min).
5. 50 µL Enzyme sample is added to blank tubes.

The reducing groups released from the PGA can be quantified by the Somogyi method ***(51)*** as follows:

1. Keeping the tubes in ice-water bath, add 500 mL of Somogyi reagent.
2. Tubes are heated for 10 min in a boiling-water bath.
3. Stop the reaction by cooling the tubes in ice-water bath (5–10 min).
4. Mix with 500 mL Nelson reagent ***(52)*** at room temperature.
5. Allow to stand 30 min at room temperature; add 4500 mL of water to each tube.
6. Measure absorbance at 660 nm against a colorimetric reaction blank, replacing the sample and the PGA solution by 500 mL of the buffer used in the enzymatic reaction (AcB or CPB).

7. From a standard curve made with D-galacturonic acid monohydrate (*see* **Subheading 2.3.6., item 6**) up to 250 mg/L, the equivalent reducing amount of samples are determined by subtracting the respective blank sample absorbance from the total absorbance.

One unit of PGase activity is the activity that liberates reducing groups equivalent to 1 μmol of D-galacturonic acid per minute, under the conditions of the assay.

3.7. Pectin Lyase Activity Determination

1. Pectin lyase activity can be determined spectrophotometrically by monitoring the absorbance increase at 235 nm of a reactive solution containing 2 mL pectin solution (*see* **Subheading 2.3.7., item 1**) with 0.5 mL dialyzed enzymatic sample (adequately diluted in the citrate-phosphate buffer used), at room temperature (23°C).
2. The absorbance change is followed for 20 min.

One pectin lyase unit is defined as the amount of enzyme needed to increase the absorbance by one unit per minute per milliliter of enzyme sample, under the conditions of the assay.

3.8. Pectinase Activity Assays

3.8.1. Pectinase Activity Determination in Liquid Extract

Pectin (polymethylgalacturonic acid [PMGA]) degrading activity (PMGase) is assayed in the same way as PGase activity (*see* **Subheading 3.6.**), using citrus pectin as substrate (**Subheading 2.3.8., item 2**).

One unit of PMGase activity is defined as the activity that liberates reducing groups corresponding to 1 μmol of D-galacturonic acid, under the conditions described.

3.8.2. Pectinase Activity Detection in Solid Medium

An indicator of pectinase activity is the diameter of the hydrolyzed zone around the microbial colony grown in agarized basal mineral medium plus 2.0 g/L pectin (*see* **Subheading 2.2.7.**). To visualize pectinase clearance zones, plates must be flooded with hexadecyl-trimethyl-ammonium bromide solution. This activity can be expressed as a percentage of surplus activity with respect to a reference strain.

3.9. Pectin Esterase Activity Determination

Pectin esterase activity is determined by measuring the amount of 0.1 *N* NaOH solution needed to keep reaction pH constant.

1. 10 mL Pectin solution (*see* **Subheading 2.3.10., item 2**) in a flask. Adjust pH to 7.5, keeping the flask at 30°C.
2. Add 1 mL of enzyme solution and again adjust the pH to 7.5.
3. Incubate for 30 min at 30°C.
4. Stop the reaction in ice-water bath.
5. Measure the amount of 0.1 *N* NaOH solution needed to bring the reaction pH to 7.5.

Enzyme activity is expressed as mg CH_3O released per milliliter of enzyme per minute, or as milliequivalent of hydrolyzed ester per milliliter of enzyme per minute.

Add 3 drops of phenol red solution (*see* **Subheading 2.3.10., item 3**) before item 5 of this protocol, if pH-meter is not available. Red color must appear at pH 7.5, and pink salmon color at pH below 7.0.

3.10. Protopectinase Activity Assays

Protopectinase (PPase) activity is determined in culture extracts of solid-state fermentations or culture filtrates of liquid fermentations.

Usually lemon or apple protopectin is used as substrate. Its preparation from albedo layers of commercial lemons was described by Cavallito et al. ***(53)***, where the authors proposed a method for PPase activity determination. Later, Contreras Esquivel et al. ***(54)*** reported some modifications to the technique.

3.10.1. Lemon Protopectin Preparation

1. Albedo layers (whitish part of lemon peel) are removed, using a cutter, from the washed fruit, and immediately submerged in ice-cold ethanol (1.0 mL/g wet albedo). Flabedo layers must be discarded.
2. This suspension is ground in a food processor and maintained at 4°C for 24 h.
3. Pass the suspension through a nylon voile.
4. Particles retained are washed several times with distilled water until the water-soluble substances that react with carbazole–sulfuric acid reagent are washed off, yielding a suspension of protopectin without free soluble pectin (*see* **steps 7–9** of this protocol).
5. Washed particles are sifted and those that pass through mesh 10 and are retained in mesh 20 sieves are kept.
6. This particle fraction is washed with acetone and dried at 30°C. The dried material is sifted and kept at 4°C until used. For dried protopectin, the recommended particle size is one that passes through mesh 50 and is retained in mesh 60.
7. Carbazole-sulfuric reaction: to 250 µL of washing water (from protopectin suspension) add 3.0 mL of chilled 32 *N* H_2SO_4 and 250 µL of carbazole solution, in ice bath.
8. The test tube is boiled in a water bath for 20 min and allowed to stand at room temperature for 20–40 min.
9. Absorbance is measured at 525 nm against a reaction blank ***(55)***.

3.10.2. PPase Activity Determination

Buffer solutions employed are AcB or CPB, maintained at 4°C (*see* **Subheading 2.3.6., item 1** and **Subheading 2.3.6., item 2**).

1. Enzymatic reaction courses in test tubes containing 20 mg of dried protopectin, 950 μL of AcB or CPB, keeping test tubes in crushed-ice bath, and 50 μL of enzyme sample.
2. Each sample needs a proper blank, with protopectin and buffer and without enzyme.
3. Test tubes are incubated at 37°C in water bath, slowly shaken.
4. Reaction time suggested is 30 min.
5. Stop the reaction by cooling the tubes in an ice-water bath.
6. Enzyme sample (50 μL) is added to the blank tubes.
7. The mixture is filtered through Whatman No. 1 filter paper (*see* **Subheading 2.3.12., item 3**).
8. Pectin released is measured as galacturonic acid in the filtrate by the *m*-hydroxy-diphenyl-sulfuric acid method ***(56)***: to 200 μL of filtrate, keeping the test tubes in crushed-ice bath, 1200 μL of sulfuric-sodium tetraborate solution (**Subheading 2.3.12., item 4**) is added, mixed in Vortex, and heated for 5 min at 100°C in boiling bath. Tubes are cooled in ice-water bath and 20 μL *m*-hydroxydiphenyl solution (**Subheading 2.3.12., item 5**) is added. Tubes are shaken and, within 5 min, absorbance at 520 nm is measured against a colorimetric reaction blank, replacing the 200 μL of filtrate by 200 μL of the buffer used in the enzymatic reaction.
9. From a standard curve of D-galacturonic acid monohydrate (**Subheading 2.3.6., item 6**) made up to 0.25 g/L, the equivalent reducing amount of the samples are determined by subtracting from the total absorbance, the respective blank sample absorbance.

One unit of PPase activity is the amount of enzyme that liberates pectic substances corresponding to 1 μmol of D-galacturonic acid per minute, under the conditions described.

Instead of the *m*-hydroxydiphenyl-sulfuric acid method, the carbazole-sulfuric method ***(55)*** can be used, replacing in **Subheading 3.10.1., step 7**, the washing water by thc filtrate obtained in **Subheading 3.10.2., step 2**.

4. Notes

1. As we have pointed out, phytopathogenic fungi produce an enzyme pool (cellulolytic, xylanolytic, ligninolytic, pectolytic, and other enzymatic activities) that degrade the plant tissue. The microorganisms mentioned above are examples of the strains used.

    a. Strains of cellulolytic fungi: *Trichoderma reesei, Botrytis cinerea.*
    b. Strains of xylanolytic fungi: *Fusarium oxysporum, Fusarium solani.*

c. Strains of ligninolytic fungi: *Phanerochaete chrysosporium*. Lacasse producer microorganisms are, e.g., *Botrytis cinerea* and *Trichophyton rubrum*.
d. Strains of pectolytic fungi: *Nectria hematococca, Fusarium oxysporum*.

2. Other carbon and nitrogen sources can be added to the basal mineral medium, according to the nutritional requirements and the enzymes to be induced. For example, sucrose, organic nitrogen, insoluble polymeric substrates such as paper and fruit pulp, lignocellulosic materials, and so on.
3. Many lignocellulosic materials (such as sugar cane bagasse pith) and others of synthetic origin (such as polyurethane foam) are used as inert support, where the liquid medium is absorbed.
4. The failure to find extracellular enzyme activity in a culture filtrate may be due to the enzyme being cell bound or substrate bound. It is also possible that an essential part of an enzyme may be cell bound, while the rest is cell free. Polymer-degrading enzymes (cell free) may be strongly absorbed by insoluble substrate and thus be undetectable in the culture filtrate, or the culture extract from a solid-state culture. If the current protocol does not yield results, a deeper study into this particular field must be done.
5. Most polymeric substrate-degrading enzymes act at acidic pH, but it is possible to find enzymatic activities at other pH values. Buffers and temperatures suggested here are only a general guide in searching for the optimum enzymatic reaction conditions.
6. The xylan used must give low turbidity to the substrate solution, and also must have an extended range of linearity between the amount of reducing sugars produced and the time of hydrolysis. It must be commercially available as well.
7. Availability of a standard enzyme preparation for comparison of the activity of the unknown enzyme preparations will minimize the errors involved on account of absence of linearity between enzyme concentration of unknown samples and xylose liberation. In the absence of a standard enzyme preparation, conditions for a linear relationship must be experimentally established before proceeding with an unknown enzyme sample assay. If a standard enzyme preparation for comparison is available, conditions for assay with respect to time, temperature, buffer molarity, and pH will be those belonging to the standard enzyme. One unit of enzyme activity is defined as the amount of enzyme producing 1 μmol of xylose equivalents per min under the given conditions, and corresponds to 16.67 nkat (μmol/s).
8. A plate assay with agarized medium may be done. Plates containing (g/L) 20 agar, 1.5 birchwood 4-*O*-methyl glucuronoxylan (covalently linked to Remazol brilliant blue R dye), 6.7 Yeast Nitrogen Base without amino acids, 2 g/L D-L-asparagine, and 5 g/L KH_2PO_4. The same methodology may be used to screen endo-1,4-β glucanase activity using as color substrate 1.5 g/L of hydroxy ethylcellulose covalently linked to Ostazin Brilliant Red H3B dye *(57)*.

References

1. Guo, W., González-Candelas, L., and Kolattukudy, P. E. (1995) Cloning of a novel constitutively expressed pectate lyase gene *pelB* from *Fusarium solani* f. sp. *pisi* (*Nectria haematococca,* Mating Type VI) and characterization of the gene product expressed in *Pichia pastoris. J. Bacteriol.* **177,** 7070–7077.
2. Schäfer, W. (1994) Molecular mechanisms of fungal pathogenicity to plants. *Annu. Rev. Phytopathol.* **32,** 461–477.
3. Freelove, A. C. J., Bolam, D. N., White, P., Hazlewood, G. P., and Gilbert, H. J. (2001) A novel carbohydrate-binding protein is a component of the plant cell wall-degrading complex of *Piromyces equi. J. Biol. Chem.* **276,** 43,010–43,017.
4. Sachslehner, A., Nidetzky, B., Kulbe, K. D., and Haltrich, D. (1998) Induction of mannanase, xylanase, and endoglucanase activities in *Sclerotium rolfsii. Appl. Environ. Microbiol.* **64,** 594–600.
5. Berry, D. R. (1975) The environmental control of the physiology of filamentous fungi, in *The Filamentous Fungi* (Smith, J. E., Berry, D. R., and Kristiansen, B., eds.), Edward Arnold, London.
6. Acuña-Argüelles, M. E., Gutiérrez-Rojas, M., Viniegra-González, G., and Favela-Torres, E. (1995) Production and properties of three pectinolytic activities produced by *Aspergillus niger* in submerged and solid-state fermentation. *Appl. Microbiol. Biotechnol.* **43,** 808–814.
7. Rogers, L. M., Kim, Y.-K., Guo, W., González-Candelas, L., Li, D., and Kolattukudy, P. E. (2000) Requirement for either a host- or pectin-induced pectate lyase for infection of *Pisum sativum* by *Nectria hematococca. PNAS* **97,** 9813–9818.
8. García-Maceira, F., Di Pietro, A., Huertas-González, M. D., Ruiz-Roldán, M. C., and Roncero, M. I. G. (2001) Molecular characterization of an endopolygalacturonase from *Fusarium oxysporum* expressed during early stages of infection. *Appl. Environ. Microbiol.* **67,** 2191–2196.
9. Di Pietro, A. and Roncero, M. I. G. (1998) Cloning, expression and role in pathogenicity of *pg1* encoding the major extracellular endopolygalacturonase of the vascular wilt pathogen *Fusarium oxysporum. Mol. Plant-Microbe Interact.* **11,** 91–98.
10. Doss, R. P. (1999) Composition and enzymatic activity of the extracellular matrix secreted by germlings of *Botrytis cinerea. Appl. Environ. Microbiol.* **65,** 404–408.
11. Doss, R. P., Potter, S. W., Soeldner, A. H., Christian, J. K., and Fukunaga, L. E. (1995) Adhesion of germlings of *Botrytis cinerea. Appl. Environ. Microbiol.* **61,** 260–265.
12. Vierheilig, H., Alt, M., Lange, J., Gutrella, M., Wiemken, A., and Boller, T. (1995) Colonization of transgenic tobacco constitutively expressing pathogenesis-related proteins by the vesicular-arbuscular mycorrhizal fungus *Glomus mosseae. Appl. Environ. Microbiol.* **61,** 3031–3034.
13. Cervone, F., De Lorenzo, G., Aracri, B., Bellincampi, D., Caprari, C., Clark, A. J., et al. (1996) The role of polygalacturonase, PGIP and pectin oligomers in fungal

infection, in *Pectins and Pectinases* (Visser, J. and Voragen, A. G. J., eds.), Elsevier Science, The Netherlands.

14. Salzman, R. A., Tikhonova, I., Bordelon, B. P., Hasegawa, P. M., and Bressan, R. A. (1998) Coordinate accumulation of antifungal proteins and hexoses constitutes a developmentally controlled defense response during fruit ripening in grape. *Plant. Physiol.* **117,** 465–472.
15. Kubicek, C. P., Mach, R. L., Peterbauer, C. K., and Lorito, M. (2001) *Trichoderma:* from genes to biocontrol. *J. Plant Pathol.* **83,** 11–23.
16. Pere, J., Puolakka, A., Nousiainen, P., and Buchert, J. (2001) Action of purified *Trichoderma reesei* cellulases on cotton fibers and yarn. *J. Biotechnol.* **89,** 247–255.
17. Mach, R. L., Seiboth, B., Myasnikov, A., González, R., Strauss, J., Harkki, A. M., et al. (1995) The *bgl1* gene of *Trichoderma reesei* QM 9414 encodes an extracellular, cellulose-inducible β-glucosidase involved in cellulase induction by sophorose. *Mol. Microbiol.* **16,** 687–697.
18. Seiboth, B., Hakola, S., Mach, R. L., Suominen, P. L., and Kubicek, C. P. (1997) Role of four major cellulases in triggering of cellulase gene expression by cellulose in *Trichoderma reesei. J. Bacteriol.* **179,** 5318–5320.
19. Sunna, A. and Antranikian, G.(1997) Xylanolytic enzymes from fungi and bacteria. *Crit. Rev. Biotechnol.* **17,** 39–67.
20. Kulkarni, N., Shendye, A., and Rao, M. (1999) Molecular and biotechnological aspects of xylanases. *FEMS Microbiol. Rev.* **23,** 411–456.
21. Sinnot, M. L. (1990) Catalytic mechanisms of enzymic glucosyl transfer. *Chem. Rev.* **90,** 1171–1202.
22. Kirk, T. K. and Farrell, R. L. (1987) Enzymatic "combustion": the microbial degradation of lignin. *Annu. Rev. Microbiol.* **41,** 465–505.
23. Caramelo, L., Martínez, M. J., and Martínez, A. T. (1999) A search for ligninolytic peroxidases in the fungus *Pleurotus eryngii* involving α-keto-γ-thiomethylbutyric acid and lignin model dimers. *Appl. Environ. Microbiol.* **65,** 916–922.
24. Xu, F., Kulys, J. J., Duke, K., Li, K., Krikstopaitis, K., Deussen, H. J., et al. (2000) Redox chemistry in laccase catalyzed oxidation of N-hydroxy compounds. *Appl. Environ. Microbiol.* **66,** 2052–2056.
25. O'Malley, D. M., Whetton, R., Bao, W., Chen, C. L., and Sederoff, R. R. (1993) The role of laccase in lignification. *Plant J.* **4,** 751–757.
26. Thurston, C. F. (1994) The structure and function of fungal laccases. *Microbiology* **140,** 19–26.
27. Slomczynski, D., Nakas, J. P., and Tanenbaum, S. W. (1995) Production and characterization of laccase from *Botrytis cinerea. Appl. Environ. Microbiol.* **61,** 907–912.
28. Saparrat, M. C. N., Martínez, M. J., Tournier, H. A., Cabello, M. N., and Arambarri, A. M. (2001) Production of ligninolytic enzymes by *Fusarium solani* strains isolated from different substrata. *World J. Microbiol. Biotechnol.* **16,** 799–803.
29. Sakai, T., Sakamoto, T., Hallaert, J., and Vandamme, E. J. (1993) Pectin, pectinase and protopectinase: production, properties and applications. *Adv. Appl. Microbiol.* **39,** 213–294.

30. Rombouts, F. M. and Pilnik, W. (1980) Pectic enzymes, in *Microbial Enzymes and Bioconversion* (Pilnik, W., ed.), Academic Press, London, UK, pp. 227–282.
31. Lang, C. and Dörnenburg, H. (2000) Perspectives in the biological function and the technological application of polygalacturonases. *Appl. Microbiol. Biotechnol.* **53,** 366–375.
32. Sakai, T. and Okushima, M. (1978) Protopectin-solubilizing enzyme from *Trichosporon penicillatum. Agric. Biol. Chem.* **42,** 2427–2429.
33. Sakai, T. and Sakamoto, T. (1990) Purification and some properties of a protopectin-solubilizing enzyme that has potent activity on sugar beet protopectin. *Agric. Biol. Chem.* **54,** 879–889.
34. Brinton, C. S., Wichmann, H. J., Willaman, J. J., Wilson, C. P., and Dore, W. H. (1927) Definitions written by the Committee on Nomenclature of Pectin of the Agriculture-Food Division. *J. Am. Chem. Soc.* **49,** 38–40.
35. Eggert, C., Temp, U., and Eriksson, K.-E. L. (1996) The ligninolytic system of the white rot fungus *Pycnoporus cinnabarinus:* purification and characterization of the laccase. *Appl. Environ. Microbiol.* **62,** 1151–1158.
36. Jung, H.-J. G., Valdez, F. R., Abad, A. R., Blanchette, R. A., and Hatfield, R. D. (1992) Effect of white rot basidiomycetes on chemical composition and in vitro digestibility of oat straw and alfalfa stems. *J. Anim. Sci.* **70,** 1928–1935.
37. Reid, I. D. (1989) Optimization of solid-state fermentation for selective delignification of aspen wood with *Phlebia tremellosa. Enzyme Microb. Technol.* **11,** 786–803.
38. Martínez, A. T., Camarero, S., Guillén, F., Gutiérrez, A., Muñoz, C., Varela, E., et al. (1994) Progress in biopulping of non-woody materials: chemical, enzymatic and ultrastructural aspects of wheat straw delignification with ligninolytic fungi from the genus *Pleurotus. FEMS Microbiol. Rev.* **13,** 265–274.
39. Biely, P. (1985) Microbial xylanolytic systems. *Trends Biotechnol.* **3,** 286–290.
40. Viikari, L., Kantelinen, A., Sundquist, J., and Linko, M. (1994) Xylanases in bleaching: from an idea to the industry. *FEMS Microbiol. Rev.* **13,** 335–350.
41. Park, J.-W., Park, K., Song, H., and Shin, H. (2002) Saccharification and adsorption characteristics of modified cellulases with hydrophilic/hydrophobic copolymers. *J. Biotechnol.* **93,** 203–208.
42. Viesturs, U., Leite, M., Treimanis, A., Eremeeva, T., Apsite, A., Eisimonte, M., et al. (1996) Production of cellulases and xylanases by *Trichoderma viride* and biological processing of lignocellulose and recycled paper fibers. *Appl. Biochem. Biotechnol.* **57/58,** 349–360.
43. Saxena, A., Kuhad, R. C., Saxena, R. K., and Gupta, R. (1994) Production and characterization of a xylanase from *Cyathus stercoreus. World J. Microbiol. Biotechnol.* **10,** 293–295.
44. Jung H., Xu, F., and Li, K. (2002) Purification and characterization of laccase from wood-degrading fungus *Trichophyton rubrum* LKY-7. *Enzym. Microb. Technol.* **30,** 161–168.
45. Miller, L. (1959) Use of dinitrosalicylic acid reagent for determination of reducing sugars. *Anal. Chem.* **31,** 426-428.

46. Ghose, T. K. (1987) Measurement of cellulase activities. *Pure Appl. Chem.* **59,** 257–268.
47. Bailey, M. J., Biely, P., and Poutanen, K. (1992) Interlaboratory testing of methods for assay of xylanase activity. *J. Biotechnol.* **23,** 257–270.
48. Ghose, T. K. and Bisaria, V. S. (1987) Measurement of hemicellulase activities. Part 1: xylanases. *Pure Appl. Chem.* **59,** 1739–1752.
49. Vares, T., Kalsi, M., and Hatakka, A. (1995) Lignin peroxidases, manganase peroxidase, and other ligninolytic enzymes produced by *Phlebia radiata* during solid-state fermentation of wheat straw. *Appl. Environ. Microbiol.* **61,** 3515–3520.
50. Machuca, A. and Ferraz, A. (2001) Hydrolytic and oxidative enzymes produced by white- and brown-rot fungi during *Eucaliptus grandis* decay in solid medium. *Enzyme Microb. Technol.* **29,** 386–391.
51. Somogyi, M. (1952) Notes on sugar determination. *J. Biol. Chem.* **195,** 19–23.
52. Nelson, N. (1944) A photometric adaptation of the Somogyi method for determination of glucose. *J. Biol. Chem.* **153,** 376–380.
53. Cavallito, S. F., Hours, R. H., and Mignone, C. F. (1997) Quantification of pectin-releasing activity of protopectinase-SE from *Geotrichum klebhanii. Biotechnol. Tech.* **11,** 331–334.
54. Contreras Esquivel, J. C., Hours, R. A., Voget, C. E., and Mignone, C. F. (1999) *Aspergillus kawachii* produces an acidic pectin releasing enzyme activity. *J. Biosc. Bioeng.* **88,** 48–52.
55. McComb, E. A. and McCready, R. M. (1952) Colorimetric determination of pectic substances. *Anal. Chem.* **24,** 1630–1632.
56. Blumenkrantz, N. and Asboe-Hansen, G. (1973) New method for quantitative determination of uronic acids. *Anal. Biochem.* **54,** 484–489.
57. Lee, H., Biely, P., Latta, R. K., Barbosa, M. F. S., and Schneider, H. (1986) Utilization of xylan by yeasts and its conversion to ethanol by *Pichia stipitis* strains. *Appl. Environ. Microbiol.* **52,** 320–324.

32

An Analysis of Microorganisms in Environments Using Denaturing Gradient Gel Electrophoresis

Paul W. Baker and Shigeaki Harayama

1. Introduction

Denaturing gradient gel electrophoresis (DGGE) has been developed as a technique to screen for mutations in a particular gene. Usually, PCR products of a particular gene in eukaryotic cells, prokaryotic cells, or clones (e.g., wild-type strain and a mutant) are amplified, and separated on a gel in which a gradient of increasing chemical denaturant is established. Under optimal conditions, PCR products that have nucleotide substitutions will migrate to a different position on DGGE in comparison to the original PCR products, and thus allow the detection of mutations in this particular gene.

A basic theory has been developed to describe the migration of DNA through DGGE and temperature gradient gel electrophoresis (TGGE). Double-stranded DNA is separated into single-stranded DNA, as the DNA migrates into an increasing concentration of a chemical denaturant. The denaturant concentration (or temperature for TGGE) at which a double-stranded DNA melts (melting temperature) is mainly determined by the G+C content: GC-rich regions melt at higher temperatures than AT-rich regions. Therefore DNA molecules that differ by only one nucleotide will have slightly different melting temperatures. When separated by electrophoresis through a gradient of increasing denaturant (chemical or temperature), the mobility of a DNA molecule will be retarded when DNA melts into single strands.

To prevent the complete separation of DNA into two single strands, a GC-rich region called a GC clamp is introduced at one end of the PCR products. This is done by using a primer pair—one of the primer pair possesses a GC clamp at the 5' end. A PCR product with a GC clamp partially melts at a particular

From: *Methods in Biotechnology: Environmental Microbiology: Methods and Protocols*
Edited by: J. F. T. Spencer and A. L. Ragout de Spencer

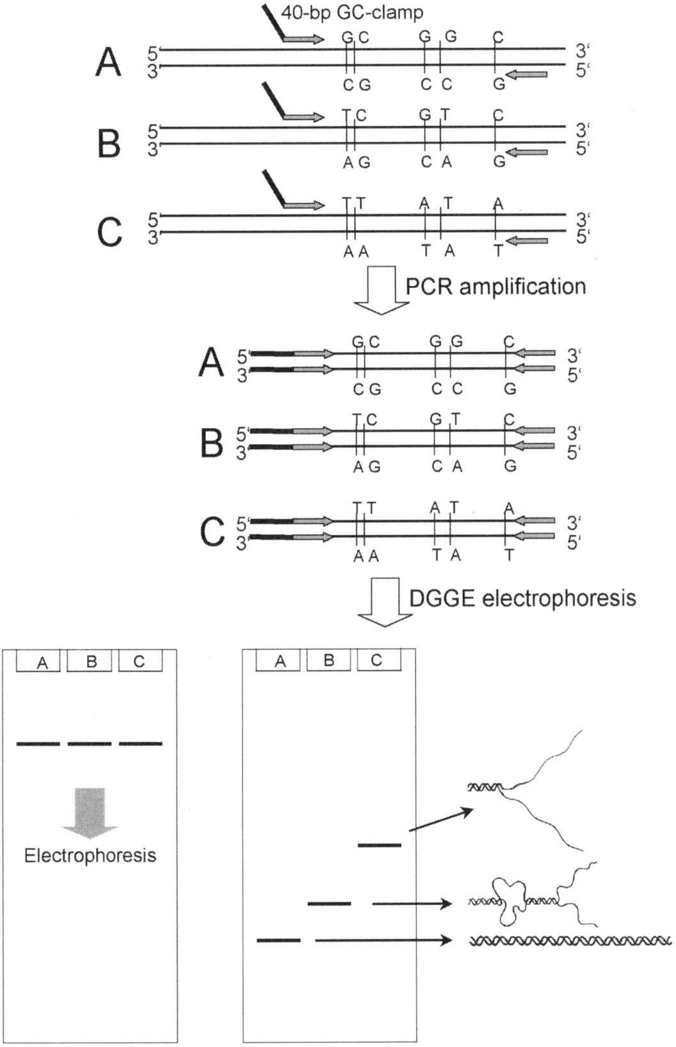

Fig. 1. Schematic diagram showing the formation of GC-clamped PCR products from genes A, B, and C, which have single base substitutions. Because PCR product A has a higher melting temperature than PCR products B and C, it migrates further in the DGGE gel. On complete melting of the gene product, a branched structure is formed, which migrates much more slowly through the DGGE gel.

position of the DGGE gel, and forms a branched structure. This structure greatly reduces the migration of the PCR products (**Fig. 1**).

DGGE is also useful in microbial ecology for determining the community structure (diversity and density of microorganisms) and population dynamics

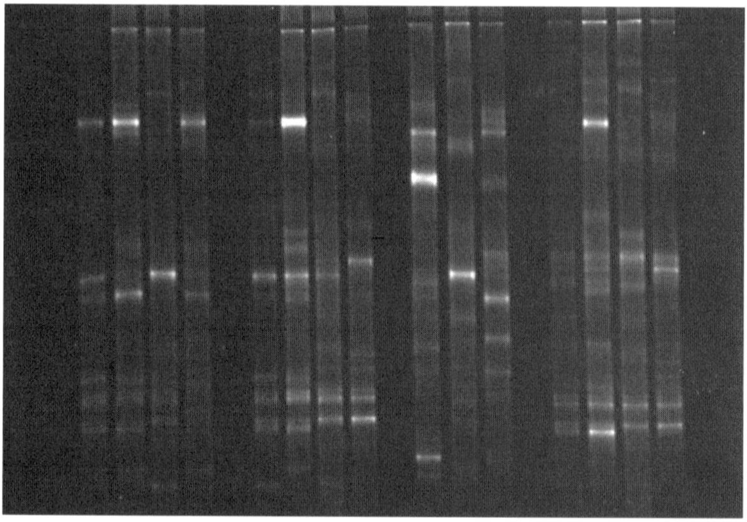

Fig. 2. 16S rDNA DGGE profiles of seawater samples. Seawater samples were collected from the Japan Sea at different locations at different seasons. The microorganisms in 2 L of the seawater samples were collected on a GV membrane (Millipore) by filtration, and DNA was extracted by the Marmur procedure. The PCR amplification of 16S rDNA was carried out using a primer set, 341fGC and 534r (**Table 1**).

(change in densities of microorganisms) in environments by analyzing short PCR products of target genes (e.g., 16S rDNA) amplified from DNA isolated from environmental samples *(1)*. In the ecological analysis, each PCR product generally contains multiple DNA species, and they are separated into multiple bands on a DGGE gel (**Fig. 2**). Already many ecological studies have been published that have incorporated DGGE as the main technique *(2)*. Such analyses allow the identification of factors that influence microbial flora. For example, the effects of variation of plants and seasons on bacterial populations in rhizospheres have been investigated *(3,4)*. DGGE can be used as a guide to isolate the prominent microorganism species in environments *(5)*. Furthermore DGGE can be used to identify multiple genes within any single microorganism *(6,7)*, or to estimate the efficiency of DNA extraction from microorganisms *(8)*.

Several potential problems exist that make the interpretation of DGGE results difficult. Primarily, there is a possibility of amplification bias during PCR amplification, but other potential problems of DGGE are comigration of sequences, heteroduplex formation, migration of single-stranded DNA, variable migration of single-sequence species, and the complete disappearance of bands during DGGE analysis *(9–11)*. Nevertheless, the advantages of quickly processing many samples far outweigh any possible disadvantages.

2. Equipment and Materials

2.1. PCR Amplification

1. Sequence data from database on microorganisms most likely to be amplified from the sample.
2. MacMelt (Bio-Rad) or Poland program (http://www.biophys.uni-duesseldorf.de/local/POLAND/poland.html).
3. DNA extract from microorganisms or environmental sample (can also use microorganisms directly).
4. PCR primers (one of the primers has a GC clamp attached).
5. *Taq* polymerase, 10X *Taq* polymerase buffer (100 m*M* Tris-HCl, pH 8.0, 500 m*M* KCl, 15 m*M* $MgCl_2$, 1% [v/v] Triton X-100).
6. 100 m*M* dGTP, 100 m*M* dATP, 100 m*M* dTTP, 100 m*M* dCTP.
7. PCR machine.

2.2. DGGE Gel Preparation

1. Dcode Universal Mutation Detection System, glass plates, 0.75-mm spacers and combs (Bio-Rad), or an equivalent electrophoresis apparatus.
2. Ethanol (70% [v/v]).
3. Acrylamide:bisacrylamide (40% [w/v], 37.5:1) solution: dissolve 100 g acrylamide and 2.7 g bisacrylamide in H_2O to a final volume of 150 mL.
4. Formamide (*see* **Note 1**).
5. Urea (electrophoresis grade).
6. 10% (w/v) ammonium persulphate.
7. *N,N,N',N'*-tetramethylenediamide (TEMED).
8. 50X TAE buffer stock: dissolve 242 g of Tris base, 57.1 mL of glacial acetic acid, and 100 mL of 0.5 *M* EDTA at pH 8 in H_2O to a final volume of 1000 mL.
9. Thin paper tissues.
10. 6X Type I gel loading buffer: 0.25% (w/v) Bromophenol Blue, 0.25% (w/v) xylene cyanol, 24% (w/v) sucrose.
11. Gradient Delivery System (Bio-Rad) or Gradient Former (Bio-Rad) and a peristaltic pump.
12. 0% Denaturant stock solution: 6% (w/v) acrylamide in 1X TAE. Dissolve 75 mL acrylamide (40X stock) and 10 mL TAE buffer (50X stock) in H_2O to a final volume of 500 mL.
13. 80% Denaturant stock solution: 6% acrylamide, 32% (v/v) formamide, 5.6 *M* urea in 1X TAE. Dissolve 75 mL acrylamide (40% stock), 170 g electrophoresis-grade urea, and 10 mL TAE (50X stock) in H_2O to a final volume of 500 mL.

2.3. Gel Staining

2.3.1. SYBR Gold Staining (or SYBR Green I or II)

1. Opaque plastic box.
2. SYBR Gold (or SYBR Green I or II).

3. 10X TBE buffer: dissolve 108 g Tris base, 55 g boric acid, and 9.3 g EDTA disodium salt in H_2O to a final volume of 1000 mL.

2.3.2. Ethidium Bromide Staining

1. Opaque plastic box.
2. Ethidium bromide stock solution (10 mg/mL).
3. 10X TBE buffer.

2.3.3. Silver Staining

Use ultrapure water for the preparation of the solutions.

1. Plastic box.
2. Fix/stop solution (10% [v/v] glacial acetic acid).
3. Staining solution: 2 g silver nitrate and 3 mL 37% formaldehyde in H_2O to a final volume of 2000 mL.
4. Developing solution: dissolve 60 g of sodium carbonate in H_2O to a final volume of 2000 mL. Chill to 4–10°C. Immediately before use, add 3 mL 37% formaldehyde and 400 µL of 5% (w/v) sodium thiosulfate.

2.4. Sequence Determination of Major Bands

1. Eppendorf tubes.
2. Knife.
3. Shaker.
4. TE: 10 m*M* Tris-HCl, 1 m*M* EDTA, pH 8.0.
5. Low-melting agarose.
6. Marker ladder.
7. 10 mg/mL Ethidium bromide.
8. Spin columns (Qiagen).

2.5. DGGE for the Identification of Isolated Microorganisms

1. PCR amplification materials (*see* **Subheading 2.1.**).
2. DGGE preparation materials (*see* **Subheading 2.2.**).
3. Sequence determination of bands (*see* **Subheadings 2.3.** and **2.4.**).

2.6. Competitive PCR and DGGE

1. Micoorganism that will be used as an internal standard.
2. Sterile distilled water, 50 m*M* phosphate buffer at pH 8.0, or saline (0.8% [w/v] NaCl) to resuspend the inoculant microorganism.
3. Phase-contrast microscope and 4'-6-diamidino-2-phenylindole (DAPI).
4. PCR amplification materials (*see* **Subheading 2.1.**).
5. DGGE preparation materials (*see* **Subheading 2.2.**).
6. GelDoc system (Bio-Rad) or a densitometer.

2.7. Hybridization of DGGE Gel Using Specific Probes

1. Denaturation solution: dissolve 87.7 g NaCl and 20 g NaOH in H_2O to a final volume of 1000 mL.
2. Neutralization buffer: dissolve 87.7 g NaCl and 5 g Tris base in H_2O; adjust pH to 7.5 and make volume up to 1000 mL.
3. Nylon membranes (Hybond-N^+).
4. Electroblotting apparatus.
5. 0.5X TAE buffer.
6. UV crosslink apparatus (e.g., Stratagene UV crosslinker).
7. Hybridization bottle (Hybaid) or plastic bag or plastic box.
8. Hybridization oven (Hybaid) or water bath.
9. Prewarmed prehybridization solution: 50% formamide, 5X SSC, 0.1% [w/v] SDS, 2% [w/v] blocking reagent for nucleic acid hybridization (Boehringer Mannheim). 1X SSC is 0.15 *M* NaCl, 0.015 *M* sodium citrate.
10. Two solutions of prewarmed 2X SSC containing 0.1% SDS.
11. Two solutions of prewarmed 0.1X SSC containing 0.1% SDS.

3. Methods

3.1. PCR Amplification

3.1.1. Primer Design

The PCR products in DGGE analysis contain a GC-rich region at one end, which basically acts as a clamp. This GC-rich region is attached to one of the primers and is incorporated during PCR amplification. This GC-rich region has been specifically designed to prevent self-reannealing, which would then create truncated PCR products. An example of a GC clamp is 5'-CGCCCGGGCGCGCCCCGGGCGGGGCGGGGCACGGGGG-3' (*see* **Note 2**). Primers should be carefully designed so that PCR products have a unit melting domain, and the GC clamp should be positioned adjacent to the highest melting domain. Several computer programs exist to determine melting profiles of PCR products. The primer pair may be universal amplifying genes from all microorganisms. Examples of universal PCR primers to amplify bacterial 16S rRNA genes are listed in **Table 1**. Group-specific or species-specific primers may also be designed. Ideally the PCR products should be between 100 and 400 bp in size. Try to avoid using degenerate primers, which may create closely migrating multiple bands on DGGE.

1. Obtain sequence data from a database. Design primers, and obtain sequences to be amplified by the primers. Add a GC-clamp sequence to one end of the sequences of the PCR products.
2. Input the sequence data into a computer program (e.g., MacMelt or Poland). The programs produce melting profiles.
3. Repeat again but with the GC clamp attached at the other end.

Table 1
PCR Primers That Can Be Used to Amplify 16S rRNA Genes of 36 Division-Level Phylogenetic Groups in the Domain *Bacteria*

Primer[a]	Position[b]	Sequence (5'–3')[c]	Reference
341f	341–357	CCTACGGGAGGCAGCAG	*25*
341fGC	341–357	GC clamp connected to the 5' end of 341f	*25*
534r	518–534	ATTACCGCGGCTGCTGG	*25*
926r	907–926	CCGTCAATTC(A/C)TTTGAGTTT	*25*
968f	968–984	AACGCGAAGAACCTTAC	*7*
968fGC	968–984	GC clamp connected to the 5' end of 968f	*7*
1401r	1358–1401	CGGTGTGTACAAGACCC	*7*
I-341f	341–356	CCTACGGGIGGCIGCA	*21*
I-341fGC	341–356	GC clamp connected to the 5' end of I-341f	*21*
I-533r	515–533	TIACCGIIICTICTGGCAC	*21*
I-926r	907–926	CCGICIATTIITTTIAGTTT	*21*
GC clamp		CGCCCGCCGCGCGCGGCGGGCGG	*25*
GC clamp		GGCGGGGGCACGGGGG	*25*

[a]f, forward primer; r, reverse primer.
[b]Corresponding to the numbering in the *Escherichia coli* sequence.
[c]I, inosine.

4. Melting profiles will be obtained for each sequence. The optimal profiles are those that comprise a single melting domain, excluding the GC clamp, and show differences between different sequences.
5. The programs also calculate the difference in melting temperature (T_C) for different PCR products. The optimal melting profiles are those which comprise a single melting domain and show differences in melting temperature between different sequences. Select primer pairs that fulfill these requirements.

3.1.2. PCR Amplification

1. In a thin-walled PCR tube, add 5 µL of 10X *Taq* polymerase buffer, 1 U of *Taq* polymerase, each deoxynucleoside triphosphate at a final concentration of 200 µ*M,* each primer at a concentration of 25 pmol, and 10–50 ng of the extracted DNA, or approximately a quarter of a 2-mm-diameter colony. Add H_2O to a final volume of 50 µL.
2. Perform touchdown PCR. After the initial hot-start at 94°C for 10 min to completely denature all DNA molecules, the program begins at an annealing temperature 10°C higher than the optimum annealing temperature for the primers without a GC clamp. For example, 94°C for 1 min, 65°C for 1 min, and 72°C for 2 min. After every second cycle, the temperature is decreased by 1°C until the optimum temperature is reached, and then 30 cycles are performed (*see* **Note 3**). One final cycle is performed using the optimum annealing temperature but a longer extension time to ensure there are no single-stranded DNA molecules.

3.2. DGGE Gel Preparation

The DGGE gel is prepared between two clean glass plates separated by plastic spacers on either side. A denaturant gradient is formed that is highest at the bottom and lowest at the top. Gradient profiles for the optimal separation depend on PCR products and are based on many different trials. The maximum and minimum denaturant concentrations are 0% and 100%. The 100% denaturant contains 42% (w/v) urea and 40% formamide (*see* **Note 4**). For PCR products of about 200 bp, 300–600 bp, and 600–1500 bp in length, prepare a 10%, 6%, and 3.5% polyacrylamide gel, respectively. Solutions to prepare 6% polyacrymide gel are given in **Subheading 2.2., items 12** and **13**. The gradient can be formed using either a Gradient Delivery System (Bio-Rad) or Gradient-Former (Bio-Rad) in combination with a peristaltic pump. The Gradient-Former consists of two columns (reservoir and mixing chamber) interconnected by a stopcock at the bottom of the columns. The mixing chamber is attached at the bottom to another tube, which passes through the peristaltic pump and connects to the top of the space between the two glass plates. The solution is introduced from the Gradient-Former to cast a gradient gel.

It is necessary to run preliminary gels to determine the optimal separation conditions. To establish the optimum time for separation of PCR products, the PCR product may be run on a time-travel gel. It may be necessary to perform a time-travel gel to determine the length of time necessary for separation of the PCR products. This is performed by loading the product into the initial lane(s) and running the gel for a short period of time. The gel is stopped and the PCR product is loaded into the next well. This is repeated until all the wells have been filled. Therefore PCR products loaded first onto the gel migrate for the longest period of time. The running time of a lane giving the best separation will thus be selected. The range of denaturants in the gel can also be adjusted to obtain the best separation but maintain sharpness of the bands.

3.2.1. Gel Preparation Using the Gradient Delivery System

1. Wash the inside of the plates with 70% ethanol.
2. Put the denaturants of high and low concentrations into syringes and displace all air bubbles from the syringes. Set up the syringes so that the high denaturant is dispensed first, followed by the second denaturant.
3. Leave a space of approx 1.5 cm at the top of the gel, so that the wells can be inserted. Leave the DGGE gel for 10 min to allow the gel to solidify.
4. Remove the top meniscus by absorption with paper tissues.
5. Overlay a 0% denaturant onto the gel and insert the comb.
6. Leave the DGGE gel for at least 1 h to completely polymerize.
7. Once the gel has polymerized, remove the comb and place the gel into a DGGE gel tank containing hot 1X TAE buffer, usually at 60°C.

8. Wash the wells using a pipet to remove unpolymerized acrylamide.
9. Add 1 vol of type I gel loading buffer to 3 vol of each PCR product sample.
10. If the PCR product originates from an environmental sample containing many microorganisms, load plenty of the product onto the gel (approx 40 μL). If the PCR product originates from a single microorganism, load a small aliquot (approx 6 μL) into each well.
11. Run the gel for a set period of time, usually several hours, and at a voltage ranging from 50 to 200 V.
12. See Model 475 Gradient Delivery System Instruction Manual (Bio-Rad catalog number 170-9042) for detailed information.

3.2.2. Gel Preparation Using the Gradient-Former

1. Wash the inside of the plates with 70% ethanol.
2. When using the Gradient-Former, set the speed of the peristaltic pump to a low desired speed.
3. Place the low denaturant into one of the columns of the Gradient-Former when the stopcock is open.
4. Close the stopcock and pipet the low denaturant from the mixing chamber into the reservoir. This step removes air bubbles that may become trapped in the section connecting the two columns around the stopcock.
5. Place the high denaturant into the mixing chamber together with a magnetic flea.
6. Put the mixing chamber over a magnetic stirrer.
7. Ensure the levels in both of the columns are the same before opening the stopcock interconnecting the two columns.
8. Switch on the magnetic stirrer and peristaltic pump to dispense the denaturants into the space between the two glass plates.
9. For subsequent steps, *see* **Subheading 3.1.1., steps 3–11**.

3.3. Gel Staining

The best method for staining DGGE gels requires using SYBR Gold. However, the gel may also be stained with ethidium bromide or silver stain. Both of these alternatives are described in this section. Silver staining has a higher resolution than ethidium bromide although the gels also have a background stain.

3.3.1. SYBR Gold Staining (or SYBR Green I or II)

1. Pipet 10 μL of 10,000X SYBR Gold stock into 100 mL of 1X TBE buffer.
2. Place the solution into a box protected from sunlight.
3. Place the gel into this solution and gently shake for 30 min.
4. View the gel under a UV transilluminator, and photograph using a yellow filter attachment.

3.3.2. Ethidium Bromide Staining

1. Alternatively, stain the gel in 100 mL of 10 mg/mL ethidium bromide.
2. View the gel under a UV transilluminator and photograph.

3.3.3. Silver Staining

1. The other alternative is silver staining. Place the gel into a plastic tray containing the fix/stop solution for 30 min.
2. Remove the solution using a pipet attached to a suction pump; do not touch the gel, otherwise staining will be impaired.
3. Rinse the gel three times with ultrapure water.
4. Add the staining solution. Completely submerge gel in the staining solution, agitate for 30 min.
5. Transfer the gel into another plastic tray containing ultrapure H_2O. Handle the gel with gloved hands. Touch only the edge of the gel. (Keep the used staining solution for silver recovery.)
6. Wash the gel briefly (less than 10 s) in ultrapure H_2O and then transfer immediately to another plastic tray containing freshly prepared ice-cold developing solution (1000 mL).
7. Agitate the gel until the bands start to develop. Change the developing solution for a fresh one. Continue until all the bands become visible. Avoid overdevelopment.
8. Add 1000 mL of fix/stop solution directly into the developing solution, and keep for 2–3 min.
9. Rinse two times with ultrapure H_2O. The gel can be viewed directly or after drying. Photograph the gel.

3.4. Sequence Determination of Major Bands

1. Clean a knife using 70% ethanol.
2. Excise the major bands from the gel using the knife, as close as possible to the band. Avoid cutting into other closely migrating bands.
3. Place the band into an Eppendorf tube with 100 µL TE and place onto a shaker overnight at room temperature.
4. Pipet the suspension into a fresh Eppendorf tube and re-amplify using the same primers as before with PCR.
5. Analyze the product on another DGGE gel alongside the PCR product from which the band was excised to check that the appropriate band has been amplified.
6. Repeat this process until one major band is obtained corresponding to the band of interest.
7. Once this has occurred, run the PCR product on a 1.5% (w/v) low-melting agarose gel.
8. Stain the gel in 100 mL of a solution of ethidium bromide (10 mg/mL).
9. Excise the band from the gel using an ethanol-washed knife.
10. Further purify using spin columns (Qiagen) as recommended by the manufacturer.

11. Sequence the product using 1 ng of each of the original primers without the GC clamp (*see* **Note 5**).

3.5. DGGE for the Identification of Isolated Microorganisms

When the PCR product amplified from an environmental sample is analyzed on DGGE, major bands may appear, indicating that these populations form a significant proportion of the total microbial population. Sequencing of any of the bands may reveal a high identity with sequences of microorganisms previously isolated. Alternatively it may show a low affiliation with the sequences of known microorganisms. In both cases, the isolation of microorganisms corresponding to major DGGE bands is of interest. The following method allows the attribution of isolated strains to DNA bands in the DGGE patterns. However, many microorganisms are yet uncultured, and microorganisms corresponding to some DGGE bands may not be isolated by conventional cultivation methods.

1. Isolate many microorganisms using an appropriate isolation method.
2. Perform PCR on DNA extracted from microorganisms or using a small sample of the microorganisms directly (*see* **Note 6**). However, the amplification by PCR directly from the microorganism may be inhibited owing to insufficient lysis of cells or by interfering cellular proteins and fatty acids.
3. Add 1 vol of type I loading dye to each 3 vol of PCR product.
4. Load a small aliquot of each PCR product amplified from the microorganism (approx 6 μL) into each well on the prepared DGGE gel (*see* **Subheadings 3.1.1.** or **3.1.2.**, **steps 1–8**).
5. Load the PCR products of the environmental sample, approx 40 μL, from which microorganisms have been isolated.
6. Run the gel for a predetermined period of time and voltage (*see* **Subheading 3.1.1.**, **step 11**).
7. Stain the gel and photograph (*see* **Subheadings 3.3.1.**, **3.3.2.**, or **3.3.3.**).
8. If a PCR product from a microorganism does migrate to the same position as the PCR product in the environmental sample, excise the band and purify.
9. Sequence the band to confirm it has the same sequence as the PCR product of interest from the environmental sample.

3.6. Competitive PCR and DGGE

There are several advantages of using competitive PCR and DGGE in comparison to using either epifluorescence *in situ* hybridization or dot blotting/ Southern blotting followed by hybridization in order to estimate bacterial populations in an environmental sample. For example many of the dominant bacterial populations in a sample can be quantified using universal ribosomal RNA primers. Moreover a bacterial population that is presumed to be small may still be quantified provided specific primers are used. Competitive PCR may be performed using a microorganism that is inoculated into the environmental samples at a known population density (*see* **Note 7**).

1. A microorganism is grown in a liquid medium to stationary phase.
2. Centrifuge and discard the supernatant.
3. Resuspend the pellet in either sterile distilled water, 50 m*M* phosphate buffer, or saline. The solution chosen depends on where the sample was obtained and its resistance to lysis.
4. Repeat the wash several times.
5. Count microorganisms directly using a phase contrast microscope, perhaps with the aid of a fluorescent dye such as DAPI.
6. Once the microbial counts have been determined, serially dilute the microorganisms.
7. Inoculate a small aliquot into each environmental sample. At each serial dilution, aliquots are inoculated into triplicate environmental samples.
8. Extract the DNA from each of the inoculated environmental samples. Also extract DNA from a sample of the microorganism that was used as the inoculant.
9. Amplify the extracted DNA by PCR using the appropriate primers.
10. Add 1 vol of type I gel loading buffer to 3 vol of each PCR product.
11. Add approx 40 μL of PCR products from each environmental sample and 6 μL of PCR product from the inoculated microorganism into each well on the prepared DGGE gel (*see* **Subheadings 3.1.1.** or **3.1.2.**, **steps 1–8**).
12. Run DGGE gel at the predetermined time period and voltage (*see* **Subheading 3.2.**, **step 11**).
13. Stain and view under a GelDoc System (Bio-Rad) (*see* **Subheadings 3.3.1.**, **3.3.2.**, or **3.3.3.**).
14. Quantify the intensities of the major bands and the inoculated microorganism. Alternatively, if the GelDoc system is not available, the gel can be viewed and photographed under a UV transilluminator. The photograph can then be examined in a densitometer and the bands quantified.

3.7. Hybridization of DGGE Gel Using Specific Probes

DGGE analysis of environmental samples using universal primers may result in a large diversity of bands. To obtain more information and reduce the level of complexity, the DGGE gels can be probed with specific primers ***(12–14)***. Furthermore, the DGGE gel can be hybridized with a specific probe to confirm the presence of specific amplification products on the DGGE gel that were amplified with group-specific primers ***(15)***.

1. Once PCR products from individual microorganisms or environmental samples have been separated by DGGE, photograph the gel.
2. Denature DNA in gel by soaking in the 1X denaturation solution for 30–60 min.

3. Neutralize the gel by soaking in the 1X neutralization solution for 30–60 min.
4. Transfer DNA from the gel to nylon membranes (Hybond-N^+) by electroblotting in 0.5X TAE buffer for 1 h at 40 mA.
5. Air-dry filter, UV crosslink with a Stratagene UV crosslinker for approx 30 s, and bake the membrane at 80°C for 2 h.
6. Place the nylon membrane into a plastic bag, plastic box, or hybridization bottle.
7. Add 50 mL prehybridization solution to the membrane and incubate with shaking for 1 h or more at 41°C. Ensure no bubbles form (otherwise a high background will result).
8. Prepare an oligonucleotide probe. Label the probe by ^{32}P, digoxigenen (DIG), or a fluorescent dye.
9. Discard the prehybridization solution and replace with a 6-mL fresh prehybridization solution containing an appropriate amount of probe, and incubate overnight at the hybridization temperature specific for the probe.
10. Discard the solution and replace with 2X SSC containing 0.1% SDS. Incubate at the hybridization temperature for 30 min (save the prehybridization solution containing the probe to reuse).
11. Replace the solution with 0.1X SSC containing 0.1% SDS. Incubate at the hybridization temperature for 30 min.
12. Use an appropriate method to detect the hybridization.

4. Notes

1. Formamide can be deionized by the following steps.

   a. Add 5 g ion exchanger (e.g., Bio-Rad mixed bed resin) to 100 mL formamide.
   b. Stir for 2 h or more.
   c. Filter through Whatman no. 1 paper.
   d. Store at –20°C.

2. An alternative to the GC clamp is the ChemiClamp, which is attached to one of the primers. Psoralen, which photoreacts with DNA to form an adduct, is covalently linked to the 5'-end of an oligonucleotide. The usual 5'-end base chosen is adenine, and psoralen is intercaled between the two strands. When exposed to long-wavelength UV light, psoralen reacts with thymidine residue to form a covalent photoadduct. Two stands become crosslinked. A restriction site is incorporated into the primer and enables the PCR product to be detached from ChemiClamp so that it can be sequenced ***(16)***. DGGE can be performed using a long GC clamp (40 bp) attached at one end of the PCR product and a short sequence of GCs (3–7 bp) at the other end ***(17)***. Also a ChemiClamp can be attached to both ends of the PCR product ***(18)***. These modifications may improve the resolution of bands observed on DGGE.
3. The touchdown PCR program ensures specific amplification of the desired amplification products starting at a stringent annealing temperature. The program may be changed to create optimum amplification conditions, provided that a higher annealing temperature is used initially.

4. The high denaturant may also contain a higher concentration of acrylamide compared to the low denaturant. This is known as double DGGE, and it allows PCR products which migrate further into the gel to become more focused *(19,20)*.
5. The PCR product may be amplified using the original primers without the GC clamp, run on low-melting agarose gel, excised, and purified through spin columns. This method removes the GC clamp so that guanine and cytosine nucleotides are not exhausted during the sequencing reaction. The products may also be amplified using specifically designed forward primer, which anneals further upstream of the sequence. Sequencing using this primer enables the entire sequence of the PCR product to be determined in one direction, which would not otherwise be possible using the original primers *(21)*. This method does not guarantee a pure sequence, because it is possible that another sequence has co-migrated with the band of interest. In such a case, cloning of the PCR products and examination of the cloned sequences would be necessary.
6. Beside screening microorganisms, this method may be used for identifying an *E. coli* clone containing a PCR product identical to the sequence of interest *(22)*.
7. Competitor DNA of a similar size but different base composition may be added to the environmental sample prior to or after DNA extraction *(23)*. Before addition, the competitor DNA is measured at 260 nm on the spectrophotometer to determine DNA concentration (ng/μL). This value is multiplied by 45.5 and by the dilution factor *(24)*. The DNA is serially diluted and inoculated into the environmental samples or DNA extracts of the environmental samples.

References

1. Muyzer, G. (1999) DGGE/TGGE a method for identifying genes from natural ecosystems. *Curr. Opin. Microbiol.* **2,** 317–322.
2. Muyzer, G. and Smalla, K. (1998) Application of denaturing gradient gel electrophoresis (DGGE) and temperature gradient gel electrophoresis (TGGE) in microbial ecology. *Antonie van Leeuwenhoek* **73,** 127–141.
3. Yang, C.-H., Crowley, D. E., Borneman, J., and Keen, N. T. (2000) Microbial phyllosphere populations are more complex than previously realized. *Proc. Natl. Acad. Sci. USA* **98,** 3889–3894.
4. Piceno, Y. M., Noble, P. A., and Lovell, C. R. (1999) Spatial and temporal assessment of diazotroph assemblage composition in vegetated salt marsh sediments using denaturing gradient gel electrophoresis analysis. *Microb. Ecol.* **38,** 157–167.
5. Watanabe, K., Teramoto, M., Futamata, H., and Harayama, S. (1998) Molecular detection, isolation, and physiological characterization of functionally dominant phenol-degrading bacteria in activated sludge. *Appl. Environ. Microbiol.* **64,** 4396–4402.
6. Rosado, A. S., Duarte, G. F., Seldin, L., and van Elsas, J. D. (1998) Genetic diversity of *nifH* gene sequences in *Paenibacillus azotofixans* strains and soil samples analyzed by denaturing gradient gel electrophoresis of PCR-amplified gene fragments. *Appl. Environ. Microbiol.* **64,** 2770–2779.

7. Nübel, U., Englen, B., Felske, A., Snaidr, J., Weishuber, A., Amann, R. I., et al. (1996) Sequence heterogeneities of genes encoding 16S rRNAs in *Paebacillus polymyxa* detected by temperature gradient gel electrophoresis. *J. Bacteriol.* **178,** 5636–5643.
8. Krsek, M. and Wellington, E. M. (1999) Comparison of different methods for the isolation and purification of total community DNA from soil. *J. Microbiol. Methods* **39,** 1–16.
9. Bruns, M. A., Hanson, J. R., Mefford, J., and Scow, K. M. (2001) Isolate PM1 populations are dominant and novel methyl tert-butyl ether-degrading bacteria in compost biofilter enrichments. *Environ. Microbiol.* **3,** 220–225.
10. Guldeberg, P., Grønæk, K., Aggerholm, A., Platz, A., thor Straten, P., Ahrenkiel, V., et al. (1998) Detection of mutations in GC-rich DNA by bisulphite denaturing gradient gel electrophoresis. *Nuc. Acids Res.* **26,** 1548–1549.
11. Weber, C. K., Shaffer, D. J., and Sidman, C. L. (1991) Unexpected behavior of H2Kb mutant DNAs in denaturing gradient gel electrophoresis. *Nuc. Acid Res.* **19,** 3331–3335.
12. Bruns, M. A., Stephen, J. R., Kowalchuk, G. A., Prosser, J. I., and Paul, E. A. (1999) Comparative diversity of ammonia oxidizer 16S rRNA gene sequences in native, tilled, and successional soils. *Appl. Environ. Microbiol.* **65,** 2994–3000.
13. Kowalchuk, G. A., Naoumenko, Z. S., Derikx, P. J., Felske, A., Stephen, J. R., and Arkhipchenko, I. A. (1999) Molecular analysis of ammonia-oxidizing bacteria of beta subdivision of the class proteobacteria in compost and composted materials. *Appl. Environ. Microbiol.* **65,** 396–403.
14. Stephen, J. R., Kowalchuk, G. A., Bruns, M.-A. V., McCraig, A. E., Phillips, C. J., Embley, M., et al. (1998) Analysis of β-subgroup proteobacterial ammonia oxidizer populations in soil by denaturing gradient gel electrophoresis analysis and hierarchical phylogenetic probing. *Appl. Environ. Microbiol.* **64,** 2958–2965.
15. Brinkhoff, T. and Muyzer, G. (1997) Increased species diversity and extended habitat range of sulfur-oxidizing *Thiomicrospira* spp. *Appl. Environ. Microbiol.* **63,** 3789–3796.
16. Costes, B., Girodon, E., Ghanem, N., Chassignol, M., Thuong, N. T., Dupret, D., et al. (1993) Psoralen-modified oligonucleotide primers improve detection of mutations by denaturing gradient gel electrophoresis and provide an alternative to GC-clamping. *Hum. Mol. Genet.* **2,** 393–397.
17. Wu, Y., Hayes, V. M., Osinga, J., Mulder, I. M., Looman, M. W. G., Buys, C. H. C. M., et al. (1998) Improvement of fragment and primer selection for mutation detection by denaturing gradient gel electrophoresis. *Nuc. Acids Res.* **26,** 5432–5440.
18. Gille, C., Gille, A., Booms, P., Robinson, P. N., and Nurnberg, P. (1998) Bipolar clamping improves the sensitivity of mutation detection by temperature gradient gel electrophoresis. *Electrophoresis* **19,** 1347–1350.
19. Gelfi, C., Righetti, S. C., Zunino, F., Della Torre, G., Pierotti, and Righetti, P. G. (1997) Detection of p553 point mutations by double-gradient, denaturing gradient gel electrophoresis. *Electrophoresis* **18,** 2921–2927.

20. Cremonesi, L., Carrera, P., Fumagallia, A., Lucchirara, S., Cardillo, E., Ferrari, M., et al. (1999) Validation of double gradient denaturing gradient gel electrophoresis through multigenic retrospective analysis. *Clin. Chem.* **45,** 35–40.
21. Watanabe, K, Kodoma, Y., and Harayama, S. (2001) Design and evaluation of PCR primers to amplify bacterial 16S ribosomal DNA fragments used for community fingerprinting. *J. Microbiol. Methods* **44,** 253–262.
22. Schabreiter-Gurtner, C., Pinar, G., Lubitz, W., and Rolleke, S. (2001) An advanced molecular strategy to identify bacterial communities on art objects. *J. Microbiol. Methods* **45,** 77–87.
23. Bruggemann, J., Stephen, J. R., Chang, Y. J., MacNaughton, S. J., Kowalchuk, G. A., Kline, E., et al. (2000) Competitive PCR-DGGE analysis of bacterial mixtures: an internal standard and an appraisal of template enumeration accuracy. *J. Microbiol. Methods* **40,** 111–123.
24. Sambrook, J. and Russell, D. W. (2001) In *Molecular Cloning—a Laboratory Manual,* Cold Spring Harbor Laboratory Press, Cold Spring Harbor, NY.
25. Muyzer, G., Hottentrager, S., Teske, A., and Wawer, C. (1996) Denaturing gradient gel electrophoresis of PCR-amplified 16S rDNA. A new molecular approach to analyze the genetic diversity of mixed microbial communities, in *Molecular Microbial Ecology Manual* (Akkermanns, A. D. L., van Elsas, J. D., and de Brujin, F. J., eds.), Kluwer Academic Publishing, Dordrecht, The Netherlands, pp. 3.4.4.1–3.4.4.22.

33

Testing for Evolutionary Correlations in Microbiology Using Phylogenetic Generalized Least Squares

Scott R. Miller

1. Introduction

Most of the metabolic, physiological, and ecological diversity on the planet is found within the bacterial and archaeal domains, and it is the adaptive limits of microorganisms that define the conditions that support life. Microbial ecologists interested in understanding the origins and distribution of this diversity often look for associations between the phenotypes of microorganisms and the environments in which they live. A researcher may be interested in whether, for example, a bacterial strain's optimal temperature has evolved in association with (and perhaps as an adaptation to) environmental temperature, or whether that trait evolved in association with other traits, such as lower temperature limit or maximal growth rate. A negative correlation in the latter case may indicate the presence of genetic constraints on phenotypic evolution that have in part delimited the distribution of microbial diversity in nature.

While answering these sorts of questions can provide insights into how historical interactions between microbial genomes and the environment have shaped present day ecological diversity, actually testing hypotheses of phenotypic evolution from comparative data poses a couple of challenges. One is how to infer historical events from data collected from extant organisms. This is a particularly critical issue for microorganisms, for which there is a paucity of information about the past available from the fossil record. Another has to do with statistical problems inherent in phenotypic data collected from organisms related to each other phylogenetically. Standard statistical models such as least squares regression and analysis of variance make several assumptions about the data being analyzed, including that they are independent, normally

From: *Methods in Biotechnology: Environmental Microbiology: Methods and Protocols*
Edited by: J. F. T. Spencer and A. L. Ragout de Spencer

distributed variables with equal variances. Phylogenetic comparative data are expected to violate at least some of these assumptions, most notably the assumption that the data are independent. This is because the phenotypes of organisms related to each other by a hierarchical pattern of shared ancestries are expected to be correlated: in general, we expect close relatives to resemble each other more than do distant ones. Treating dependent data as if they are independent leads us to overestimate both the degrees of freedom of our sample and, as a result, our confidence during hypothesis testing.

For the past 20 yr, the development of statistical methods for analyzing comparative phenotypic data has been an active area of research in evolutionary biology (for a review, *see* **ref. *1***). Phylogenetic comparative methods have been devised for addressing a variety of questions, most commonly either to test for correlated trait evolution or to estimate ancestral phenotypes. These methods require both a statistical framework that can take into account the phylogenetic relationships inherent in the dataset and a model of how phenotypes evolve along a phylogeny in order to estimate the expected similarity among samples. This chapter describes one approach, in which phylogenetic data are incorporated into a generalized least squares (GLS) model ***(2,3)*** in order to test for a correlation between a phenotypic trait and one or more other variables (which may be other traits or components of the environment). The method is illustrated using the software package COMPARE (**ref. *4***; http://compare.bio.indiana.edu).

1.1. The GLS Model

How does one incorporate phylogeny into the statistical analysis of comparative data? One approach is to use GLS, which relaxes some of the assumptions of the simple least squares regression model (*see* **Note 1**) to allow pairs of datapoints to be correlated (i.e., not independent) and individual samples to have different variances. These possibilities are specified in the variance-covariance matrix of the model's error term. When two samples (i.e., data points) are not independent because of shared evolutionary history, the off-diagonal element of this matrix describing the covariance between them will not be equal to zero. If the two samples have different variances, the diagonal elements for each sample will not be equal to each other. Because all that is assumed by the model is that this error term is normally distributed with mean zero and known variance and covariance structure, GLS is very flexible ***(3,5)***. Not only can it be used in correlation analysis, but models can also be developed to reconstruct ancestral traits and estimate rates of phenotypic evolution. COMPARE implements some of these models, including one which tests for correlations between characters (*see* **Subheading 3.**).

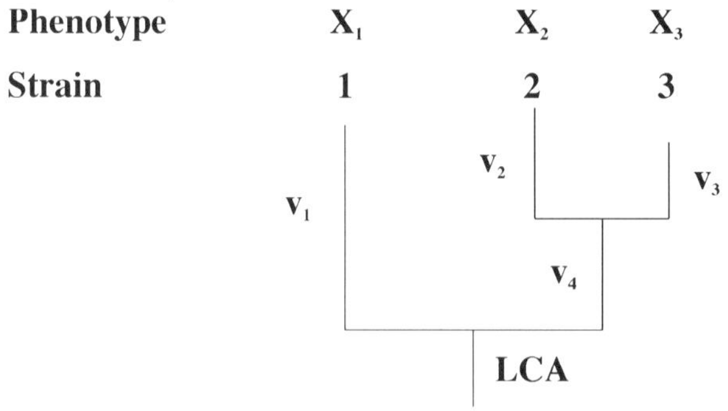

Fig. 1. Hypothetical phylogeny reconstructed from DNA sequence data for three bacterial strains and their phenotypic values for a hypothetical trait. Branch lengths indicate the amount of sequence divergence in units of nucleotide substitutions per site. LCA, last common ancestor of the strains.

1.2. Models of Phenotypic Evolution

In order to actually specify the variance-covariance structure of the GLS error term, we need to transform branch-length data from a phylogeny (usually in units of molecular sequence divergence) into units of expected phenotypic similarity by applying a model of how traits evolve over time. This is similar to phylogeny reconstruction methods, which also make explicit and/or implicit assumptions about the evolutionary process, in this case to determine the relatedness of organisms. The most widely applied microevolutionary model used in phylogenetic comparative methods is Brownian motion, the random walk in continuous time used to model a variety of stochastic processes, including neutral evolution in the field of population genetics *(6)*. In a Brownian-motion model of phenotypic evolution along a phylogeny, the change in mean phenotype over time is independent, normally distributed, and expected to be zero. However, the expected *variance* of the mean does change and is proportional to the branch lengths of the phylogeny in units of sequence divergence. It is this property of the model that allows us to convert phylogenetic relationships into the expected phenotypic similarities among organisms due to their shared evolutionary histories.

To illustrate, strains 1, 2, and 3, with phenotypic values X_1, X_2, and X_3 for a hypothetical continuously distributed trait, are related to each other by the phylogeny shown in **Fig. 1**. The change in mean phenotype along each branch of the phylogeny is expected to be zero according to the Brownian-motion model. The variance in the mean for each strain is expected to be proportional to the total amount of sequence divergence since that strain diverged from the last common ancestor of all three taxa. For strain 1 this variance is v_1, while the variances for strains 2 and 3 are $v_2 + v_4$ and $v_3 + v_4$, respectively. The shared term in the variance component for strains 2 and 3, v_4, is the expected covariance, or dependence, between the two strains. That is, if phenotypes are evolving by Brownian motion, then the covariance between two strains is simply proportional to the sum of the lengths of the branches shared between them by common ancestry, or the fraction of the amount of time from the root of the phylogeny until the present that the organisms evolved together prior to divergence from their most recent common ancestor. The model also assumes that the phenotypes of strains 2 and 3 have evolved independently, with variances v_3 and v_4 following their divergence from their last common ancestor. Similarly, because there has been no shared evolutionary history between strain 1 and the other strains since they last shared an ancestor, its phenotype has evolved entirely independently of the others, i.e., with zero covariance. We can now specify the elements of the variance-covariance matrix, with variances on the diagonal and covariances off-diagonal:

$$\begin{bmatrix} v_1 & 0 & 0 \\ 0 & v_2 + v_4 & v_4 \\ 0 & v_4 & v_3 + v_4 \end{bmatrix}$$

The GLS model under Brownian motion is statistically equivalent to Felsenstein's independent contrasts *(7,8)*, a recursive algorithm for estimating continuous character correlations and the most popular phylogenetic comparative method in use today.

Of course, we usually do not know whether the traits in our dataset are really evolving by Brownian motion. As noted previously, in Brownian-motion models, phenotypic change is a drift-like process. Often, however, we are interested in a particular system precisely because we believe that the patterns of change that we observe are not owing solely to random drift but are being effected by natural selection. What, then, are the consequences of mis-specifying the evolutionary model? Hansen and Martins *(9)* demonstrated that the expected correlation between phenotypes following a speciation event decreases linearly with increasing phylogenetic distance under not only Brownian motion but also a number of other evolutionary scenarios. The latter include directional selection when the direction of selection fluctuates, stabilizing selection with a fluctuating optimum, and bursts of evolution followed by long periods of phenotypic stasis. Therefore, although we may not know exactly the extent to which different evolutionary forces are contributing to the patterns of phenotypic diversity in our data, Brownian motion may capture its expected phylogenetic dependence.

The exception to this linearity is when phenotypic evolution is constrained, as is the case, for example, when traits are under stabilizing selection for a fixed optimum. Here, phenotypic values that differ greatly from the optimum are more likely to evolve toward rather than away from it. Stabilizing selection is typically modeled by an Ornstein-Uhlenbeck "rubber-band" process in which phenotypes evolve in a drift-like manner but are pulled toward an optimal value by a restraining force, alpha, which increases in strength the further the phenotype moves away from that optimum ***(9,10)***. The constraint on phenotypic evolution imposed by alpha leads to an exponential decay of phenotypic similarity with phylogenetic distance ***(9)***. Therefore, the phenotypic similarity of relatives is less than would be expected if traits were evolving by, for example, random drift, and is an exponentially declining function of both the evolutionary distance (branch lengths) separating them and the strength of the restraining force. Specifying the elements of the variance–covariance matrix under the exponential model thus requires not only phylogenetic data but also an estimate for the alpha parameter. This can be achieved by performing the regression analysis for different values of alpha and optimizing the estimate as the value that maximizes the likelihood of the GLS model. Brownian motion is a special case of this process in which the strength of the restraining force is zero.

Evaluations by both computer simulation and experimental evolution of the relative abilities of the Brownian-motion and exponential models to accurately estimate regression parameters have both shed light on their performance when their assumptions have been violated (i.e., the model of phenotypic evolution is wrong) and also point out the dangers of ignoring phylogeny during correlation analysis. GLS under a Brownian-motion model greatly outperformed the standard nonphylogenetic approach of calculating Pearson product–moment correlations under almost all simulation conditions, except for the case when there are extremely strong phenotypic constraints ***(11,12)***. This makes sense, because it is under such conditions that most historical information will be quickly erased, thereby making the data approach the latter model's assumption of independence. Similarly, Felsenstein's independent contrasts ***(7)*** was unbiased and more accurate than Pearson correlation analysis when applied to real phenotypic data collected for bacteriophage taxa related by a known, experimentally generated phylogeny ***(13)***. On the other hand, inclusion of the alpha parameter in the GLS model resulted in good statistical performance under all computer simulation conditions ***(12)***, regardless of the true evolutionary process underlying phenotypic change. This indicates the general utility of the exponential model for analyzing real data sets.

2. Materials

The following will be required to estimate correlations by phylogenetic GLS: phenotypic data for a group of organisms (and possibly environmental data from their habitats), phylogenetic data indicating how those organisms are related to each other and the software package COMPARE *(4)*, downloadable from http://compare.bio.indiana.edu.

The type of phenotypic data collected for the study organisms, e.g., laboratory strains, will of course depend upon the hypothesis being tested. For correlation analysis, data for two or more traits and/or environmental parameters are required. Many traits, such as growth rate, are continuously distributed, and most phylogenetic comparative methods are designed to deal with such data. One advantage of a GLS approach is that it can also accommodate discrete characters, e.g., binary data such as the ability to metabolize a substrate. While the example in **Subheading 3.** below uses data collected from laboratory isolates, there is no reason in principle why phenotypic data could not be collected from microorganisms directly in the environment, provided that one can unambiguously link a phenotype to a specific genotype (for use in phylogeny building).

In order to account for the potential nonindependence of samples in the phenotypic dataset, one must also have information about the phylogenetic relationships of the organisms from which the data were collected. Phylogenies are typically inferred from aligned DNA sequence data, and many methods for phylogenetic reconstruction have been developed (for a detailed review of existing methods and information on phylogenetic analysis software, *see* **ref.** ***14***). To avoid circularity, it is advisable to avoid inferring your phylogeny using sequence data for a gene that you have reason to believe may underlie the phenotypes you are analyzing.

Phylogenies should be denoted in the Newick format used by most phylogenetic software. Pairs of sister taxa are grouped in parentheses and separated by commas. Branch lengths for each taxon or clade follow a colon, and a semicolon indicates the end of the tree. Thus, in the Newick format, the phylogeny in **Fig. 1** is "$([2:v_2,3:v_3]:v_4,1:v_1)$." Most comparative methods assume that both the topology and the branch lengths of the phylogeny are known without error. Phylogenies, however, are rarely known with complete certainty. Strategies for dealing with uncertainty in your phylogeny are discussed below (*see* **Note 2**).

3. Methods: Correlation Between rRNA Copy Number and Ecology

Bacterial genomes exhibit great variation in ribosomal RNA (rRNA) operon copy number. Because previous studies had not established a clear link between an organism's operon copy number and its "ecological strategy," Klappenbach et al. *(15)* were interested in testing whether there is a positive association between a strain's rRNA operon copy number and how quickly it could respond to a favorable change in resource availability (coded as early or late colony appearance on a plate). For the 24 strains examined (**Table 1**), the authors found that early-appearing strains tended to have more operon copies per genome (3–10 copies) than did late-appearing strains (1–3 copies), but they did not perform a statistical test to evaluate whether there was a positive correlation between these two traits because of the likely non-independence of their data. Below we use GLS to test whether there is a significant relationship between these traits:

1. Go to the COMPARE website (http://compare.bio.indiana.edu), run COMPARE and then open the main window.
2. To perform correlation analysis, select "PGLS-Relationships" as the method.
3. Enter the names of the organisms in your dataset as a column in the "Taxon Names" text box, either directly or by copying-and-pasting from another application. In our example, these are the 24 strains listed in **Table 1**.
4. Enter the phenotypic data in the "Taxon Means" text box. Data for each trait should be entered as a separate column followed by an estimate of within-taxon variation (standard error), enclosed in triangle brackets <>. If you have an estimate of this variation (e.g., the standard error of measurements obtained for sub-clones of a strain), it can be incorporated by COMPARE into the variance components (diagonal elements) of the error term of the GLS model ***(3)***. This capacity for combining different sources of error into the analysis highlights the flexibility of GLS.

   Unlike traits such as growth rate or enzymatic activity, the data of Klappenbach et al. ***(15)*** are not continuously distributed, although they are discrete for different reasons. While a strain's number of rRNA operons must take an integer value, its response to replenished resources is an underlying continuous variable (growth rate) coded as one of two categories (early/late). GLS can accommodate discrete data because, although they cannot evolve by Brownian motion like continuously distributed traits, their error terms can, and that is what matters for the analysis ***(1)***. While data for continuous traits can simply be entered as is, dummy variables must be created before discrete traits can be incorporated into the GLS model. In our example, let Z_1 be a dummy variable representing growth response, with values $Z_1 = 0$ if the observation is from a late colony former and $Z_1 = 1$ if from an early colony former (**Table 1**). For rRNA operon copy number, 12 of the strains have either 1 or 2 copies, while the other 12 have greater than 2. We assign the following values to dummy variable Z_2: $Z_2 = 0$ if the strain has 1 or 2 copies; $Z_2 = 1$ if the strain has more than 2 copies (*see* **Note 3**). We assume that there is no within-taxon variation in these traits. The data for growth response and operon number should therefore be entered in the "Taxon Means" box as shown below:

   1 <0> 1 <0>
   0 <0> 0 <0>
   0 <0> 0 <0>
   0 <0> 0 <0>
   0 <0> 0 <0>
   0 <0> 1 <0>
   and so on.

Table 1
Discrete Comparative Data Coded as Dummy Variables for Correlation Analysis

Strain	Z_1	Z_2
HF3-S21027	1	1
HS5-S24542	0	0
KBS-LC2	0	0
HS7-S24561	0	0
KBS-LC8	0	0
KBS-LC1	0	1
HF2-S21012	1	1
KBS-EC4	1	1
HF5-S21125	1	1
HS2-S24520	0	0
HS6-S24556	0	0
HS1-S23321	0	0
HS3-S24538	0	0
HS4-S24541	0	0
KBS-EC2	1	1
KBS-LC6	0	0
KBS-LC4	0	0
KBS-LC9	0	0
KBS-EC3	1	1
HF1-S21003	1	1
KBS-EC5	1	1
HF6-S22243	1	1
KBS-EC1	1	1
HF4-S21032	1	1

Adapted with permission from Klappenbach et al. ***(15)***.

Enter the number of taxa and traits for your dataset in the appropriate text boxes. In our example these are 24 and 2, respectively.

5. Enter the number of phylogenies to be used in your analysis in the "Number of phylogenies" box. In our example this is 1. Partial 16S rRNA gene sequences (approx 470 bp) for the 24 strains were obtained from GenBank, and a maximum likelihood phylogeny was reconstructed for the aligned sequences using a general time-reversible model with *Aquifex pyrophilus* as an outgroup. Enter the phylogeny in the text box under "Enter phylogeny below":

((((KBS-EC3:0.0,KBS-EC5:0.002635):0.00000001,HF1-S21003:0.0):0.029073,HF6-S22243:0.043198):0.066069,((KBS-EC1:0.072848,HF4-S21032:0.053723):0.122946,((((HS1-S23321:0.00000001,(HS6-S24556:0.002806,HS2-S24520:

0.014190):0.008496):0.058124,(HS3-S24538:0.047107,HS4-S24541:0.080642): 0.041662):0.047884,(HF3-S21027:0.138372,(((KBS-EC4:0.015445,HF2-S21012: 0.0):0.021123,HF5-S21125:0.002105):0.073557,(HS5-S24542:0.049863,((KBS-LC1:0.008264,KBS-LC2:0.079047):0.015862,(KBS-LC8:0.046871,HS7-S24561: 0.028210):0.009285):0.060324):0.034712):0.081412):0.048808):0.034106,((KBS-EC2:0.099980,KBS-LC9:0.068839):0.059236,(KBS-LC4:0,KBS-LC6:0.002484): 0.095889):0.212395):0.076194):0.071234);

While this phylogeny is generally well supported by bootstrap analysis and agrees well with trees reconstructed with other methods, it is still a hypothesis which is not known with complete certainty. For starters, note that there were two unresolved polytomies (multifurcations with more than two branches radiating from a single internal node) that were "resolved" by setting the affected branch lengths equal to 10^{-8}. We do not know whether these multifurcations were due to insufficient data (i.e., a soft polytomy) or historical reality (i.e., a hard polytomy). Regardless, comparative methods, including phylogenetic GLS, generally require that your phylogeny be completely resolved into a bifurcating tree. If there are polytomies in your tree, one way to get around this problem is to set internal branches estimated to have a length equal to zero to a positive but very small number. Although this solution is arbitrary, it does not affect either the variance–covariance matrix or the parameter estimates of the model ***(8,16)***. The more uncertainty there is in your phylogeny, the more important it becomes to consider incorporating this source of error into your model. A number of approaches for dealing with phylogenetic uncertainty have been proposed, but which strategy is best will depend upon the quality of your phylogenetic information. This is discussed in more detail below (*see* **Note 2**).

6. Choose "Execute." To implement a Brownian-motion model of evolution, select "use alpha" and set it equal to zero.

The output from COMPARE includes a summary of the phylogeny and comparative data followed by statistics for the GLS regression. These include estimates of the intercept and slope and their errors, an estimate of how much of the variation in the data is explained by the model (R^2), its log likelihood, and the estimate of the correlation coefficient (R). For the example data, the slope of the regression of growth response on operon copy number was significantly different from zero (with a mean ± standard error of 0.40 ± 0.111), which provides statistical support that the two traits are evolving in a correlated fashion. The estimated correlation coefficient was 0.61.

Redo the analysis, this time implementing an exponential model by estimating the value of the restraining force alpha. Although we generally do not know whether a linear or an exponential model is more appropriate for describing the phenotypic dependence of our dataset, virtually all evolutionary models can be described by one or the other (but *see* **Note 4** for discussion of an alternative model). For this reason I suggest you run both models. However, as mentioned under **Subheading 1.2.** above, the exponential model does appear to be a better general purpose model in that it is quite robust to violations of its assumptions about the evolutionary process underlying phenotypic change. Therefore, if you obtain markedly different results with the respective models, I would favor the results of the exponential model. That said, parameter estimates appear to be more accurate under the Brownian-motion model when its assumptions are not seriously violated ***(12)***, so one can achieve better estimates with this model provided that the value of alpha is estimated to be low. Because alpha is scaled to tree length, however, what constitutes a low value of this parameter is phylogeny dependent. If the sum of all branch lengths is equal to one, for example, then alpha can be removed from the model if its estimated value is lower than approx 2.5.

For the regression of growth response on operon copy number, the COMPARE output shows that the maximum likelihood estimate of alpha was 7.8, a moderate constraint on phenotypic evolution. Like the Brownian-motion model, the slope of the regression was significantly different from zero (0.75 ± 0.100), and there was a strong positive correlation between the two traits (0.85). One reason why the exponential model may provide a good fit is because the data are categorical and are constrained from diverging freely. Under a common model of binary trait evolution, in which there is a constant probability of reversibly changing from one state to the other over any time interval, phenotypic similarity indeed declines exponentially with increasing evolutionary distance ***(17)***.

4. Notes

1. Consider the familiar simple least squares regression model:

$$Y_i = \beta_0 + \beta_1 X_i + \varepsilon_i \qquad i = 1, \ldots, n$$

In matrix form, this is:

$$\mathbf{Y} = \mathbf{X}\boldsymbol{\beta} + \boldsymbol{\varepsilon}$$

where **Y** is the vector of the n observations of the dependent variable Y_i:

$$\begin{bmatrix} Y_1 \\ Y_2 \\ \vdots \\ Y_n \end{bmatrix}$$

β is the vector of regression coefficients:

$$\begin{bmatrix} \beta_0 \\ \beta_1 \end{bmatrix}$$

X is the matrix:

$$\begin{bmatrix} 1 & X_1 \\ 1 & X_2 \\ \cdot & \cdot \\ \cdot & \cdot \\ \cdot & \cdot \\ \cdot & \cdot \\ 1 & X_n \end{bmatrix}$$

which, when multiplied by **β**, yields the expected values of $Y_i = \beta_0 + \beta_1 X_i$. The coefficients of the model are fit by minimizing the sum of the squared deviations of the data points from the regression line.

ε is the error term vector, with ε_i expected to be zero for all observations. The standard least squares regression model also assumes that these error terms are normally distributed, independent, and have equal variance σ^2. These latter two assumptions are captured in the variance-covariance matrix, for which diagonal elements of the matrix (the variances) are σ^2 and the off-diagonal elements (the covariances) are equal to zero (because of the model's assumption that the errors of different observations are independent). This model is a special case of the generalized least squares (GLS) model in which (1) error terms of different observations may have different variances and therefore have different values along the diagonal; and (2) pairs of error terms may be correlated, resulting in off-diagonal elements greater than zero.

2. For the worst-case scenario of having no phylogenetic information for your study organisms, randomized phylogenies can be generated by computer simulation ***(18)***. Topologies are simulated according to a Markovian branching process of speciation (e.g., **ref.** ***19***), with branch lengths specified by either the branching process or a coalescent model ***(19)***. GLS analysis of the comparative data can then be conducted on all of the sample trees to obtain a distribution of estimated correlation coefficients. The spread of this distribution is a measure of confidence in the mean that takes into account the uncertainty in the phylogeny. To simulate random trees in COMPARE, select the "Generate trees or data" button in the main window. Choose some number of trees to generate (e.g., 1000) without constraint (the default option) for the number of taxa in your dataset by filling in the appropriate text boxes. Select "Generate" to initiate the simulation and then transfer the results to the main window for analysis using the "Replace" button.

   In the event that your phylogeny is partly resolved (e.g., you know that a subset of the taxa in your dataset form a clade), a distribution based on completely randomized trees will overestimate the degree of phylogenetic uncertainty and yield confidence intervals for your parameter estimate that are too conservative. COMPARE allows you to constrain the random sampling of phylogenies to a subset of ordered topologies according the algorithm of Housworth and Martins ***(20)***. To do this, choose "Generate trees or data." Select "With constraint," at which point COMPARE will prompt you to enter your available topological information in the standard Newick format in the "Constraints" text box. Proceed as above to generate a random sample of phylogenies from this subset of all possible phylogenies.

If, on the other hand, your phylogeny is mostly resolved but lacking, say, in strong bootstrap support at a minority of nodes, then another approach that can be used is to run a correlation analysis on each of the trees inferred from the bootstrapped molecular data and then obtain confidence intervals from the resultant distribution of coefficient estimates ***(21)***. This would be another way to resolve the polytomies in the example phylogeny.

3. One is not restricted to two values when using dummy variables to introduce discrete data into regression models. In our example, there are $N = 7$ strain groups based on number of rRNA operons ***(15)***. Rather than code the data as two groups one could choose to incorporate the exact number of operons a strain possessed into the model by creating multiple dichotomous dummy variables. For $N = 7$ groups, this requires $N - 1 = 6$ dummy variables Z_2, Z_3, Z_4, Z_5, Z_6, and Z_7. Strains with one copy would be coded (Z_2, Z_3, Z_4, Z_5, Z_6, Z_7) = (0,0,0,0,0,0), those with two copies (1,0,0,0,0,0), those with three (0,1,0,0,0,0), and so on. Care should be taken, however, when choosing how to code the data, particularly when datasets are not large, as is often the case with comparative data. This is because more complex models can lead to poor parameter estimation and misleading results.
4. Consider the possibility that acquiring an additional rRNA operon via a tandem duplication event immediately allows the variant to respond more quickly to favorable changes in the environment than its ancestor. This would reflect a model of evolution in which trait changes occur at "speciation events." In such a case the expected phenotypic correlation between taxa decreases with the proportion of shared speciation events ***(9)***. To analyze your data with this model, proceed as before, but this time enter "0" in the main window box following "Raise all branch lengths to the power of," thereby setting all branch lengths of the phylogeny equal to 1. For our example, we again find evidence for a highly significant positive correlation between the two traits ($R = 0.83$).

Acknowledgments

I would like to thank Emília Martins for introducing me to phylogenetic comparative methods and for her helpful comments on a draft of this manuscript.

References

1. Martins, E. P. and Hansen, T. H. (1996) The statistical analysis of interspecific data: a review and evaluation of phylogenetic comparative methods, in *Phylogenies*

and the Comparative Method in Animal Behavior (Martins, E. P., ed.), Oxford University Press, Oxford, England, pp. 22–75.

2. Grafen, A. (1989) The phylogenetic regression. *Phil. Trans. R. Soc. Lond. B* **326,** 119–157.
3. Martins, E. P. and Hansen, T. F. (1997) Phylogenies and the comparative method: a general approach to incorporating phylogenetic information into the analysis of interspecific data. *Am. Nat.* **149,** 646–667.
4. Martins, E. P. (2001) COMPARE: Statistical analysis of comparative data. Computer programs distributed by the author via http://compare.bio.indiana.edu.
5. Grafen, A. (1992) The uniqueness of the phylogenetic regression. *J. Theor. Biol.* **156,** 405–423.
6. Edwards, A. F. W. and Cavalli-Sforza, L. L. (1964) Reconstruction of evolutionary trees, in *Phenetic and Phylogenetic Classification* (Heywood, W. H. and McNeill, J., eds.), *Syst. Assoc. Publ. No. 6,* London, pp. 67–76.
7. Felsenstein, J. (1985) Phylogenies and the comparative method. *Am. Nat.* **125,** 1–15.
8. Rohlf, F. J. (2001) Comparative methods for the analysis of continuous variables: geometric interpretations. *Evolution* **55,** 2143–2160.
9. Hansen, T. F. and Martins, E. P. (1996) Translating between microevolutionary process and macroevolutionary patterns: the correlation structure of interspecific data. *Evolution* **50,** 1404–1417.
10. Lande, R. (1976) Natural selection and random genetic drift in phenotypic evolution. *Evolution* **30,** 314–334.
11. Martins, E. P. and Garland, T., Jr. (1991) Phylogenetic analyses of the correlated evolution of continuous characters: a simulation study. *Evolution* **45,** 534–557.
12. Martins, E. P., Diniz-Filho, A. F., and Housworth, E. A. (2002) Adaptive constraints and the phylogenetic comparative method: a computer simulation test. *Evolution* **56,** 1–13.
13. Oakley, T. H. and Cunningham, C. W. (2000) Independent contrasts succeed where ancestor reconstruction fails in a known bacteriophage phylogeny. *Evolution* **54,** 397–405.
14. Swofford, D. L., Olsen, G. J., Waddell, P. J., and Hillis, D. M. (1996) Phylogenetic inference, in *Molecular Systematics* (Hillis, D. M., Moritz, C., and Mable, B. K., eds.), Sinauer Associates, Sunderland, MA, pp. 407–514.
15. Klappenbach, J. A., Dunbar, J. M., and Schmidt, T. M. (2000) rRNA operon copy number reflects ecological strategies of bacteria. *Appl. Environ. Microbiol.* **66,** 1328–1333.
16. Purvis, A. and Garland, T., Jr. (1993) Polytomies in comparative analyses of continuous characters. *Syst. Biol.* **42,** 569–575.
17. Martins, E. P. and Hansen, T. F. (1996) A microevolutionary link between phylogenies and comparative data, in *New Uses for New Phylogenies* (Harvey, P. H., Leigh Brown, A. J., Maynard Smith, J., and Nee, S., eds.), Oxford Univ. Press, Oxford, pp. 273–288.
18. Martins, E. P. (1996) Conducting phylogenetic comparative studies when the phylogeny is not known. *Evolution* **50,** 12–22.

19. Kingman, J. F. C. (1982) The coalescent. *Stochastic Processes and their Applications* **13,** 235–248.
20. Housworth, E. A. and Martins, E. P. (2001) Random sampling of constrained phylogenies: conducting phylogenetic analyses when the phylogeny is partially known. *Syst. Biol.* **50,** 628–639.
21. Felsenstein, J. (1988) Phylogenies and quantitative characters. *Ann. Rev. Ecol. Syst.* **19,** 445–471.

34

Conjugated Linoleic Acid Detection Produced by Dairy Starter Cultures

Carina Van Nieuwenhove, Silvia N. González, and Alicia Bardón

1. Introduction

Conjugated linoleic acids (CLA) are a group of octadecadecenoic acids containing conjugated double bonds that are separated by one single bond. CLA has gained considerable attention because it is a natural component in foods with anticarcinogenic and antioxidative properties. Other properties (antidiabetic and immunomodulating) have been attributed to CLA.

These isomers of linoleic acid (LA) are formed by rumen bacteria *(1)*, of which *Butyrivibrio fibrisolvens* *(2)* is the most widely known. Some of these bacteria have a linoleic acid isomerase, which catalyzes the isomerization of linoleic acid into CLA. Of the various isomers of CLA, the *cis-9, trans-11* and the *trans-10, cis-12* isomers are considered to be the most effective. The conversion reaction has been studied both in growing bacterial cultures and washed-cell suspensions *(3)*. CLA can be found naturally in dairy products and meat of ruminant animals. Many authors report higher levels of CLA in milk products than in unprocessed milk *(4,5)*. Different dairy starter cultures have been tested for CLA formation for use in the dairy industry. In order to examine the isomerizing effects of bacteria on LA, many methods of detection have been tested, but most are difficult to carry out. In this chapter, we describe a simple method available for extraction and detection of CLA produced by different dairy starter cultures. The procedure used for lipid extraction has a very good yield for long-chain fatty acids.

From: *Methods in Biotechnology: Environmental Microbiology: Methods and Protocols*
Edited by: J. F. T. Spencer and A. L. Ragout de Spencer

2. Materials

1. Chloroform.
2. Methanol containing no more than 0.5% w/w water.
3. Pure HCl.
4. NaOH.
5. *n*-Hexane.
6. Anhydrous Na_2SO_4.
7. LAPTg: 15 g/L peptone, 10 g/L triptone, 10 g/L yeast extract, 10 g/L glucose, 1.08 g/L Tween-80; pH 6.5 ± 0.2. Sterilize in autoclave at 121°C for 15 min ***(6)***.
8. Tween 80 (polyoxyethylene sorbitan monooleate, Sigma).
9. Linoleic acid.
10. Conjugated linoleic acid.
11. LA supplementation: different LA concentrations (0–1000 μg/mL) were added to aqueous solution in 1% Tween-80 (*see* **Note 1**).
12. Saline buffer: 0.8% w/v NaCl, 0.121% K_2HPO_4, and 0.034% KH_2PO_4, pH 7.2 (*see* **Note 2**).

3. Methods

3.1. Bacterial Strains

Propionibacteria, Lactobacilli, and Lactococci strains were grown in LAPTg medium (*see* **Note 3**) with linoleic acid supplementation. Each test organism was inoculated to a density of 10^6 organisms/mL. All strains were grown until stationary phase.

3.2. Medium Preparation

1. 10-mL Aliquots of medium with added LA (0–1000 μg/mL) were autoclaved in test tubes at 121°C for 15 min.
2. After inoculation with 1% of culture (v/v), the culture was incubated under conditions appropriate for each specific bacterium until stationary phase, i.e., at 37°C for lactobacilli (20 h), and 20°C for lactococci (72 h) and propionibacteria (72 h). Then isomerase activity was measured by CLA detection using gas chromatography (*see* **Note 4**).

3.3. Lipid Extraction

1. Samples of bacterial suspensions were centrifuged at 2500*g*. The bacterial pellets were washed twice with saline buffer. The bacterial cells and supernatant fluid were analyzed for extracellular and intracellular fatty acid composition, respectively (*see* **Note 5**).
2. Following incubation, the lipids (*see* **Note 6**) of 5 mL of medium or bacteria suspension were extracted with 15 mL of chloroform/methanol 2:1 (v/v) mixture (*see* **Note 7**).
3. The lower layer was collected and evaporated with a rotary evaporator at 35°C for 15 min.

3.4. Lipid Hydrolysis

The residue was saponified with 6 mL of 0.9% NaOH (*see* **Note 8**) in methanol at 100°C for 20 min, then cooled to room temperature.

3.5. Preparation of Fatty Acid Methyl Esters

1. Free fatty acids were methylated with 5 mL of 4% chloridric acid in methanol at 60°C for 30 min *(7)*.
2. The methylated sample was mixed with 4 mL of hexane:water 1:1 v/v (*see* **Note 7**) and centrifuged at 500*g* for 10 min at 4°C.
3. The hexane layer was dried with anhydrous Na_2SO_4 and then kept under nitrogen at room temperature. The residue was redissolved in 1 mL of hexane for further quantification.

3.6. Standards Methylation

LA and CLA standards (99% pure; Sigma) were methylated as described above (*see* **Notes 10** and **11**).

3.7. Gas Chromatography (GC) Analysis

1. GC analysis was carried out by using an Agilent Technologies Model 6890N gas chromatograph equipped with a flame ionization detector and a model 7683 automatic injector with a 10 µL syringe.
2. 1 µL of fatty acid preparation was injected into a HP 5 column (30 m × 0.25 mm id × 0.25 µm of thickness). GC conditions: injector temperature 250°C; initial oven temperature 50°C (increased to 150°C at 20°C/min and held for 50 min, then increased to 225°C at 10°C/min and held for 20 min); and detector temperature, 250°C. Nitrogen was used as the carrier gas.
3. Fatty acids were identified by a comparison of the retention times with the methylated standards. The areas of LA and CLA peaks were calculated as mg CLA or LA per 10 mL medium.

4. Notes

1. It is not recommended to add more than 1000 µg/mL of LA to the medium due to its growth-inhibiting effect.
2. All solutions must be prepared with distilled or demineralized water or water of equivalent purity.
3. Other media can be used, depending on selected strain.
4. The major production of CLA is observed during growing phase.
5. Most of the CLA (98%) is detected in supernatant fluid.
6. This method of lipid extraction is for long-chain fatty acids (C 12–C 20).
7. All solvents used were HPLC grade.
8. 0.9% NaOH in methanol can be replaced by 0.9% KOH in methanol.
9. Special precautions must be taken when handling organic solvents.
10. Keep fatty acids out of light and air to prevent degradation.

11. Using this method, LA and CLA recovery percentages are 97% and 98%, respectively.

References

1. Harfoot, C. G. and Hazlewood, G. P. (1997) Lipid metabolism in the rumen, in *The Rumen Microbial Ecosystem* (Hobson, P. N. and Stewart, C. S., eds.), Blackie Academic and Professional Press, London, pp. 383–426.
2. Kepler, C. R. and Tove, S. B. (1967) Biohydrogenation of unsatured fatty acids. *J. Biol. Chem.* **242,** 5686–5692.
3. Rainio, A., Vahvaselkä, M., Suomalainen, T., and Laakso, S. (2001) Reduction of linoleic acid by *Propionobacterium freudenreichii* ssp. *Shermanii. Can. J. Microbiol.* **47,** 735–740.
4. Ha, Y. L., Grimm, N. K., and Pariza, M. W. (1989) Newly recognized anticarcinogenic fatty acids: identification and quantification in natural and processed cheeses. *J. Agric. Food Chem.* **37,** 75–81
5. Aneja, R. P and Murthi, T. N. (1991) Beneficial effects of ghee. *Nature* **350,** 280.
6. Raibaud, P., Caulet, M., Galpin, J. V., and Mocquot, G. (1961) Studies on the bacterial flora of the alimentary tract of pigs II. Streptococci: selective enumeration and differentiation of the dominant group. *J. Appl. Bacteriol.* **24,** 285–291.
7. Chin, S. F., Liu, W., Storkson, J. M, Ha, Y., and Pariza, M. W. (1992) Dietary sources of conjugated dienoic isomers of linoleic acid, a newly recognized class of anticarcinogens. *J. Food Compos. Anal.* **5,** 185–197.

IV

Reviews

35

Techniques for Manipulating the Bacterial Endophyte *Bacillus mojavensis*

Charles W. Bacon and Dorothy M. Hinton

1. Introduction

Bacterial endophytes actively colonize plant tissues and establish long-term associations, actually lifelong natural associations. The endophyte presumably benefits from inhabiting the plant interior because it is a protected niche in which there is relatively little competition from other microorganisms for a reliable source of nutrition. Thus, an endophyte is a microbe that lives in association with plants, forming a symptomless biotrophic association (**Fig. 1**). However, not all endophytes are exclusively intercellular, as some form intracellular associations as well, e.g., inside phloem and/or xylem tissues. Endophytes may colonize plants either exclusively below ground or above ground or both. Endophytic bacteria are distinguished from transient visitors, usually dormant or latent bacterial infections, which form associations as happenstances that will not survive long. The term *bacterial endophytes,* as used here, encompasses the broader use of the narrowly defined term of Kado *(1)* to include that of Quispen *(2)*, which defines bacterial endophytes as those intercellular bacteria that impart an ecological benefit to the plant without doing substantive harm to the plant. While this definition is restrictive, it conveys the concepts useful in defining fungal endophytes *(3)*, and will serve as the primary focus for readers of this review as that of using endophytes as biocontrol agents. Inherent in the definition of endophytic bacteria is the concept of a symptomless, nondisease-producing infection that results in a series of interactions that range from no effects on the hosts (neutralism), to beneficial to the hosts and bacteria (mutualism), or to benefiting only one member (commensalism). We base these interactions on the terminology used for fungal endo-

From: *Methods in Biotechnology: Environmental Microbiology: Methods and Protocols*
Edited by: J. F. T. Spencer and A. L. Ragout de Spencer

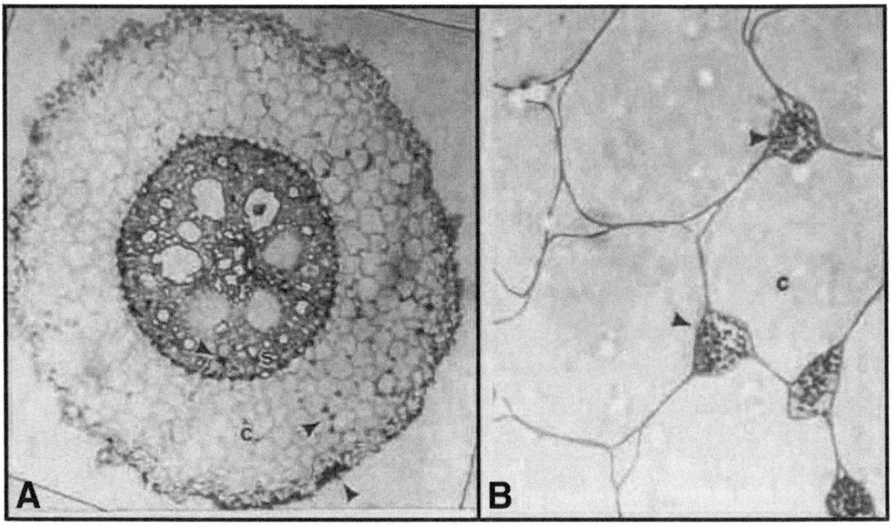

Fig. 1. Endophytic habit of *Bacillus mojavensis* in corn roots. **A,** Light micrograph (40X) showing a cross-section of root cells showing the distribution of bacteria both within the intercellular spaces of the cortex, c, and the steel, s. **B,** Light micrograph (100X) of corn plant showing the distribution of aniline blue stained bacterial cells of *B. mojavensis* within the intercellular spaces of root tissue (arrows), c, cell.

phytes and their quantitative and qualitative relationships with grasses ***(3)***. However, the specific interactions that would apply to *B. mojavensis* specifically, and other bacterial endophytes in general, have not yet been established.

The definition of endophytic bacteria used here purposely excludes latent and/or pathogenic bacteria and the resulting negative pathological, epidemiological, and physiological effects that these bacteria produce in plants. Also excluded are those bacteria whose interactions with plants are restricted, i.e., epiphytic and rhizoplane or phyllosphere colonizers and dormant or latent stages of endophytic colonizers, as well as diazotrophic bacteria that form root and stem nodules, such as species of *Rhizobium, Bradyrhizobium,* and so on. However, the diazotrophic endophytic species that are included are those that do not form nodules, and these include the species of *Azorhizobium, Azospirillum,* and others.

Bacterial endophytes are rapidly becoming a distinct group of biocontrol organisms, as indicated by the recent increases in publication numbers that reflect an interest in their potential benefits to agriculture. However, endophytic bacteria that did not produce diseases were described well over 50 yr ago ***(4,5)***.

Endophytic bacteria show a tremendous amount of diversity not only in plant hosts but also in bacterial taxa ***(6,7)***. Usually, endophytic bacteria are represented by specific strains, and these include Gram-positive and -negative species. However, there are some species that apparently have evolved along with plants, exploiting the endophytic habit. We have discovered one such species of desert-dwelling bacteria, *Bacillus mojavensis*. All isolates of this species tested to date readily form endophytic associations with plants, and several of these have been reported to be successful in preventing disease development ***(8–10)***. One isolate of *B. mojavensis* has been used to demonstrate for the first time that a bacterial endophyte can be used to reduce the *in planta* accumulation of a mycotoxin produced by an endophytic fungus ***(11)***. Endophytic bacteria can also be used as transformed vectors for other useful products that can be expressed *in planta,* augmenting either the natural vitamin and other value-added components of food crops, as well as a natural pesticide vector. Such surrogate-transformed plants might have greater value and public acceptability than transgenic plants.

Endophytic bacteria intended for use in biotechnology must have specific natural requirements relative to plants and potential uses, which include biocontrol agents, delivery systems for specific pest-protecting genes or their products, and providing plant growth–enhancing traits. The first requirement is that the endophyte must not have the potential to induce plant disease. A second requirement is that the endophyte must be capable of being distributed within the plant, especially at locations where pests are expected to reside. Equally important, and in the absence of a uniform distribution, is that the delivery system of the substance being used to control or deter pests must be translocated from the bacterium through the plant and to the target. The third requirement is that the organism must be equally transformable and easily manipulated genetically.

Since the discovery of the *B. mojavensis* group of organisms is relatively recent, there is scant knowledge concerning the specifics for genetic manipulations, and molecular modifications of a natural group of bacteria that is a self-contained, plant protected and nourished, and self-regulated system. Thus, isolates of this group fulfill the above three requirements for usefulness in plant health and protection. The objective of this chapter is to provide both a brief review of bacterial endophytes, and present useful methodologies and basic tools that can be used to exploit *B. mojavensis* at the molecular level for a greater impact on emerging technologies using an endophytic bacterium as a biocontrol, delivery system, plant-growth promoter, and plant-health enhancer. In general, some of the techniques presented will also prove suitable for other species of endophytic bacteria.

2. Materials: The Bacterium

The species of bacteria that have been established as endophytic are numerous, and include species within the genera *Pseudomonas* (fluorescent pseudomonads), *Burkholderia, Bacillus, Erwinia, Clavibacter, Xanthomonas, Agrobacterium, Serratia, Stenotrophomonas, Phyllobacterium, Enterobacter, Corynebacterium, Micrococcus, Flavobacterium,* and others ***(6,12–16)***. These bacterial endophytes have been used as biological controls for pathogens ***(6,17–20)***, medicinal qualities ***(21)***, other pathogenic and mycotoxic fungi ***(10,22)***, and for the control of bacterial pathogens such as the use of *Pseudomonas fluorescens* in the control of several pathogens of cotton, cucumber, and pea other offer bacterial pathogens ***(23–25)***. Additional uses of bacterial endophytes include control of plant-parasitic cyst and root-knot nematodes ***(7,26)***. Further, most of the genera indicated previously have a restricted endophytic colonization that is usually limited to the roots, or subcuticle areas of roots, and are not distributed into the root cortex or foliage and stem areas. True bacterial endophytes that would offer the most as biocontrol agents should colonize root, stem, and leaves, although their distribution among these organs may not be the same.

Outstanding strains among the bacterial species used as biological controls, and not necessarily as endophytes, include several strains of *Bacillus subtilis. Bacillus subtilis* (Ehrenberg) Cohn is the oldest, and the nomenclatural type for the Bacillaceae and the genus *Bacillus*. This Gram-positive, spore-forming bacterium has proven safe over many years as a nonpathogenic species, and is consumed in ton quantities in several human food preparations. The bacterium is widely distributed in nature and has been isolated from several botanical environments, primarily the soil, where it has been shown to have antibiotic properties and biocontrol potentials. A strain of *B. subtilis* was recently reported ***(27)*** to be endophytic, intercellular, and nonpathogenic, to enhance plant growth, and to protect plants against fungi ***(22)***. This strain was patented (Patent Number 5,994,117; ATCC 55732) as a biocontrol for diseases of maize caused by fungi ***(28)***. This isolate was subsequently found to belong to the closely related *B. subtilis*-like phenotypes that were recently described as *Bacillus mojavensis* ***(29)*** (**Fig. 2**). The discovery of the patented isolate initiated the search for additional endophytic strains and spurred experimental studies of the species in general.

The type and other strains of *B. mojavensis* were isolated from soil samples from the Mojave Desert in California, and other major deserts of the world, respectively (**Table 1**). This bacterium is distributed throughout the axis of the corn plant, although it is found in the roots in a much greater density than other parts of the plants, and found in equal frequencies in leaves and stems ***(9)***. A similar distribution is observed for wheat, and it is assumed that *B. mojavensis* is distributed in most plant organs. All isolates of this species were subsequently shown to be endophytic, nonpathogenic to plants, antagonistic to fungi, and to enhance plant growth ***(9)***. Thus, we have discovered a species with a

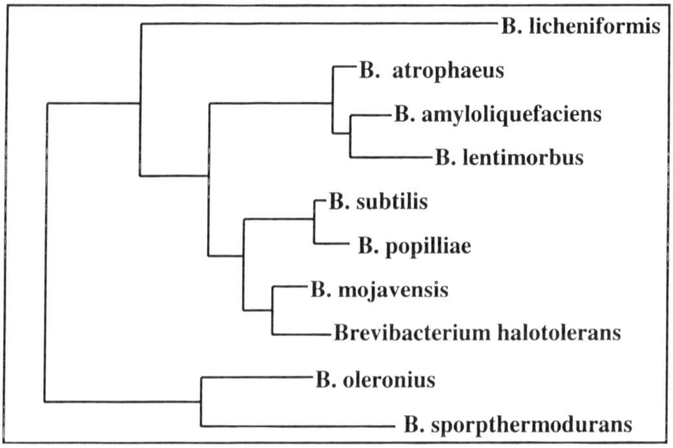

Fig. 2. Neighbor joining tree for species within the *Bacillus* group, indicating that they are very closely related to each other by 16S rRNA gene sequence analysis *(29)*.

natural unique biocontrol habit. Its occurrence in desert soil suggests that it is a natural endophyte of desert plants, since very little growth and colonization can occur in the desert soils used to isolate these strains, which were isolated from near-surface soil depths of 10 to 20 cm *(29)*.

2.1. Culture Media

This bacterium is easy to culture on Difco (Detroit, MI) nutrient agar, which can also be used to isolate the bacterium from tissue, and to demonstrate fungal antagonism (see below). Other agar media found suitable for the isolation of *B. mojavensis* include Difco tryptose blood agar base *(29)*, J-agar (5 g tryptone, 15 g yeast extract, 3 g K_2HPO_4, 20 g agar, 2 g glucose, 1000 mL distilled water, pH 7.3 to 7.5) *(37)*. The J-agar is recommended for culturing the fastidious *Bacillus* species used as biocontrols for insects, and it produces rapid growth of *B. mojavensis* strains. Several strains of *B. mojavensis* are apparently nutritionally fastidious, since they grow slowly on other media, especially nutrient broth. Nevertheless, bioassays designed to determine fungal antagonism should be conducted on nutrient agar, as all isolates produce antagonistic substances on it, and because the zone of inhibition is far larger than that observed on other media. This suggests that there is more production of inhibitory substances on nutrient agar than on other media.

2.2. Initial Isolation

A working definition for bacterial endophytes includes bacteria that can be isolated from surface disinfested plant tissue that shows no evidence of pathogenicity. Such symptomless infections formed the basis for the early observa-

Table 1
A Listing of the World's Strains and Mutants of *Bacillus mojavensis* According to Roberts et al. *(29)*

Bacillus mojavensis strain*	Desert or collection site	Collector
NRRL B-14698^T	Mojave Desert, CA	F. M. Cohan
RRC 101 (ATCC 55732)	Northern Italy, maize kernel	C. W. Bacon
RRC 111	Transformant of RRC 101	C. W. Bacon
RRC 113	UV nonantagonistic mutant	C. W. Bacon
RRC 114	UV nonantagonistic mutant	C. W. Bacon
NRRL B-14699	Mojave Desert, CA	F. M. Cohan
NRRL B-14700	Mojave Desert, CA	F. M. Cohan
NRRL B-14701	Mojave Desert, CA	F. M. Cohan
NRRL B-14702	Mojave Desert, CA	F. M. Cohan
NRRL B-14703	Gobi Desert, People's Republic of China	M. S. Roberts
NRRL B-14704	Gobi Desert, People's Republic of China	F. M. Cohan
NRRL B-14705	Gobi Desert, People's Republic of China	F. M. Cohan
NRRL B-14706	Gobi Desert, People's Republic of China	M. S. Roberts
NRRL B-14707	Gobi Desert, People's Republic of China	F. M. Cohan
NRRL B-14708	Gobi Desert, People's Republic of China	F. M. Cohan
NRRL B-14709	Gobi Desert, People's Republic of China	M. S. Roberts
NRRL B-14710	Gobi Desert, People's Republic of China	F. M. Cohan
NRRL B-14711	Gobi Desert, People's Republic of China	F. M. Cohan
NRRL B-14713	Gobi Desert, People's Republic of China	M. S. Roberts
NRRL B-14714	Sahara Desert, Nefta, Tunisia	M. S. Roberts
NRRL B-14712	Gobi Desert, People's Republic of China	F. M. Cohan
NRRL B-14715	Sahara Desert, Nefta, Tunisia	M. S. Roberts
NRRL B-14716	Sahara Desert, Nefta, Tunisia	M. S. Roberts
NRRL B-14718	Sahara Desert, Nefta, Tunisia	F. M. Cohan
NRRL B-14719	Sahara Desert, Nefta, Tunisia	M. S. Roberts
NRRL B-14817	Tumamoc Hill, AZ	K. E. Duncan
NRRL B-14818	Tumamoc Hill, AZ	K. E. Duncan
NRRL B-14824	Sahara Desert, Nefta, Tunisia	M. S. Roberts

*Note: Strain RRC 111 is an endophytic, antagonistic, auxotrophic *(trpC2, thr5)* transformant that was produced by transforming the nonendophytic, nonantagonistic, auxotrophic *(trpC2, thr5)* *B. subtilis* BD 170 (ATCC 33608) with DNA from the endophytic antagonistic *B. mojavensis* RRC101 ***(9)***. Two other strains, RRC 113 and RRC 114, were derived from RRC 111 as random UV-resistant mutants, selected on the basis of not being nonantagonistic to *F. moniliforme* and other fungi.

tions of Pasteur in 1870, as reviewed by Smith ***(30)***, that there were bacteria residing inside plant tissues that were symptomless. Most of the early observations on bacterial endophytes were based on isolations from seeds, ovules, and tubers ***(4,5,30)***. Henning and Villforth ***(31)*** made the early observations that endophytic bacteria resided in stems and leaves, and Mundt and Hinkle ***(32)*** made a survey of plants that demonstrated the diversity of plant species harboring bacterial endophytes. This survey has been extended to include well over 54 genera, representing 129 bacterial species from both monocotyledonous and dicotyledonous plants ***(6,7,14,15,33,34)***.

2.2.1. Plant Tissue Isolation

Demonstration that a bacterial isolate originated from internal organs of plants is exceedingly important according to the definition of endophytic bacteria. The methods used must involve one or more highly concentrated disinfectants. The commonly used disinfectants include sodium hypochlorite (supplied either in commercial bleach or the dry reagent), mercuric chloride, ethyl and methyl alcohols, sulfuric acid, hydrochloric acid, and hydrogen peroxide ***(14,35,36)***. Two or more of these disinfectants are used in sequence, with one or more distilled-water rinses between each treatment. The time period used for each depends on both the nature of the material and the degree of external contamination observed. However, some disinfectants are used for specific plant parts. Roots can be disinfected with commercial bleach only if it is known for certain that the endophyte is present, and if live endophyte recovery is not important. For example, a 1% chloramine-T aqueous solution is recommended to disinfect roots ***(35)***. The plant materials used for recovery of bacteria consisted of roots that were surface disinfested with 1% chloramine-T for 30 min and shoots that were surfaced disinfested with commercial bleach, full strength (5.25% sodium hypochlorite), for 5 min as described above ***(14,35,36)***. Roots disinfected with bleach would readily take up the sodium hypochlorite and become sterilized internally, yielding no bacteria. Disinfected plant parts were cut into smaller (0.5 to 1 cm) sections, placed on rifampin-augmented nutrient agar medium, and incubated for 5–14 d at 25°C.

The presence of bacteria in roots should be used only as a tentative indicator of endophytic infection, because the bacteria can colonize roots through cracks in broken roots and abrasions present on root surfaces and remain in these locations. Thus, bacteria found in roots might not represent true endophytic colonization. Additional criteria for endophytic infection activity are listed in **Subheading 3.**

2.2.2. Soil Isolation

The isolates of *B. mojavensis* were obtained from soil samples collected near the surface of deserts at depth ranging from 10 to 20 cm ***(29)***. Because all except one of the known world's isolates ***(28)*** were collected by Roberts et al. ***(29)***, their procedure is listed here in some detail. All collections were made under aseptic conditions, and each 150-g desert soil sample is heat treated at

80°C for 10 min to kill vegetative cells and induce sporulation *(29)*. Appropriate dilutions are made of the heated soil samples, each is plated out on tryptose blood agar, and all emergent colonies streaked three times for isolation. Preliminary screening is conducted on the isolates as described above for the ability to hydrolyze gelatin, utilize citrate, reduce nitrate, and ferment L-arabinose and mannitol. Any isolate that is positive for all five is grouped into the *B. subtilis* group *(29)*. Additional studies concerning antagonism, molecular typing, and endophytic colonization are performed as described in this chapter.

In addition to soil, it should be noted that additional isolates are possible from desert plants, either fresh or dried material, and the procedure used for isolation from plant material following inoculation is described in **Subheading 3.1.1.** There has not been any survey of desert plants for these bacteria, so their biological role and distribution in desert and other habitats have not been defined.

2.3. Identification

The identification of *Bacillus* species is difficult, and molecular methods *(29,38–40)* are the basis for distinguishing species within this very large, complex genus (**Fig. 2**). Isolates may be identified tentatively to the genus *Bacillus* if they have the ability to hydrolyze gelatin, utilize citrate, reduce nitrate, and ferment l-arabinose and mannitol. Any isolate that is positive for all five of these phenotypic tests probably belongs to the *Bacillus subtilis* group of closely related species (**Fig. 1**), this may include *B. mojavensis* isolates, but additional analysis may be required for proper identification, including molecular analysis. It is important to remember that all isolates of *B. mojavensis* are identical to the biochemical and morphological phenotype of *B. subtilis*.

Strains of *B. mojavensis* can be distinguished from the *B. subtilis* group only by differences in whole-cell fatty-acid composition, resistance to genetic transformation between taxa, and PCR amplification of polymorphic DNA *(29)*. However, we have established that ribotyping of *B. mojavensis* strains with *Eco*RI can also be used both for genetic typing and for species and strain identification (***10***, and unpublished; *see* below). Ribotyping is based on restriction digestion of bacterial chromosomal DNA, in this case with *Eco*RI, followed by Southern hybridization with a ribosomal operon probe, producing a genetic fingerprint, or riboprint pattern, which is automatically identified and characterized by an analytical software from patterns stored in a data bank *(40)*. Thus, a large number of isolates can be screened and patterns compared in a relatively short time. For identification, patterns are compared to a set of reference patterns in an identification database. Sample patterns are characterized by clustering mathematically indistinguishable or identical patterns. Each cluster of sample patterns is referred to as a "ribogroup." Ribogroups are categorized by placement into specific libraries that can be utilized to compare specific ribogroups within a species to each other (**Fig. 3**).

The ribotyping procedure is simple, since all aspects of it are performed with the automated ribotyping system and software that is commercially available (RiboPrinter, Microbial Characterization System, Qualicon/DuPont, Wilmington, DE). As starting material, a few cells from a single colony of *B. mojavensis* or cells directly isolated from plant tissue are sufficient to generate a ribotype pattern. The cells are then subjected to a heat treatment to inactivate nucleases and kill cells. DNA preparation involves lysing cells and cutting or digesting the DNA with a restriction endonuclease. The restriction fragments of DNA are separated by electrophoresis and the DNA immobilized on a membrane. The DNA is hybridized with a labeled rRNA *E. coli* operon probe and processed on the membrane, and the pattern is detected and analyzed with specific statistical software and compared with identification database software resulting in delineation of species or ribogroups (**Fig. 3**).

3. Methods: Bioassay for Endophytism

Successful isolation of *B. mojavensis* from plants is only one criterion used to determine its endophytic status. All suspect isolates must be reintroduced into plants, usually of the same host species, which had no prior infection. The inoculated plants are examined microscopically for the presence of the bacterium, and the bacterium reisolated on media (**Subheading 2.1.**). Noninfected plants can be produced from seeds that have been treated with disinfectants like those indicated above, and the seedling grown under aseptic conditions, which can best be achieved under control culture conditions such as a plant growth room *(28)*. Additional procedures are described in other references designed to study the relationships of endophytic bacteria with plants *(14,22,33,36,42,43)*.

3.1. Endophytic Colonization and Generating Mutants

All experiments to evaluate the ability of *B. mojavensis* strains to endophytically colonize plants should be based on the recovery of a wild type and its spontaneous antibiotic-resistant mutant derived from the parent isolate. The most popular antibiotic used for this purpose is rifampin (or rifampicin), which readily produces spontaneous rifampin-resistant mutants (Rif^r, encoded by *rpo*B). Rifampin is used at a concentration of 100 or 50 mg/L of medium. Rifampin resistance has been combined with other marker genes such as the *lac*ZY genes in the fluorescent pseudomonads, and with chromosomally inserted kanamycin resistance in several *Erwinia* species. These combinations have worked for most soil bacteria, but as discussed below, there might be

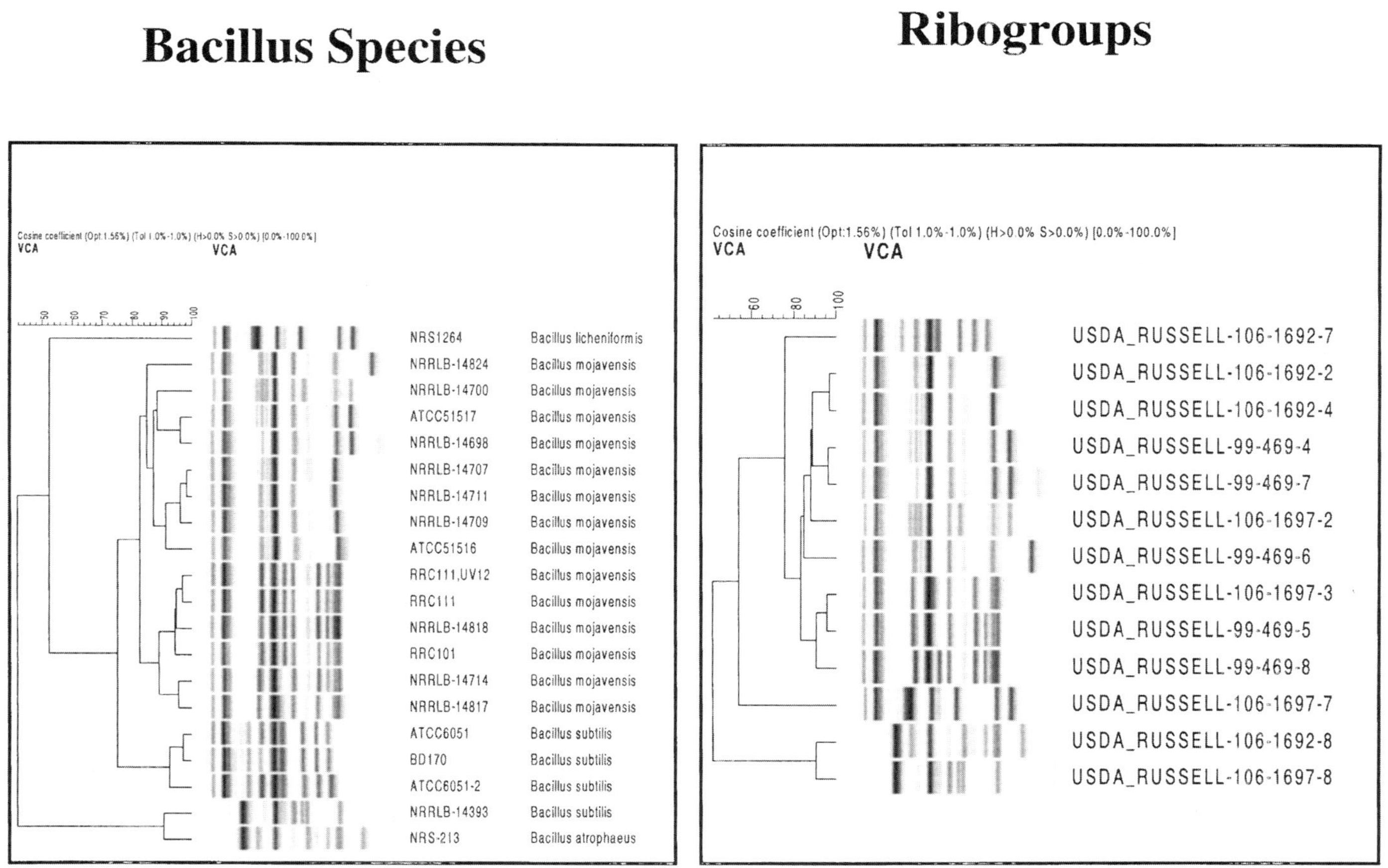

Fig. 3. Ribotypying of *Bacillus* species and the ribogroups produced by *B. mojavensis* strains. Sizes of *Eco*R1 fragment containing rDNA sequences derived from *Bacillus* species and strains of *B. mojavensis*. Ribotyping was performed using RiboPrinter Microbial Characterization System (Qualicon/DuPont, Wilmington, DE).

minor problems, probably strain-specific, which are related to a reduction in growth and/or competitiveness of the mutant test isolate.

Careful observation is necessary, since approx 1% of all rifampin mutations of *B. mojavensis* lose the ability to antagonize fungi ***(22)***. This and other deviations from the wild type have been observed for rifampin-resistant mutants of several other bacterial species ***(23)***. Spontaneous chromosomal rifampin-resistant mutants (Rif ^{r}) of each test strain are generated on nutrient agar containing rifampin (50 or 100 mg/L, Sigma Chemical Co., St. Louis, MO). Several single rifampin-resistant colonies are selected from each isolate and streaked once more onto rifampin-augmented nutrient agar. To determine the stability of mutants, each isolate is carried through 10 serial passages on nutrient agar without rifampin. Rifampin resistance of each isolate is confirmed following the tenth serial passage by plating onto nutrient agar augmented with rifampin (100 μg/mL). The working stock cultures of mutants are maintained on nutrient agar containing 50 μg/mL rifampin and stored at 4°C until used. Stock cultures of all strains are stored on silica gel at –30°C.

3.1.1. Plant Inoculation and Culture

It is important to emphasize that this endophytic species readily colonizes plants following the topical application to seed. It is not necessary to force, inject, or reapply bacterial inoculum to plants once the seed has received the bacterium. We have no information on the endophytic colonization of plants following the application of inocula to leaves and stems of mature plants. The maize seedling assay ***(44)*** is used as the test for endophytic colonization by the bacterial strains. The seeds are surface and internally sterilized before use, rinsed in sterile distilled water, soaked in water for 4 h, heat treated in a water bath at 60°C for 5 min, and rinsed in sterile cool water ***(44)***. Bacterial inoculum (0.01 mL, 2×10^6 cfu/mL), prepared from a wild type and its Rif ^{r} mutants, is placed on sterilized seeds and incubated on moist filter paper at 25°C for 48 h. These kernels are planted in a sterile synthetic soil mix (Redi-Earth, pH 5.5–6.0) in 6-in plastic pots (thinned to 12 plants per pot). Plants are watered daily, and after 2 wk, they are fertilized at weekly intervals with a liquid 20-20-20 (N-P-K) fertilizer containing micronutrients. Uninoculated seeds are used as controls and planted as described for the treatment groups. Companion treatment groups consisting of non-Rif ^{r} mutants are also used in both preliminary experiments and in the Rif ^{r} mutant recovery experiments to confirm that the mutation did not alter the endophytic effects of the parental strains. Additional plants to use include bush bean, oat, rye, and wheat. The colonizations of seedlings of these plants are identical to that of corn seedlings sterilized, inoculated and grown as described above.

3.1.2. Endophyte Recovery

Surface disinfected 2- to 4-wk-old plant material was used for recovery of bacteria or biocontrol activity. This time period is the standard time used to determine endophytic colonization by the patented strain and any *in planta* biochemical control assessment, i.e., mycotoxin reduction or disease protection ***(22,45)***. All plant materials are disinfected as described above. However, there is a phenomenon referred to as antibiotic masking, which prevents the rifampin-resistant endophytic mutants to grow upon recovery from plants when tissue is first placed onto rifampin-augmented media ***(43)***. This phenomenon has been observed for field-grown corn infected with *B. mojavensis,* but not necessarily so with plants grown under light-room conditions. Possibly there is either a time period for this expression which is longer than the culture period under light-room conditions, or the added microbial competition under field conditions induces this effect. To be sure of recovery under any condition, it is recommended that Rif^r mutants of *B. mojavensis*-infected tissue be first placed on nutrient agar without rifampin, and colonies picked and then inoculated onto nutrient agar augmented with rifampin, and treated as described above for isolation of the bacterium from plant tissues. Dilution plating (selective plating), using spread plate, poured plate, or droplet plating on media, allows enumeration of populations as required. Enrichment techniques may be used, especially for detecting *B. mojavensis* in seed, since it is present in very low levels, or when the results are negative, and the colonies enumerated by the most probable number method.

3.2. In Planta Detection and Microscopy

Microscopic examinations are performed on all bacteria-colonized plants that yield bacteria from leaves. *In planta* visualization is accomplished with plant materials cut into thin sections by hand with a razor blade and stained first with 2,3,5-triphenyltetrazolium chloride stain ***(46)*** followed by a second stain for 1 min in an aqueous solution of 0.1% aniline blue ***(35)***. Both stains are prepared as sterile solutions. The double-stained material is observed as a wet mount under oil at 100X. The bacterial cells are located in the intercellular spaces of plant tissue as stained, blue-purple, rod-shaped cells. If the whole bacteria-infected coleoptile from a 2–3-d-old seedling is subjected to this staining procedure, the region containing the bacteria is stained darkly, providing quick macroscopic evidence of bacterial endophytic colonization.

3.3. Assay for Antagonism and Plant Responses

3.3.1. In Vitro Test

Inhibition of fungal growth by strains of *B. mojavensis* is determined on nutrient agar. Bacterial inocula are prepared from strains grown on nutrient

Table 2
In Vitro Assay for Type of Inhibition to *Fusarium moniliforme* by Selected Strains of *Bacillus mojavensis (9)*

Bacillus mojavensis strains	Type of fungal antagonism[a]
RRC 101	+++
RRC 111	++++
RRC 51516	CCC
ATCC 51517	C
NRRL B-14714	+
NRRL B-14711	+
NRRL B-14707	C
NRRL B-14709	C
NRRL B-14698[T]	C*
NRRL B-14818	++
NRRL B-14817	+
NRRL B-14700	–
RRC 113 (UV-12)	–
RRC 114 (UV-50)	–

[a] –, no inhibition; +, weak (<3 mm) in vitro inhibition; ++, moderate (> 3–9 mm) in vitro inhibition; +++, strong (>9–18 mm) in vitro inhibition; ++++, very strong (>18 mm) in vitro inhibition; C, a contact inhibition (no zone of inhibition but dead hyphae of *F. moniliforme*); C*, very strong contact kill.

agar for 7 to 10 d. The fungus used in this test is *Fusarium moniliforme (=F. verticillioides),* RRC PATGUS, a strain rated as a strong pathogen of maize seedlings ***(47)*** and transformed ***(48)*** with both the *gusA* reporter gene, which codes for β-glucoronidase (gus), and the *hph* gene, which codes for hygromycin resistance. This marked fungal isolate is also suitable for studies of plant-fungus-*B. mojavensis* interactions. Inoculum for the test fungus is prepared with 14-d-old cultures grown on potato dextrose agar plates (PDA). Control plates consist of fungi or bacteria placed on nutrient agar plates alone as described previously. A plug of *F. moniliforme* is placed on one side of a nutrient agar plate, and a loop of bacteria is streaked down the opposite side of the plate. The plates are incubated in the dark at 25–27°C until the fungi on the control plates have grown together. Antagonism is expressed within strains of this species as either a zone of inhibition, due to a diffusible toxin in the medium, or to the death of hyphae upon contact with the zone of bacteria (**Table 2; Fig. 4A–D**). The first measure of antifungal activity from the bacteria is the size of the zone of inhibition formed between the radial growth on reverse, measured from the edge of the colony of *F. moniliforme* to the edge of a bacterial colony on the plate of nutrient agar (**Fig. 4C**). The second measure of

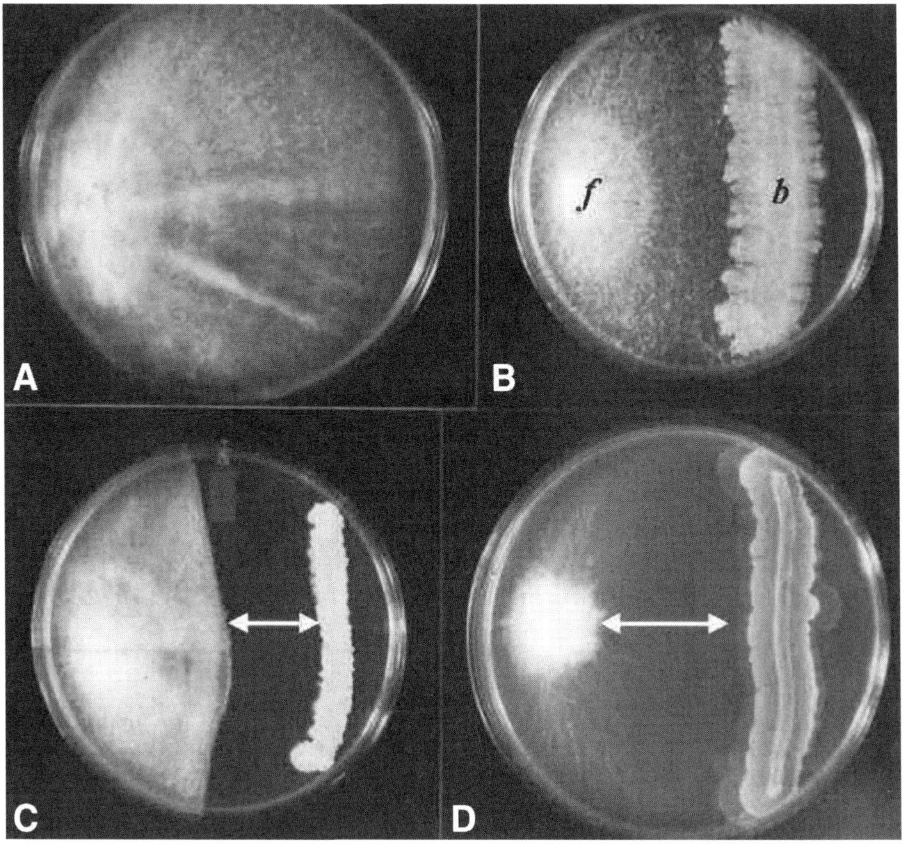

Fig. 4. Forms of antagonisms showed by strains of *B. mojavensis* to *Fusarium moniliforme*. **A**, and **B**, *F. moniliforme* control, and *F. moniliforme* and a noninhibitory bacterium, respectively. **C**, antagonism owing to death from a diffusable toxin into the medium, producing a clear zone (arrow). **D**, Death owing to direct contact of hyphae with cells of *B. mojavensis*, gradually producing a dying zone of appressed hyphae back to the initial inoculum location (arrow).

antifungal activity, in the absence of an inhibition zone, is the appearance of hyphae in contact with the bacterial colony (**Fig. 4D**). The hyphal walls in contact with the bacterial colony are characterized by lysis and necrosis along the bacterial zone of contact with hyphae.

Other fungi used in this bioassay and found to be antagonized by *B. mojavensis* include *Fusarium graminearum, F. oxysporum, F. proliferatum, Helminthosporium maydis, H. carbonum, Clabdsporium herbarum, Colletotrichum graminicola, Diplodia zeae, Aspergillus flavus, A. parasiticus, Penicillium chryogeum,* and *Phythium* sp. Indeed, the list of fungi that are inhibited by *B. mojavensis* is expected to be extensive, since the fungi listed here

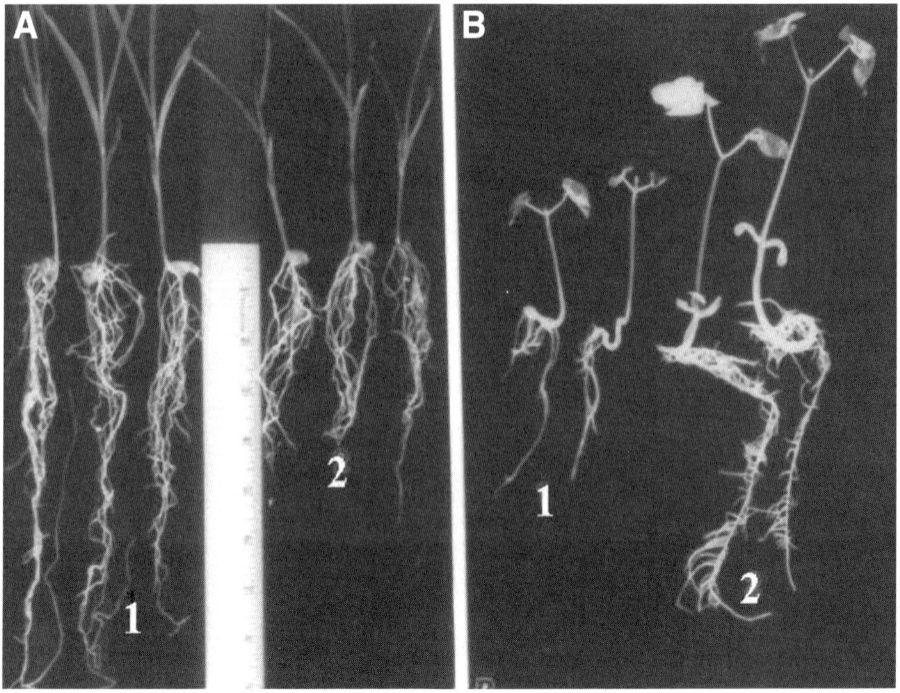

Fig. 5. Effects of *Bacillus mojavensis* RRC101 on plant growth. **A**, Effects of *Bacillus mojavensis* RRC101 on 3-wk-old corn seedling root growth, with (1) and without (2) the bacterium. **B**, Effects on bean seedling growth, with (1) and without (2) the bacterium ***(9)***.

were the species available for testing in this laboratory and all were inhibited. Endophytic bacteria are also indicated as controls for viruses, nematodes, and pathogenic bacteria ***(6)***, but the effects of *B. mojavensis* on these pathogens are unknown.

3.3.2. In Vivo Effects of B. mojavensis *on Plant Performance*

Plant materials used to determine the effects of *B. mojavensis* on plant growth include several commonly used vegetables, including the maize cultivar "Truckers Favorite," the bush bean cultivar "Blue Lake," and the Durham wheat cultivar "Ingot." Seeds of these plants are subjected to the double sterilization procedure described in **Subheading 3.1.1.** ***(44)*** and then planted in a sterilized synthetic soil mix in 6-in plastic sterile pots. An aqueous suspension of washed bacterial inoculum (10^6 cfu/mL), prepared from a 48-h shake nutrient broth culture, is placed on the disinfected seeds, which are air-dried for approx 12 h. The bacterium may also be applied to the seed with various seed coatings, without any deleterious effects. Seed dried as described may be stored under aseptic conditions under refrigeration (4°C) for at least 6 wk without germination being affected (data not shown). The inoculated seeds are planted in sterile soil. All plants are grown under aseptic conditions in a plant growth light room at 32°C under a 16-h light (cool-white, high-output fluorescent tubes, an average of 254 µmol • m^2 • s) and an 8-h dark regime at 29°C for 14 to 21 d. Treated plants are harvested, washed, and separated into roots and stems. The soil growth medium should be monitored throughout the experiment for sterility to both non–*B. mojavensis* bacteria and fungi. The lengths of roots, secondary root development, and shoot growth are used to measure the effects of specific strains of *B. mojavensis* on plant growth (**Fig. 5A,B**).

Acknowledgments

We thank L. Nakamura, NCAUR, Peoria, IL, for the strains of *Bacillus mojavensis* and other species of *Bacillus* used in this study. We express special appreciation to Gustafson, Inc., for early financial support on seedling growth and disease protection, and to Lance Bolton, Qualicon/DuPont, Wilmington, DE, for assistance in the ribotyping analyses, which were performed with the RiboPrinter Microbial Characterization System.

References

1. Kado, C. I. (1992) Plant pathogenic bacteria, in *The Prokaryotes* (Balows, A., Truper, H. G., Dworkin, M., Harder, W., and Schleifer, K.-H., eds.), Springer-Verlag, New York, pp. 660–662.
2. Quispel, A. (1992) A search for signals in endophytic microorganisms, in *Molecular Signals in Plant-Microbe Communications* (Verma, D. P. S., ed.), CRC Press, Boca Raton, FL, pp. 471–490.
3. Stone, J. K., Bacon, C. W., and White, J. F., Jr. (2000) An overview of endophytic microbes: endophytism defined, in *Microbial Endophytes* (Bacon, C. W. and White, J. F., Jr., eds.), Marcel Dekker, New York, pp. 3–30.
4. Tervet, I. W. and Hollis, J. P. (1948) Bacteria in the storage organs of healthy plants. *Phytopathology* **38,** 960–967.
5. Hollis, J. P. (1951) Bacteria in healthy tissue. *Phytopathology* **41,** 350–367.
6. McInroy, J. A. and Kloepper, J. W. (1995) Survey of indigenous bacterial endophytes from cotton and sweet corn. *Plant Soil* **173,** 337–342.
7. Hallmann, J. (2001) Plant interactions with endophytic bacteria, in *Biotic Interactions in Plant-Pathogen Associations* (Jeger, M. J. and Spence, N. J., eds.), CABI Publishing, New York, pp. 87–119.
8. Bacon, C. W., Yates, I. E., Hinton, D. M., and Meredith, F. (2001) Biological control of *Fusarium moniliforme* in maize. *Environ. Health Perspect.* **109,** 325–332.
9. Bacon, C. W. and Hinton, D. M. (2002) Endophytic and biological control potential of *Bacillus mojavensis* and related species. *Biol. Control* **23,** 274–284.
10. Bacon, C. W. and Hinton D. M. (2001) Control of seedling blight in wheat by *Bacillus mojavensis. Phytopathology* **91** (Suppl), S4.

11. Bacon, C. W. and Hinton, D. M. (1996) Symptomless endophytic colonization of maize by *Fusarium moniliforme. Can. J. Bot.* **74,** 1195–1202.
12. Sturz, A. V., Christie, B. R., and Nowak, J. (2000) Bacterial endophytes: potential role in developing sustainable systems of crop production. *Crit. Rev. Plant Sci.* **19,** 1–30.
13. Chanway, C. P. (1996) Endophytes: they're not just fungi. *Can. J. Bot.* **74,** 321–322.
14. Hallmann, J., Quadt-Hallmann, A., Mahaffee, W. F., and Kloepper, J. W. (1997) Bacterial endophytes in agricultural crops. *Can. J. Microbiol.* **43,** 895–914.
15. Kobayashi, D. Y. and Palumbo, J. D. (2000) Bacterial endophytes and their effects on plants and uses in agriculture, in *Microbial Endophytes* (Bacon, C. W. and White, J. F., Jr., eds.), Marcel Dekker, New York, pp. 199–233.
16. McCully, M. E. (2001) Niches for bacterial endophytes in crop plants: a plant biologist's view. *Aust. J. Plant Physiol.* **28,** 983–990.
17. Sturz, A. V., Christie, B. R., and Matheson, B. G. (1998) Associations of bacterial endophyte populations from red clover and potato crops with potential for beneficial allelopathy. *Can. J. Microbiol.* **44,** 162–167.
18. Bent, E. and Chanway, C. P. (1998) The growth-promoting effects of a bacterial endophyte on lodgepole pine are partially inhibited by the presence of other rhizobacteria. *Can. J. Microbiol.* **44,** 980–988.
19. Pleban, S., Ingel, F., and Chet I. (1995) Control of *Hizoctonia solani* and *Sclerotium rolfsii* in the greenhouse using endophytic bacillus spp. *Eur. J. Plant Pathol.* **101,** 665–672.
20. Chanway, C. P. (1998) Bacterial endophytes: ecological and practical implications. *Sydowia* **50,** 149–170.
21. Sokmen, A., Jones, B. M., and Erturk, M. (1999) The in vitro antibacterial activity of Turkish medicinal plants. *J. Ethnopharmacol.* **67,** 79–86.
22. Bacon, C. W., Yates, I. E., Hinton, D.M., and Meredith, F. (2001) Biological control of *Fusarium moniliforme* in maize. *Environ. Health Perspect.* **109,** 325–332.
23. Compeau, G., Al-Achi, B. J., Platsouka, E., and Levy, S. B. (1988) Survival of rifampin-resistant mutants of *Pseudomonas fluorescens* and *Pseudomonas putida* in soil systems. *Appl. Environ. Microbiol.* **54,** 2432–2438.
24. Raaidmakers, J. M., Leeman, M., van Oorschot, M. M. P., van der Sluis, I., Schipprs, B., and Bakker, P. A. H. M. (1992) Dose-response relationships in biological control of *Fusarium* wilt of radish by *Pseudomonas* spp. *Phytopathology* **85,** 1075–1081.
25. Liu, L., Kloepper, J. W., and Tuzun, S. (1995) Induction of systemic resistance in cucumber against bacterial angular leaf spot by plant growth-promoting rhizobacteria. *Phytopathology* **85,** 843–847.
26. Hallmann, J., Quadt-Hallmann, A., Miller, W. G., Sikora, R. A., and Lindow, S. E. (2001) Endophytic colonization of plants by the biocontrol agent *Rhizobium etli* G12 in relation to *Meloidogyne incognita* infection. *Phytopathology* **91,** 415–422.
27. Hinton, D. M., and Bacon, C. W. (1995) *Enterobacter cloacae* is an endophytic symbiont of corn. *Mycopathologia* **129,** 117–125.

28. Bacon, C. W. and Hinton, D. M. (1999) Use of *Bacillus subtilis* as an endophyte for the control of diseases caused by fungi. US Patent & Trademark Office, November 30, 1999.
29. Roberts, M. S., Nakamura, L. K., and Cohan, F. M. (1994) *Bacillus mojavensis* sp. nov., distinguishable from *Bacillus subtilis* by sexual isolation, divergence in DNA sequence, and differences in fatty acid composition. *Internat. J. Systemat. Bact.* **44,** 256–264.
30. Smith, E. F. (1911) *Bacteria in Relation to Plant Diseases.* Carnegie Institution of Washington, Washington, DC.
31. Hennig, K. and Villforth, F. (1940) Experimentelle untersuchungen zur frage der bakteriensymbiose in hoheren pflanzen und ihrer beeinflussung durch leitelemente. *Biochem. Z.* **305,** 299–309.
32. Mundt, J. O. and Hinkle, N. F. (1976) Bacteria within ovules and seeds. *Appl. Environ. Microbiol.* **32,** 694–698.
33. McInroy, J. A. and Kloepper, J. W. (1995) Population dynamics of endophytic bacteria in field-grown sweet corn and cotton. *Can. J. Microbiol.* **41,** 895–901.
34. Brooks, D. S., Gonzales, C. F., Appel, D. N., and Filer, T. H. (1994) Evaluation of endophytic bacteria as potential biological control agents for oak wilt. *Biol. Control* **4,** 373–381.
35. Bacon, C. W. and Hinton, D. M. (1997) Isolation and culture of endophytic bacteria and fungi, in *Manual of Environmental Microbiology* (Knudsen, G. R., McInerney, M. J., Stetzenbach, L. D., and Walter, M. V., eds.), ASM Press, Washington, DC, pp. 413–421.
36. Assi, S. M. P., da Silveria, E. B., de Mariano, L. R., and Menezes, D. (1998) Endophytic bacteria—method for isolation and antagonistic potential against cabbage black rot. *Summa Phytopathol.* **24,** 216–220.
37. Gordon, R. E., Hayward, M. D., and Pang, C. H. (1973) *The Genus Bacillus.* Agriculture Handbook No. 427, U.S. Depart. Agriculture, Washington, DC.
38. Ash, C., Farrow, A. E., Wallmanks, S., and Collins, M. D. (1991) Phylogenetic heterogeneity of the genus *Bacillus* revealed by comparative analysis of small-subunit ribosomal RNA sequences. *Let. Appl. Microbiol.* **13,** 202–206.
39. Nakamura, L. K., Roberts, M. S., and Cohan, F. M. (1999) Relationship of *Bacillus subtilis* clades associated with strains 168 and W23: a proposal for *Bacillus subtilis* subsp. *subtilis* subsp. nov. and *Bacillus subtilis* subsp *spizizenii* subsp. nov. *Int. J. Syst. Bacteriol.* **49,** 1211–1215.
40. Bruce, J. (1996) Automated system rapidly identifies and characterizes microorganisms in food. *Food Technol.* **1996,** 77–81.
41. Logan, N. A. and Berkeley, C. W. (1984) Identification of *Bacillus* strains using the API system. *J. Gen. Microbiol.* **130,** 1871–1882.
42. Quadt-Hallmann, A., Hallmann, J., and Kloepper, J. W. (1997) Bacterial endophytes in cotton: location and interaction with other plant-associated bacteria. *Can. J. Microbiol.* **43,** 254–259.
43. McInroy, J. A., Musson, G., Wei, G., and Kloepper, J. W. (1996) Masking of antibiotic-resistance upon recovery of endophytic bacteria. *Plant Soil* **186,** 213–218.

44. Bacon, C. W., Hinton, D. M., and Richardson, M. D. (1994) A corn seedling test for resistance to *Fusarium moniliforme*. *Plant Dis.* **78,** 302–305.
45. Bacon, C. W. and Hinton, D. M. (1996) Fusaric acid and pathogenic interactions of corn and non-corn isolates of *Fusarium moniliforme,* a nonobligate pathogen of corn. *Adv. Exp. Med. Biol.* **392,** 175–191.
46. Patriquin, D. G. and Dobereiner, J. (1978) Light microscopy observations of tetrazolium-reducing bacteria in the endorhizosphere of maize and other grasses in Brazil. *Can. J. Microbiol.* **24,** 734–742.
47. Bacon, C. W., Bennett, R. M., Hinton, D. M., and Voss, K. A. (1992) Scanning electron microscopy of *Fusarium moniliforme* within asymptomatic corn kernels and kernels associated with equine leukoencephalomalacia. *Plant Dis.* **76,** 144–148.
48. Yates, I. E., Hiett, K. L., Kapczynski, D. R., Smart, W., Glenn, A. E., Hinton, D. M., et al. (1999) GUS transformation of the maize fungal endophyte *Fusarium moniliforme*. *Mycol. Res.* **103,** 129–136.

36

Engineering of Bacteria Using the *Vitreoscilla* Hemoglobin Gene to Enhance Bioremediation of Aromatic Compounds

Benjamin C. Stark, Dale A. Webster, and Krishna R. Pagilla

1. Introduction

The hemoglobin (VHb) from the Gram-negative bacterium *Vitreoscilla* was originally discovered in 1966 ***(1)***. It was initially thought to be a soluble form of the terminal oxidase of the respiratory chain (cytochrome *bo*) until its correct identification as a hemoglobin in 1986 ***(2)***. It was originally proposed that the role of VHb in *Vitreoscilla* is to help provide oxygen to the cell, particularly under conditions of oxygen shortage, to enhance respiratory ATP production. This idea fits well with the large induction in VHb levels that occurs at low oxygen concentrations ***(3)***. It was further proposed that VHb may deliver oxygen directly to cytochrome *bo* ***(3)***; supportive evidence, but not proof, exists for this idea ***(4–6)***.

Shortly after the identification of VHb, its gene, *vgb,* was cloned ***(7,8)***. This enabled experiments investigating the molecular details of the control of VHb expression, in which it was found that the *vgb* promoter contains one or more elements responsive to oxygen levels (including an Fnr site), one or more CAP-binding sites (one of which may also be an ArcA-binding site), and at least one additional putative control site of unknown function ***(9–13)***. It also allowed the transformation of heterologous microbial hosts with *vgb* to try to enhance their growth and productivity by supplying them with additional oxygen for respiration and thus ATP production. Such genetic-engineering experiments initially focused on enhancement of production of proteins ***(14–16)*** and antibiotics ***(17,18)*** by bacteria and fungi. Enhancement of fermentation product production was also investigated ***(19,20)***.

From: *Methods in Biotechnology: Environmental Microbiology: Methods and Protocols*
Edited by: J. F. T. Spencer and A. L. Ragout de Spencer

We also had the idea that this approach might be successful in enhancing bioremediation of aromatic chemicals by bacteria. There were two reasons for this. First, many of the potential bioremediating organisms themselves are obligate aerobes; supplying them with extra oxygen using *vgb*/VHb may aid their growth (as indicated above), particularly in oxygen-poor environments. Second, the pathways of aromatic-compound catabolism in such organisms invariably require addition of molecular oxygen to the aromatic ring ***(21)***; it is possible that the extra supply of oxygen afforded by *vgb*/VHb may stimulate these reactions directly.

2. Experimental Approach and Results

2.1. Initial Work With Plasmid-Borne vgb

Our attempts to enhance degradation of aromatics began with work on transforming bioremediating organisms with *vgb* and producing VHb in such cells. The first three species used for this purpose were *Pseudomonas aeruginosa, Pseudomonas putida,* and *Xanthomonas* (originally *Pseudomonas*) *maltophilia.* Transformation into these strains was done with a construct (pSC160) in which *vgb* was placed into the broad host range vector pKT230; pSC160 was stably maintained in all three species and VHb expressed at levels of about 7–13 nmol/g wet weight of cells, about one-fourth to one-third that of the induced level in *Vitreoscilla* ***(7)***. The presence of *vgb*/VHb was correlated with a modest increase in viability and cell mass following stationary phase for *P. aeruginosa* and *X. maltophilia,* but a greater decrease in the same parameters for *P. putida.* Additional experiments confirmed that proper functioning of the *vgb* promoter occurs in pseudomonads ***(10,11)***.

We investigated the ability of VHb to enhance degradation of benzoate as a model aromatic compound in *X. maltophilia.* The results, however, paralleled those of the growth experiments, with no enhancement occurring when cultures were grown starting with small inocula. Only when large amounts of cells in log phase were harvested and transferred to fresh medium to continue log-phase growth did the *vgb*-bearing strain degrade benzoate faster (by approx 15%) than the untransformed strain ***(22)***.

More promising results were obtained with the 2,4-dinitrotoluene (DNT) degrading strain isolated in the laboratory of Jim Spain ***(23,24)***. This bacterium was isolated from a DNT-contaminated site and originally designated *Pseudomonas* strain DNT. Typing by us using the API strip system identified it as a strain of *Pseudomonas cepacia.* It was subsequently reclassified in the genus *Burkholderia;* we refer to it as *B. cepacia,* even though that designation is not certain ***(25)***. *B. cepacia* strain DNT could also be readily transformed with plasmid pSC160, which could be stably maintained at about 15 copies per cell and resulted in a VHb concentration of approx 8–10 nmol/g wet weight of cells. In addition, in this system the presence of *vgb*/VHb was correlated with an increase in growth in rich medium (12% with respect to maximum OD_{600nm} achieved) and rate of DNT degradation during growth in minimal medium containing 200 ppm DNT and 200 ppm yeast extract. The degradation enhancement (as measured by DNT disappearance after one day's growth) was greatest (approx 60%) when inocula had been grown in minimal medium containing 200 ppm DNT, 200 ppm yeast extract, and 0.02 *M* succinate ***(26)***.

Further experiments were done to assess the advantage, if any, afforded by VHb regarding DNT degradation in the *B. cepacia* system over a much wider range of medium compositions, DNT concentrations, and aeration rates (and combinations thereof) ***(27)***. In general, the disappearance of DNT was very modest (due probably to poor growth) when the cosubstrate was 200 ppm yeast extract. Amounts of succinate, amino acids, or yeast extract 5–10 times higher than this, as well as tryptic-soy broth as cosubstrates, greatly enhanced both growth and DNT disappearance. The *vgb*-bearing strain degraded DNT faster than the control strain for almost every combination of conditions. For the bulk of the conditions tested the increases ranged from approx 10–75%, but the greatest advantages (up to 3.5-fold) occurred under conditions of limited aeration.

2.2. Biochemical Studies

The Spain laboratory had also cloned the genes encoding three of the enzymes in the DNT-degrading pathway ***(24)***. They supplied us with the clone (in *E. coli* on plasmid pJS39) of the operon which encodes the first of these (DntA). This plasmid, as well as plasmid pHG1 ***(28)***, which contains *vgb* cloned into a compatible vector (pVK102), were cloned simultaneously into *E. coli* JM109 ***(29)***. Expression of VHb was very high in this strain (maximum of 234 nmol/g wet weight of cells). Comparison of its DNT-degrading ability with that of the matched strain bearing pJS39 alone allowed examination of the effects of VHb on the first step in DNT degradation, without the complication of the following steps (as was the case with our original system).

Using either whole cells or whole cell lysates as enzyme preparations, the presence of VHb was determined to increase the specific activity of DntA (with respect to DNT) by three- to fourfold. It is as yet unclear whether this is due to enhancement of production of DntA, enhancement of DntA activity (perhaps by direct delivery by VHb of O_2 to DntA), or a combination of these effects (*see* **Subheading 2.5.**).

2.3. Integration of vgb into the Host Cell Chromosome

Although plasmid pSC160 was quite stable in our strains, we also worked to produce strains in which *vgb* had been integrated stably into the host chromosome. These strains would be potentially more useful in actual field applications due to increased stability of *vgb,* including absence of the need for antibiotics to guarantee its presence, as well as the lowered risk of horizontal transfer of both *vgb* and the antibiotic-resistance genes. The conjugation-transposon-based vectors developed in the laboratory of K. N. Timmis ***(30,31)*** were used for this purpose, and *vgb* successfully integrated into the chromosomes of both *P. aeruginosa* and *B. cepacia* (the *B. cepacia* used here was not strain DNT, but was obtained from the USDA collection in Peoria, IL) ***(32)***. In *P. aeruginosa,* VHb was expressed at a level of 8.8 nmol/g wet weight of cells and afforded little advantage in growth but a sizable advantage (roughly 30–70% increase) in benzoate degradation compared with the *vgb*-free control strain. In *B. cepacia,* the VHb level was only approx 0.8 nmol/g wet weight and there was little or no advantage in either growth or DNT degradation compared with the *vgb*-free control strain. It is as yet unknown why the (presumably single-copy *vgb*) *P. aeruginosa* strain has about the same level of VHb as the strain bearing pSC160 (presumably with *vgb* at about 15 copies per cell), while VHb levels in the single-copy and multicopy *B. cepacia* strains are scaled to the *vgb* copy number. In any event, it seems that there may be a threshold level of VHb below which no enhancement of bioremediation by such engineered strains occurs.

2.4. Biochemical Engineering Studies

We have also investigated bioremediation of DNT in the *B. cepacia*-pSC160 system described above (with plasmid-borne *vgb*) in sand columns and fermentors. In the first study ***(33)***, two identical fixed-bed bioreactor columns containing sand as attached growth medium were constructed. One was inoculated with *B. cepacia*-bearing pSC160, while the other was inoculated with plasmid-free *B. cepacia.* Each column was run as a continuous reactor, with feed pumped in at one end while effluent was collected at the same rate from the other end. Over a 5-mo period both the DNT and oxygen concentrations in the feed were varied (100–214 mg/L DNT and 3.1–6.8 mg/L DO) while effluent was assayed for remaining DNT. The *vgb*-bearing strain was superior in DNT removal (rate, amount removed, or both) under all conditions, being able to keep the effluent DNT concentration at or below 1 mg/L at all loading rates and DO concentrations, except during start-up and periods of loading-rate change. Under identical conditions, the effluent from the *vgb*-free *B. cepacia* bioreactor contained at least 5 mg/L DNT and as much as 22 mg/L DNT (the latter for 214 mg/L DNT and 3.1 mg/L DO in the feed).

The same two strains were also tested for ability to degrade DNT during growth in a two-phase bioreactor ***(34)***. One phase was 1.5 L of the aqueous growth medium, and the other was 100 mL oleyl alcohol into which 2 g of DNT was dissolved. Preliminary experiments showed that oleyl alcohol actually stimulated the growth of *B. cepacia.* The low solubility of DNT in water (approx 200 ppm) limits the rate at which its biodegradation can occur. Our setup simulated a system in which large amounts of DNT could be solubilized from a contaminated site with oleyl alcohol (into which DNT can be dissolved to levels at least 110 times that in water) and mixed with a culture; the DNT would continuously diffuse into the aqueous phase and be degraded. The two-phase reactor can degrade at least two times as much DNT per volume of culture as a single-phase aqueous reactor. The *vgb*-bearing *B. cepacia* degrades DNT 20–30% faster than the *vgb*-free strain under these conditions.

2.5. Current Work

As part of ongoing structure-function studies of VHb we have produced about two dozen site-directed mutants of the bacterial hemoglobin ***(35,36)***. Some of these are in the oxygen-binding site and the rest in a region (the D region) that appears disordered in the VHb crystal structure ***(37)***; all have been tested for their oxygen-binding affinities. Some have affinities higher, some the same, and some lower than that of wild-type VHb (K_d of about 7 μ*M*) ***(35)***. Aside from the importance of these data in determining how VHb works, they have also produced a collection of hemoglobins with K_ds for oxygen that range from approx 2 to 19 μ*M* that may be of use in bioremediation of aromatic molecules.

We presume that in order for VHb to be useful in the enhancement of bioremediation of aromatics, it must first be able to deliver oxygen efficiently to the terminal respiratory oxidase. As the K_d for oxygen of the oxidase (in *E. coli*) is approx 2 μ*M*, the wild type and mutant VHbs would be able to do this. The same would have to be true of the oxygenases involved in catabolism of the aromatics. In order for VHb to be able to deliver oxygen to such enzymes, its K_d for oxygen cannot be too low or it would be unable to unload oxygen to the oxygenase. There is a paucity of information about the K_ds for oxygen of these enzymes, but we anticipate that they will vary somewhat. Thus, our collection of mutant VHbs may be able to serve as a source of "custom fitting VHbs," with the particular mutant chosen so as to "match" the K_d of the oxygenase(s) of each bioremediating strain/pathway. We are currently engaged in testing this plan in both *P. aeruginosa* and *B. cepacia* (benzoate and DNT degradation, respectively) with both plasmid-borne and chromosomally integrated *vgb*s.

In a recent study, two *vgb*s with single mutations in the heme binding pocket were integrated into the host chromosomes of *B. cepacia* and *P. aeruginosa* ***(38)***. In the case of *B. cepacia* these strains had 10–20% faster degradation of DNT under limited aeration conditions than the matched strain containing wild-type *vgb*. In contrast to our initial results ***(32)***, in which the chromosomally integrated *vgb* did not enhance DNT degradation under such conditions, there was enhancement in this parameter for all *vgb*-bearing strains (20–30%, which is about half that seen with strains carrying *vgb* on pSC160). This variation in our own results needs to be investigated further, but we have some indication that the *B. cepacia* strain with chromosomally integrated *vgb* may, in fact, be useful.

In related work we have used the two-hybrid screen technique to demonstrate direct interaction between VHb and the flavin-binding subunit of DNT dioxygenase (DntA; *see* **Subheading 2.2.**). Further, use of site-directed mutants of VHb in the D region has implicated this part of VHb as a possible contact point between VHb and such oxygenases ***(36)***.

Our bioremediation work is continuing and being extended to additional aromatic compounds, the biodegradation of which can be enhanced by VHb ***(39)***. *B. cepacia* bearing pSC160 was shown to have enhanced growth and degradation of 2-chlorobenzoic acid (at an initial concentration of 0.5 *M*) compared with *vgb*-free *B. cepacia* under three different aeration conditions. Based on the disappearance of 2-chlorobenzoic acid in the first day of growth, the VHb enhancement was as much as fivefold. The mineralization of 2-chlorobenzoic acid by *vgb*-bearing *B. cepacia* is being investigated with chloride balance experiments, along with confirmation of the pathway described by Arensdorf and Focht ***(40)***. Comparison of chromosomally integrated and pSC160 carried *vgb*-bearing *B. cepacia* is continuing in order to determine the optimum strains for further bioremediation experiments. As indicated previously for DNT degradation, we do not have definitive answers yet; in addition, we might expect to find different results for each aromatic compound.

Finally, we are working with transforming *vgb* into *Gordonia amarae,* in an attempt to enhance production of biosurfactants by this species ***(41)***. Such enhancement could be useful in increasing the solubility of aromatic compounds so that they may be taken up more readily by bioremediating species. Coculturing of *vgb*-bearing *G. amarae* and *P. aeruginosa* or *B. cepacia,* for example, could result in synergistic effects in this regard.

3. Summary and Future Plans

Substantial data have been collected to support the proposition that genetic engineering with *vgb*/VHb can enhance bioremediation of aromatic compounds. This includes relatively simple shake flask experiments as well as experiments designed to help make this technology of practical use. The latter include molecular work (chromosomal integration) as well as biochemical engineering experiments (fermentor and soil-column work). At this point, it can probably be said that VHb technology is not a panacea for bioremediation. It does, however, seem to show promise that it could lead to significant increases in the rate and level of bioremediation, particularly in oxygen-limited conditions. Perhaps the most useful aspect of this technology will be bioremediation *in situ* of sites (or locations within sites) in which oxygen is limiting (for example, those far below the surface), where a biological solution will be able to augment physical solutions such as excavation and aeration.

Acknowledgments

This work was supported by NSF Grants BES-9309759, MCB-9910356, and INT-9503400; AFOSR Grant F49620-95-1-0325; and an ERIF grant from the Illinois Institute of Technology.

References

1. Webster, D. A. and Hackett, D. P. (1966) The purification and properties of cytochrome *o* from *Vitreoscilla. J. Biol. Chem.* **241,** 3308–3315.
2. Wakabayashi, S., Matsubara, H., and Webster, D. A. (1986) Primary sequence of a dimeric bacterial haemoglobin from *Vitreoscilla. Nature* **322,** 481–483.
3. Webster, D. A. (1987) Structure and function of bacterial hemoglobin and related proteins, in *Advances in Inorganic Biochemistry, Vol. 7, Heme Proteins* (Eichhorn, G. L. and Marzilli, L., eds.), Elsevier Science Publishing Co., New York, pp. 245–265.
4. Tsai, P. S., Nageli, M., and Bailey, J. E. (1996) Intracellular expression of *Vitreoscilla* hemoglobin modifies microaerobic *Escherichia coli* metabolism through elevated concentration and specific activity of cytochrome *o. Biotechnol. Bioengin.* **49,** 151–160.
5. Ramandeep, Hwang, K. W., Raje, M., Kim, K. J., Stark, B. C., Dikshit, K. L., and Webster, D. A. (2001) *Vitreoscilla* hemoglobin: intracellular localization and binding to membranes. *J. Biol. Chem.* **276,** 24,781–24,789.
6. Park, K. W., Kim, K. J., Howard, A. J., Stark, B. C., and Webster, D. A. (2002) *Vitreoscilla* hemoglobin binds to subunit I of cytochrome *bo* ubiquinol oxidases. *J. Biol. Chem.* **277,** 33,334–33,337.
7. Dikshit, K. L. and Webster, D. A. (1988) Cloning, characterization and expression of the bacterial globin gene from *Vitreoscilla* in *Escherichia coli. Gene* **70,** 377–386.
8. Khosla, C. and Bailey, J. E. (1988) The *Vitreoscilla* hemoglobin gene: molecular cloning, nucleotide sequencing and genetic expression in *E. coli. Mol. Gen. Genet.* **214,** 158–161.
9. Khosla, C. and Bailey, J. E. (1989) Characterization of the oxygen-dependent promoter of the *Vitreoscilla* hemoglobin gene in *Escherichia coli. J. Bacteriol.* **171,** 5995–6004.
10. Dikshit, K. L., Dikshit, R. P., and Webster, D. A. (1990) Study of *Vitreoscilla* globin *(vgb)* gene expression and promoter activity in *E. coli* through transcriptional fusion. *Nucl. Acid. Res.* **18,** 4149–4155.
11. Dikshit, K. L., Dikshit, R. P., and Webster, D. A. (1991) Transcriptional control of *Vitreoscilla* hemoglobin synthesis, in *Structure and Function of Invertebrate*

Oxygen Carriers (Vinogradov, S. N. and Kapp, O. H., eds.), Springer-Verlag, New York, pp. 313–321.

12. Joshi, M. and Dikshit, K. L. (1994) Oxygen dependent regulation of *Vitreoscilla* globin gene: evidence for positive regulation by FNR. *Biochem. Biophys. Res. Comm.* **202,** 535–542.
13. Yang, J. G. (2002) Study of *Vitreoscilla* globin gene *(vgb)* expression using mutations in *vgb* promoter and chromosomal *fnr* and *arcA* genes. Ph.D. thesis, Illinois Institute of Technology, Chicago, IL.
14. Kallio, P. T. and Bailey, J. E. (1996) Intracellular expression of *Vitreoscilla* hemoglobin (VHb) enhances total protein secretion and improves the production of α-amylase and neutral protease in *Bacillus subtilis. Biotechnol. Prog.* **12,** 31–39.
15. Khosravi, M., Webster, D. A., and Stark, B. C. (1990) Presence of the bacterial hemoglobin gene improves α-amylase production of a recombinant *Escherichia coli* strain. *Plasmid* **24,** 1–5.
16. Buddenhagen, R. E., Webster, D. A., and Stark, B. C. (1996) Enhancement by bacterial hemoglobin of amylase production in recombinant *E. coli* occurs under conditions of low O_2. *Biotechnol. Lett.* **102,** 695–700.
17. Magnolo, S. K., Leenutaphong, D. L., DeModena, J. A., Curtis, J. E., Bailey, J. E., Galazzo, J. L., et al. (1991) Actinorhodin production by *Streptomyces coelicolor* and growth of *Streptomyces lividans* are improved by the expression of a bacterial hemoglobin. *Bio/Technology* **9,** 473–476.
18. DeModena, J. A., Gutierrez, S., Velasco, J., Fernandez, F. J., Fachini, R. A., Galazzo, J. L., et al. (1993) The production of cephalosporin C by *Acremonium chrysogenum* is improved by the intracellular expression of a bacterial hemoglobin. *Bio/Technology* **11,** 926–929.
19. Chen, W., Hughes, D. E., and Bailey, J. E. (1994) Intracellular expression of *Vitreoscilla* hemoglobin alters the aerobic metabolism of *Saccharomyces cerevisiae. Biotechnol. Prog.* **10,** 308–313.
20. Wei, M. L., Webster, D. A., and Stark, B. C. (1998) Metabolic engineering of *Serratia marcescens* with the bacterial hemoglobin gene: alterations in fermentation pathways. *Biotechnol. Bioeng.* **59,** 640–646.
21. Moat, A. G. and Foster, J. W. (1988) *Microbial Physiology,* second edition. John Wiley and Sons, New York.
22. Liu, S. C., Webster, D. A., Wei, M. L., and Stark, B. C. (1996) Genetic engineering to contain the *Vitreoscilla* hemoglobin gene enhances degradation of benzoic acid by *Xanthomonas maltophilia, Biotechnol. Bioengin.* **49,** 101–105.
23. Spanggord, R. J., Spain, J. C., Nishino, S. F., and Mortelmans, K. E. (1991) Biodegradation of 2,4-dinitrotoluene by a *Pseudomonas* sp. *Appl. Environ. Microbiol.* **57,** 3200–3205.
24. Suen, W. C. and Spain, J. C. (1993) Cloning and characterization of *Pseudomonas* sp. strain DNT genes for 2,4-dinitrotoluene degradation. *J. Bacteriol.* **175,** 1831–1837.
25. Johnson, G. R., Jain, R. K., and Spain, J. C. (2000) Properties of the trihydroxytoluene oxygenase from *Burkholderia cepacia* R34: an extradiol dioxygenase from the 2,4-dinitrotoluene pathway. *Arch. Microbiol.* **173,** 86–90.

26. Patel, S. M., Stark, B. C., Hwang, K. W., Dikshit, K. L., and Webster, D. A. (2000) Cloning and expression of *Vitreoscilla* hemoglobin gene in *Burkholderia* sp. strain DNT for enhancement of 2,4-dinitrotoluene degradation. *Biotechnol. Prog.* **16,** 26–30.
27. Nasr, M. A., Hwang, K. W., Akbas, M., Webster, D. A., and Stark, B. C. (2001) Effects of culture conditions on enhancement of 2,4-dinitrotoluene degradation by *Burkholderia* engineered with the *Vitreoscilla* hemoglobin gene. *Biotechnol. Prog.* **17,** 359–361.
28. Geckil, H., Stark, B. C., and Webster, D. A. (2001) Cell growth and oxygen uptake of *Escherichia coli* and *Pseudomonas aeruginosa* are differently affected by the genetically engineered *Vitreoscilla* hemoglobin gene. *J. Biotechnol.* **85,** 57–66.
29. Fish, P. A., Webster, D. A., and Stark, B. C. (2000) *Vitreoscilla* hemoglobin enhances the first step in 2,4-dinitrotoluene degradation in vitro and at low aeration in vivo. *J. Molec. Catal. B: Enzymatic* **9,** 75–82.
30. Herrero, M., de Lorenzo, V., and Timmis, K. N. (1990) Transposon vectors containing non-antibiotic resistance selection markers for cloning and stable chromosomal insertion of foreign genes in Gram-negative bacteria. *J. Bacteriol.* **172,** 6557–6567.
31. De Lorenzo, V., Herrero, M., Jakubzik, U., and Timmis, K. N. (1990) Mini-Tn*5* transposon derivatives for insertion mutagenesis, promoter probing, and chromosomal insertion of cloned DNA in Gram-negative eubacteria. *J. Bacteriol.* **172,** 6568–6572.
32. Chung, J. W., Webster, D. A., Pagilla, K. R., and Stark, B. C. (2001) Chromosomal integration of the *Vitreoscilla* hemoglobin gene in *Burkholderia* and *Pseudomonas* for the purpose of producing stable engineered strains with enhanced bioremediating ability. *J. Indust. Microbiol. Biotechnol.* **27,** 27–33.
33. So, J., Webster, D. A., Stark, B. C., and Pagilla, K. R. (2004) Enhancement of 2-4-dinitrotoluene biodegradation by *Burkholderia* sp. in sand bioreactors using bacterial hemoglobin technology. *Biodegradation*, in press.
34. Lin, J. M., Stark, B. C., and Webster, D. A. (2003) Effects of *Vitreoscilla* hemoglobin on the 2,4-dinitrotoluene (2,4-DNT) dioxygenase activity of *Burkholderia* and on 2,4-DNT degradation in two-phase bioreactors. *J. Ind. Microbiol. Biotechnol.* **30,** 362–368.
35. Dikshit, K. L., Orii, Y., Navani, N., Patel, S., Huang, H. Y., Stark, B. C., et al. (1998) Site-directed mutagenesis of bacterial hemoglobin: the role of glutamine (E7) in oxygen-binding in the distal heme pocket. *Arch. Biochem. Biophys.* **349,** 161–166.
36. Lee, S. Y., Stark, B. C., and Webster, D. A. (2004) Structure-function studies of the *Vitreoscilla* hemoglobin D-region. *Biochem. Biophys. Res. Commun.*, in press.
37. Tarricone, C., Galizzi, A., Coda, A., Ascenzi, P., and Bolognesi, M. (1997) Unusual structure of the oxygen-binding site in the dimeric bacterial hemoglobin from *Vitreoscilla* sp. *Structure* **5,** 497–507.
38. Kim, Y. (2002) Evaluation of genetic engineering with mutant bacterial hemoglobin genes as a method to enhance bioremediation of aromatic compounds and

survival of pUC-based plasmids in *Psuedomonas*. PhD thesis, Illinois Institute of Technology, Chicago, IL.

39. Urgun-Demirtas, M., Pagilla, K. R., Stark, B. C., and Webster, D. A. (2003) Biodegradation of 2-chlorobenzoate by recombinant *Burkholderia cepacia* expressing *Vitreoscilla* hemoglobin under variable levels of oxygen availability. *Biodegradation*, **14,** 357–365.
40. Arensdorf, J. J. and Focht, D. D. (1994) Formation of chlorocatechol meta cleavage products by a pseudomonad during metabolism of monochlorobiphenyls. *Appl. Environ. Microbiol.* **60,** 2884–2889.
41. Dogan, I. (2002) Effect of *Vitreoscilla* hemoglobin gene on the production of biosurfactant in *Gordonia* amarae. PhD thesis, Illinois Institute of Technology, Chicago, IL.

37

The Use of Chemical Shift Reagents and ^{23}Na NMR to Study Sodium Gradients in Microorganisms

Anne-Marie Delort, Genevieve Gaudet, and Evelyne Forano

1. Introduction

The microbiology of extreme habitats is a fast-moving field with increasing isolation of new organisms *(1)*. In particular, studies on halophilic Archaea or bacteria inhabiting diverse environments such as salt lakes, saline soils, or salted food products are developing in order to understand molecular factors that impart stability to halophilic biomolecules *(2)*. Most species are able to keep an intracellular ionic environment at low levels, and complex mechanisms of adjustment of the intracellular environments enable rapid adaptation to changes in the salt concentration of the environment. To study the physiology of these microorganisms and their mechanisms of adaptation, accurate estimation of the intracellular sodium concentration, and thus of the sodium transmembrane gradient, is crucial.

Over the past 15 yr, a new methodology has been developed to measure sodium gradients directly on living cells. It is based on in vivo ^{23}Na NMR and the application of specific techniques to distinguish the resonances of intracellular and extracellular sodium. The first technique is based on the difference of relaxation properties of Na^+_{in} and Na^+_{ex} which was exploited in multiple quantum filtered NMR experiments *(3–4)*. The second approach, more generally applied, uses shift reagents, which are anionic complexes of lanthanides (Dy^{3+}, Tm^{3+}, Tb^{3+}). The paramagnetic properties of these complexes, when exchanging with external sodium, induce a chemical shift of external Na^+ resonance (**Fig. 1**). These reagents usually do not cross the cytoplasmic membrane, so the intracellular sodium is not shifted. The more currently used chemical-shift reagents are dysprosium(III) tripolyphosphate ($Dy(PPPi)_2^{7-}$) *(5)*; thulium(III)

From: *Methods in Biotechnology: Environmental Microbiology: Methods and Protocols*
Edited by: J. F. T. Spencer and A. L. Ragout de Spencer

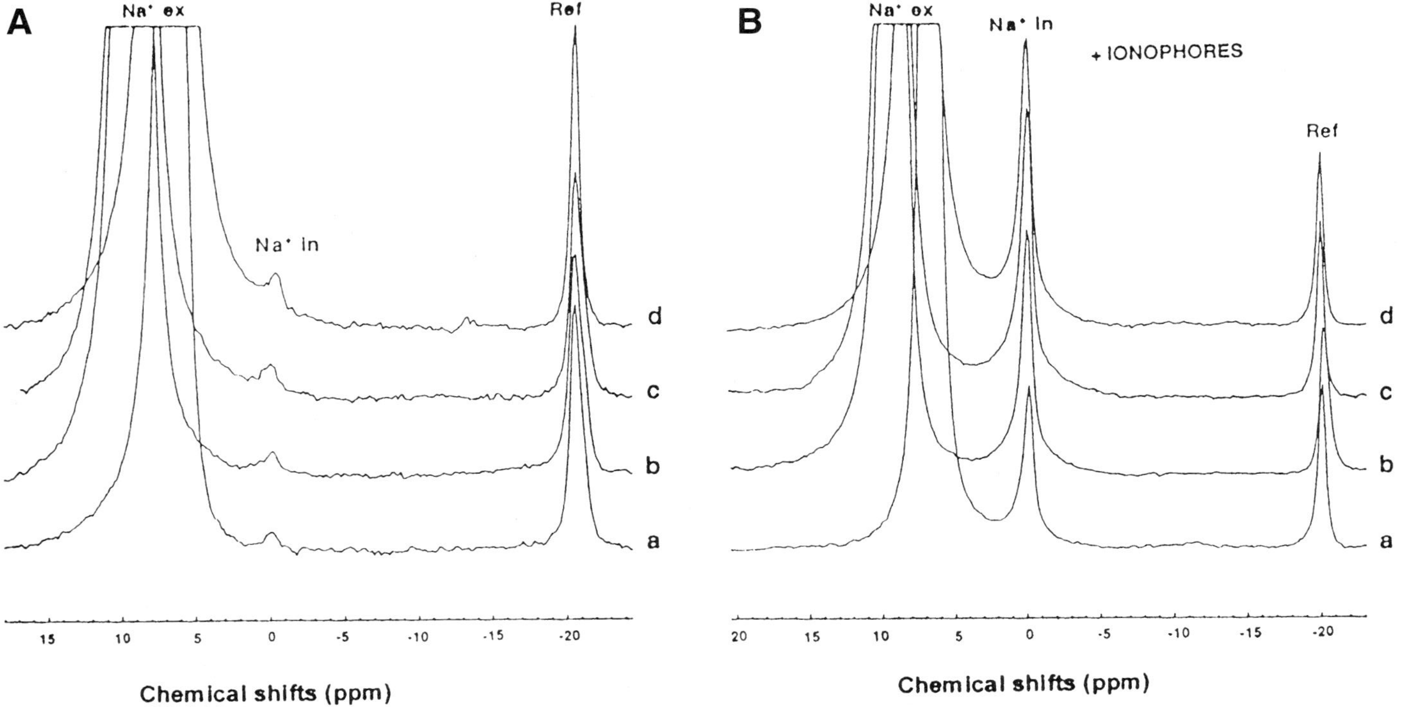

Fig. 1. In vivo ^{23}Na NMR spectra of *F. succinogenes* S85 (from ***18***) *F. succinogenes* S85 resting cells (10 mg protein/mL) washed in a 75 m*M* Na^+-containing buffer were incubated in buffers containing 50 m*M* Na^+ and 5 m*M* $Tm(DOTP)^{5-}$ (a), 75 m*M* Na^+ and 6 m*M* $Tm(DOTP)^{5-}$ (b), 100 m*M* Na^+ and 6 m*M* $Tm(DOTP)^{5-}$ (c), 150 m*M* Na^+ and 6 m*M* $Tm(DOTP)^{5-}$ (d). In vivo ^{23}Na NMR spectra collected after 15 min in the absence of ionophore (**A**); spectra collected after a further 15-min incubation of the same samples in the presence of ionophores (monensin 90 μ*M*, valinomycin 30 μ*M*, and CCCP 30 μ*M*) (**B**).

Na^+in: intracellular sodium; Na^+ex: extracellular sodium; ref: Na^+ in the reference capillary containing $Na_7\,Dy(PPPi)_2^{7-}$.

1,4,7,10-tetraazacyclododecane -1,4,7,10-tétrakis (metylenephosphate) ($Tm(DOTP)^{5-}$) ***(6–7)***; and dysprosium(III) triethylenetetramine hexaacetate ($DyTTHA^{3-}$) ***(8)***. The first two of these reagents induce an upfield shift of ^{23}Na NMR resonances, while ($DyTTHA^{3-}$) induces a downfield shift.

These reagents have been used on various biological systems, but only a few studies have been performed with microorganisms. $Dy(PPPi)_2^{7-}$ was used to study Na^+ gradients in the bacteria *Escherichia coli* ***(9–11)***, *Brevibacterium* sp. ***(11)***, and *Vibrio costicola* ***(12)***. $Dy(TTHA)^{3-}$ was used in the case of *Streptococcus faecalis (Enterococcus hirae)* ***(13–17)***. $Tm(DOTP)^{5-}$ was recently used to measure sodium gradients in *Fibrobacter succinogenes* S85 ***(18)***. A few data were also reported on eucaryotic microorganisms: *Saccharomyces cerevisiae* ***(19–21)*** and *Neurospora crassa* ***(22)***.

In this review, we present various technical and practical aspects related to the use of these three reagents and ^{23}Na NMR spectroscopy to quantify sodium gradients in microorganisms.

2. Materials

1. $Dy(TTHA)^{3-}$: This reagent is prepared according to the method of Chu et al. ***(8)***. A stock solution is prepared (typically 100 m*M*) because it will be added to the extracellular medium of the cell suspension. $DyCl_3{\bullet}6H_2O$ (Aldrich), and H_6TTHA (Fluka) are mixed in equimolar amounts in water. This chemical reaction produces $Dy(TTHA)^{3-}$ $3H^+$and $3H^+Cl^-$; the resulting pH of the solution is thus acidic and is neutralized by addition of a base. Depending on the base, one can choose the nature of the conterion of the reagent according to the experiment to carry out. The most commonly used bases are KOH, NaOH, or choline hydroxide ***(13)***; they will generate ($Dy(TTHA)^{3-}$, $3K^+$), ($Dy(TTHA)^{3-}$, $3Na^+$), or ($Dy(TTHA)^{3-}$, 3choline$^+$).
2. $Dy(PPPi)_2^{7-}$: This reagent is prepared as follows: to a 100 m*M* solution of $DyCl_3{\bullet}6H_2O$ (Aldrich) in water under agitation with a magnetic stirrer is added drop by drop a solution of $Na_5(PPPi)$ (Sigma). The theoretical molar ratio is 1:2, leading to the formation of ($7Na^+$, $Dy(PPPi)_2^{7-}$) and $3Na^+,Cl^-$; however, it is recommended to assess the formation of the complex by observation: the right complex is obtained when the mixture turns from a white (undissolved 1:1 complex) to a limpid solution (1:2 ratio, soluble complex). Note that in this reagent solution 10 Na^+ are present. As $Na_5(PPPi)$ is the only commercially available salt, if one wants to change the nature of the conterion Na^+, it should be exchanged by passing the solution through an ionic exchange resin: for instance, for K^+, one can use an AG50W-X8 resin (Bio-Rad) containing the potassium form ***(9)***; for choline$^+$ the procedure is more complex: a H^+ resin should be used (Chelex 100, 50-100 Mesh, Bio-Rad) and H^+ must be exchanged by choline$^+$ by passing choline hydroxide (Aldrich, 2*N*). The pH is controlled and the exchange reaction is stopped when it is neutral; the last step consists in passing $Na_5(PPPi)$ over this column ***(13)*** (*see* **Note 1**).

3. $Tm(DOTP)^{5-}$: $Tm(DOTP)^{5-}$ $5Na^+$ is commercially available from Macrocyclics (Dallas, TX) and is only in Na^+ form; it can be used directly without further preparation (*see* **Note 2**).

3. Methods

3.1. ^{23}Na NMR Experiment Setup

3.1.1. Cell Preparation

After cultivation under appropriate conditions, the microorganism under study has to be harvested by centrifugation. After washings in the appropriate buffer, the cells must be finally resuspended as concentrated as possible, and certainly at approx 10 mg protein/mL in the case of bacteria. If it is desired to monitor the intracellular Na^+ concentration variations according to the external ones (regulation of the Na^+ transmembrane gradient), the cells are resuspended in several buffers containing various Na^+ concentrations. The cell suspension is then transferred to a 10-mm-diameter NMR tube. The shift reagent is generally added just before the NMR tube is placed in the spectrometer.

3.1.2. Calibration of the Internal Reference Capillary

In order to quantify the intracellular and extracellular sodium concentrations, a reference capillary (Bruker) (*see* **Note 3**) is centered in the 10-mm-diameter NMR tube (*see* **Fig. 2**). As the capillary sodium area is used to quantify the extracellular sodium concentrations, its ^{23}Na NMR resonance must be shifted both from the intracellular and extracellular ^{23}Na NMR resonances. Therefore a shift reagent must be used; usually Na_7 $Dy(PPPi)_2^{7-}$ complex is used as it is easier to get a large chemical shift (*see* **Fig. 1**). It is recommended to dissolve the shift reagent in ^2H_2O, as the deuterium signal will be used later to shim the NMR spectrometer. In order to perfectly calibrate the sodium concentrations, series of ^{23}Na NMR spectra are collected on 10-mm-diameter NMR tubes filled with NaCl solutions, the concentration of which is perfectly known, and containing the capillary. A linear plot is obtained between ^{23}Na NMR signals area and NaCl concentrations, and this plot is used as a reference curve to calculate the sodium concentration in the capillary (*see* **Note 4**).

3.1.3. ^{23}Na NMR Parameters

^{23}Na NMR parameters are dependent on the spectrometer used. However, some general comments can be made: (1) T1 is very short, from 1 to 60 ms ***(23)***, so that the relaxation delay can be also quite short; (2) ^{23}Na NMR receptivity is rather high (9.27) ***(23)***, so that a few scans can be registered to get an acceptable signal to noise ratio. NMR receptivity is a measurement of the sensitivity of the NMR signal of ^{23}Na nucleus, present at natural abundance (which

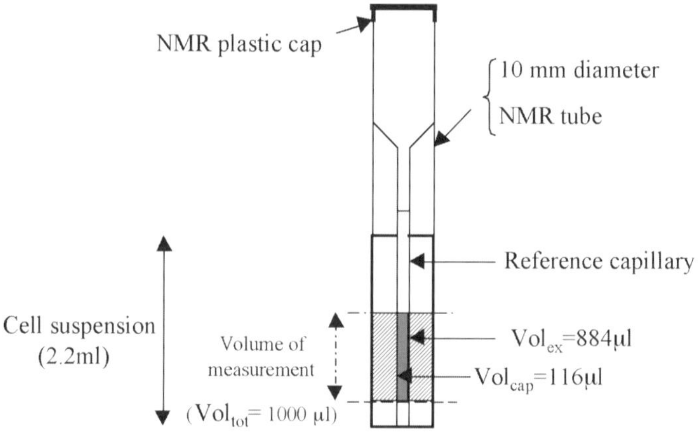

Fig. 2. NMR tube and capillary for ^{23}Na NMR experiments.

is 100%), relative to that of ^{1}H, which is assigned the value 100. Typically, when using a DSX 300 Bruker spectrometer, ^{23}Na spectra were collected at 79.39 MHz every 5 min (90° pulse:12.75 μs, repetition time 360 ms, 2K, AQ = 230 ms, 500 scans), but good quality spectra can be obtained within 2.5 min using 250 scans. A direct Fourier transformation is applied to FIDs and a linefeed correction is made using Bruker software. Integration of ^{23}Na NMR signals is made using a deconvolution routine in the Bruker XWIN NMR program.

3.2. Shift Reagents Efficiency to Separate Internal and External ^{23}Na NMR Resonances

Some general considerations should be made first to understand what factors may influence the shift efficiency of these shift reagents, which are anionic chelates of lanthanides (SR^{x-}). The anionic nature of these chelates (PPPi, TTHA, or DOPT) explains the interaction of the complexes with extracellular Na^+ according to the following equilibrium:

$$Na^+ + SR^{x-1} \rightleftarrows (Na)SR^{(x-1)-}$$

This exchange is very rapid on the time scale of the NMR chemical shift, and the observed resonance is an average of the ^{23}Na NMR resonance of free Na^+ (defined as $\delta - 0$) and the Na^+ complexed with the shift reagent. The paramagnetic nature of the shift reagent containing lanthanide ions causes the shift in resonance of the extracellular sodium; this phenomenon is due mainly to a dipolar (pseudocontact) mechanism. In some cases the difference in bulk magnetic susceptibility can also produce a shift between intra- and extracellular resonance.

Any factor with an influence on this equilibrium will affect the shift efficiency of the shift reagent *(8,23)*: (1) first the ratio Na^+:SR^{x-1}, as it will affect the resulting $(Na)SR^{(x-1)-}$ concentration—in other words, the ^{23}Na resonance is the most shifted when this ratio is the lowest; (2) temperature and ionic strength, as they can change the equilibrium situation; (3) the pH of the solution (H^+ cations) and the presence of K^+, Ca^{2+}, Mg^{2+}, as these cations can compete with Na^+ for interaction with SR^{x-1}; (4) the nature of the anionic chelate, as it is responsible not only for the strength of the SR complex but also of the interaction with sodium cation; (v) the nature of the lanthanide cation, which can influence the shift efficiency.

According to these various factors, the efficiency of the three most frequently used chemical shift reagents to separate internal and external ^{23}Na NMR resonances can be compared.

3.2.1. $Dy(TTHA)^{3-}$

$Dy(TTHA)^{3-}$ presents some advantages: it does not carry Na^+ conterion to the buffer; it is quite insensitive to pH variations, at least in the range of pH 5.5–12, and to the presence of divalent cations (Ca^{2+}, Mg^{2+}, and so on) *(8,24)*; also the linewidth of the external sodium resonance is not enlarged. However, its shifting performance is usually very low. Therefore its application is limited to experimental conditions where external sodium concentration is low (typically 25 m*M*) and the $Dy(TTHA)^{3-}$ concentration rather high (typically 10 m*M*). For instance in *F. succinogenes* *(25)*, when 10 m*M* $Dy(TTHA)^{3-}$ was used in the presence of 25 m*M* extracellular sodium, the following values were measured $\delta = 2.2$ ppm, $\Delta\nu_{1/2} = 35$ Hz. This rather weak performance was also observed in *S. faecalis* *(13)* and *E. coli* *(9)*. Simor et al. *(26)* developed a selective inversion recovery (SIR) pulse sequence to improve spectral resolution in isolated perfused rat hearts. SIR provides a very useful tool to monitor small changes in Na_{in} signal intensity when using Dy(TTHA) that induces a small shift (at least equal to 2.5 ppm).

3.2.2. $Dy(PPPi)_2^{7-}$

$Dy(PPPi)_2^{7-}$ is more efficient than $Dy(TTHA)^{3-}$ for shifting extracellular sodium signal. For instance, in the presence of 6 m*M* $Dy(PPPi)_2^{7-}$ and 60 m*M* Na^+, an upfield shift of about –6 ppm was observed for *F. succinogenes* *(25)*; the same results were reported for *E. coli* ($\delta = -7$ ppm) under rather similar conditions *(9–10)*. In *N. crassa,* a 5 m*M* $Dy(PPPi)_2^{7}$ induced a $\delta = -5$ ppm when the extracellular Na^+ concentration was 25 m*M* *(22)*; in *S. cerevisiae,* a $\delta = -8$ ppm was observed in the presence of a 10 m*M* extracellular Na^+ and 1 m*M* $Dy(PPPi)_2^{7}$ *(19)*. Finally Gilboa et al. *(12)* showed that it was possible to separate intra- and extracellular sodium NMR signals in situations where extracellular sodium was extremely concentrated (2 *M*) when studying the halophilic *V. costicola* in the presence of 50 m*M* $Dy(PPPi)_2^{7-}$. However, the use of this reagent is not always convenient for various reasons:

First, one molecule of $Dy(PPPi)_2^{7-}$ brings 10 Na^+, so changes in buffer composition are usually necessary—namely, replacing Na^+ by choline, to keep extracellular sodium concentrations in a physiological range in the case of nonhalophilic microorganisms such as *E. coli* or *F. succinogenes*. Even with these changes, 60 m*M* Na^+_{ex} is usually the lower limit for a good resolution of intra- and extracellular ^{23}Na resonances, as 6 m*M* of reagent is usually necessary. Another alternative is to exchange sodium ions present in the $Dy(PPPi)_2^{7-}$ solution for choline$^+$ ions, passing the complex through an ion exchange column, but, as largely explained in **Subheading 2.2.**, this procedure is time consuming.

Second, the linewidth of the external Na^+ resonance is very large in the presence of this shift reagent; the following values were reported in the literature: $\Delta\nu_{1/2} = 90$ Hz for *E. coli* ***(9)***, $\Delta\nu_{1/2} = 110$ Hz for *F. succinogenes* ***(25)***. In eucaryotic cells, this enlargement was observed but to a lower extent: $\Delta\nu_{1/2} = 50$ Hz in *N. crassa* ***(22)***, $\Delta\nu_{1/2} = 60$ Hz in *S. cerevisiae* ***(19)***.

Finally, $Dy(PPPi)_2^{7-}$ is very sensitive to external environment, particularly to pH ***(8,24)***; consequently, its efficiency to separate internal and external ^{23}Na NMR resonances is rapidly decreased when the external medium becomes more acidic. This pH dependence made its use difficult for studying lactic-acid-producing bacteria (*E. hirae,* unpublished results) or for studies involving the effect of ionophores on pH gradients (*F. succinogenes,* **ref. *25***).

3.2.3. $Tm(DOTP)^{5-}$

$TmDOTP^{5-}$ was introduced more recently for ^{23}Na NMR experiments but was only tested on *F. succinogenes* as far as bacteria are concerned ***(18,25)***. Its shifting abilities were very good: it could be used at a concentration ≤6 m*M* in buffers containing from 30 to 200 m*M* external sodium concentrations (*see* **Fig. 1**); the observed chemical shifts ranged from 6.5 to 8 ppm. In *N. crassa,* the addition of 2.5 m*M* $TmDOTP^{5-}$ induced a $\delta = 2.5$ ppm when the extracelluar sodium was 25.4 m*M*. Extracellular ^{23}Na NMR resonance was more or less constant when pH changed. In the literature, it has been reported that linewidth of ^{23}Na NMR signals was narrower than that observed in the presence of the other reagents ***(7,27)***. This was attributed to the lower magnetic susceptibility of Tm(III) compared to that of Dy(III) ***(7)***. However, under the experimental conditions used for studying *F. succinogenes,* the same linewidth of Na_{ex} signals was observed in the presence of $TmDOTP^{5-}$ and $Dy(TTHA)^{3-}$; in both cases it was much narrower (one third) than in the presence of $Dy(PPPi)_2^{7-}$ ***(25)***. In *N. crassa* the observed linewidth of Na_{ex} signals was the same as that of *F. succinogenes* ($\Delta\nu_{1/2} = 35$ Hz) when $TmDOTP^{5-}$ was used ***(22)***.

3.3. Control of the Toxicity of Shift Reagents

One of the major points when using chemical-shift reagents is to check that these compounds are not toxic for the microorganisms under study. Although the chelates used are not toxic per se, these reagents can become toxic during ^{23}Na experiments, for two main reasons: First, toxicity can result from interactions of the chelate with cellular cations (Ca^{2+}, Mg^{2+}), disturbing metabolic functions; as a consequence, Ca^{2+} infusions were necessary for in vivo studies in animals ***(28)***. Second, complex stability can be destroyed under in vivo conditions—for instance, it was observed that PPPi can be degraded by extracellular phosphatases, resulting in a release of free Dy^{3+} that can enter the cells and be toxic. Various approaches can be used to assess shift reagents toxicity; however, controls are usually made on the overall metabolism: carbon metabolism is monitored by ^{13}C NMR, energy metabolism (pH, ATP, and so on) by ^{31}P NMR or biochemical assays.

3.3.1. Control by ^{13}C NMR

The utilization of [1-^{13}C] glucose (or another ^{13}C-labeled substrate) by cells incubated in the presence or the absence of shift reagent can be monitored by in vivo ^{13}C NMR (*see* **Note 5**): In the case of *S. faecalis (E. hirae),* $choline_7Dy(PPPi)_2$ (at 2.5 m*M*) did not modify the carbon metabolism of the bacterium ***(13)*** (*see* **Note 6**). The same result was obtained with $Dy(TTHA)^{3-}$ ***(13)*** in this bacterial system. In contrast, it was shown that $Dy(TTHA)^{3-}$ was toxic for *F. succinogenes* cells, as it inhibited glucose degradation in parallel to metabolite synthesis ***(25)***. Finally, $TmDOTP^{5-}$ and $Dy(PPPi)_2{}^{7-}$ were also tested on glucose utilization by *F. succinogenes* and were shown to have no effect on this metabolism ***(25)***.

3.3.2. Control by ^{31}P NMR

^{31}P NMR is also a powerful tool to monitor various parameters of microbial metabolism: First, it can be used to determine intracellular and extracellular pH directly in vivo, just by measuring the chemical shift of Pi, which is pH dependent. Second, concentrations of nucleotide di- and triphosphates such as ADP and ATP can be quantified on ^{31}P NMR spectra ***(29)***. These two parameters give insight into the regulation of the energy metabolism. Also, other phosphorylated metabolites can be monitored, including sugar-phosphates, polyphosphates, 2,3-diphospho-glycerophosphate, and so on, reflecting the whole metabolism. Finally, in the case of $Dy(PPPi)_2{}^{7-}$, the degradation of the chelate into Pi can be checked by measuring the time course of Pi integral. In parallel to the chemical shift of integrals of the various phosphorylated entities detectable on ^{31}P NMR spectra, their linewidth can be measured. Any enlargement of intracellular signals can be related to an effect on membrane permeability and thus to a dysfunction of cell metabolism.

Considering these different aspects, $Dy(PPPi)_2^{7-}$ was shown to be nontoxic towards *E. coli* ***(9)*** and eukaryotic cells such as red cells ***(29–32)***. On the contrary, this reagent was hydrolyzed by phosphatases in *S. saccharomyces* ***(19)***, *N. crassa* ***(22)***, or *Dictyostelium discoideum* ***(33)***. $Dy(TTHA)^{3-}$ induced enlargement of ^{31}P signals in the latter organism, while it was shown to have no effect on *S. faecalis* ***(13)***. No data were reported on phosphorus metabolism in microorganisms using $TmDOTP^{5-}$; however, it did not alter the cellular energy state of rat liver ***(7)*** nor that of subcutaneous implanted gliosarcoma in rats ***(34)***, and it did not induce any cell damage in *N. crassa* ***(22)***.

Complementary data can be collected about the energy status of the cell by using classical biochemical techniques such as ATP assay by the luciferine-luciferase method or the measurement of membrane potential by using $[^3H]TPP^+$ ***(9)***.

3.4. Calculation of Sodium Gradients

Sodium transmembrane gradient value (ΔpNa) can be calculated from the following equation:

$$\Delta pNa = 2.3 \times RT/ZF \times \log C_{in}/C_{ex} \text{ (in mV)}$$

$R = 8.32$ J/Kmol; T = temperature (in K); $F = 96{,}500$; $Z = 1$ (cation charge)
$\Delta pNa = 62 \times \log C_{in}/C_{ex}$ (in mV); C_{in} = intracellular sodium concentration (m*M*); C_{ex} = extracellular sodium concentration (m*M*); C_{ex} is usually fixed in the experimental procedure. C_{in} can be measured from the integrals of ^{23}Na resonances on NMR spectra as follows:

$$C_{in} = [(k \times I^{eff}_{in})/I_{cap}]/Vol_{in}$$

Vol_{in} = intracellular volume (µL); I_{cap} = integral of ^{23}Na NMR signal of the reference capillary; k = calibration factor of capillary (µmol)

$$I^{eff}_{in} = I_{in} \times Vis_{in}$$

where I_{in} = integral of intracellular ^{23}Na NMR signal; Vis_{in} = visibility factor of Na^+_{in}

Finally,

$$C_{in} = [(k \times I_{in} \times Vis_{in})/I_{cap}]/Vol_{in}] \quad (1)$$

In this equation, two sets of data are difficult to measure, mainly because bacteria are small organisms and represent less than 10% of the total volume of the cell suspension: first, the intracellular volume (Vol_{in}); second, "NMR visibility factor" (Vis_{in}), due to ^{23}Na properties.

3.4.1. Intracellular Volume Measurement

Various techniques can be applied that are based on the use of two isotope tracers: one molecule is supposed to stay in the extracellular compartment while the other one diffuses passively across the cytoplasmic membrane and is thus present in both compartments. These markers can be radioisotopes or NMR-detectable isotopes. Two examples will be given:

1. Radioisotopes: Internal volume estimates for the cells can be made on the basis of the sucrose-impermeable space, using ^{14}C-sucrose and $^{3}H_2O$ tracers as initially described by Stock et al. ***(35)***. Other molecules used as extracytoplasmic markers include inulin, polyethylene glycol, or taurine ***(36)***. In the case of *F. succinogenes,* the experimental conditions were as follows: Cells prepared as for the NMR experiments (10 mg protein/mL) were incubated 15 min anaerobically with 0.5 μCi of ^{14}C-sucrose and $^{3}H_2O$ (3.3 μCi/mL each). Samples were then centrifuged in 1.5-mL tubes through oil to separate cells and extracellular fluid, as previously described ***(36)***. Silicon oil (Rhodorsil 508V70) was used in our experiments, but other oils such as bromododecane are preferred by some authors ***(36)***. 50-μL samples of the supernatants were dissolved in scintillation liquid and counted. The bottom of the tubes containing the cell pellet was cut and cells were dispersed by vigorous vortexing in the scintillation cocktail before ^{3}H and ^{14}C counting.

   Intracellular space (in μL/mg protein) was calculated by subtracting the sucrose-accessible volume from the water-accessible volume.
2. NMR detectable isotopes: The intracellular volume of de-energized cells of *Streptococcus faecalis* was determined indirectly by NMR ***(13)***. The extracellular volume was given by the ^{23}Na signal of external Na^{+} that was unable to enter the cells under these experimental conditions. The total volume was assessed by ^{2}H-NMR based on the $^{2}H_2O$ resonance present in the whole sample. This technique was derived from that described by Shinar and Navon ***(37)***, replacing ^{59}Co by ^{23}Na NMR. In their work they used a ^{59}C complex ($K_3Co(CN)_6$) for measuring the external volume, assuming that this complex stayed outside the cells. In the case of *S. faecalis,* the experimental conditions were as follows: De-energized bacterial cells were resuspended in a choline buffer (100 m*M* choline chloride, 2 m*M* acid EDTA, 40 m*M* MES, 5 m*M* $MgSO_4$, 200 μ*M* phosphoric acid/L, pH adjusted to 7.3 by choline hydroxide) to a final concentration of 20 mg protein/mL. 75 μL of H_2O and 30 μL of 1 m*M* NaCl were added to 3 mL of cell suspension. ^{2}H and ^{23}Na NMR spectra were collected (*see* **Note 7**) and then the suspension was centrifuged at 15,000*g*. A known volume of the supernatant (Δ) was again analyzed by ^{2}H and ^{23}Na NMR and NMR integrals measured. If I_C^D is the deuterium integral of the cell suspension, I_S^D the deuterium integral of the supernatant, I_C^{Na} the sodium integral of the supernatant, Vol_{tot} the total volume, and Vol_{ex} the external volume. The volume of the capillary in the measurement volume (Vol_{cap}) and the intracellular volume (Vol_{in}) can be determined as follows:

$$Vol_{in} = Vol_{tot} - Vol_{ex} - Vol_{cap} = (I_C^D \Delta\ I_S^D) - (I_C^{Na} \Delta\ I_C^{Na}) - Vol_{cap}$$

3.4.2. Sodium NMR "Visibility"

The "visibility" factor is related to the fact that intracellular sodium was shown not to be always 100% detectable ("visible"). This is due to static and/or dynamic quadrupolar effects caused by microenvironments of the Na^+ ion in the cell ***(23,38)***. NMR visibility has been shown to vary—40% in yeast ***(19)***, 60% in *S. faecalis* ***(13)***, 100% in red cells ***(29,39)***; however it is sometimes difficult to measure accurately, and some contradictory results were obtained for similar cells ***(23,38)***. Two main approaches can be used to assess sodium visibility NMR factor: (i) comparison between intracellular and extracellular ^{23}Na signals during the time course of sodium influx or efflux, and (ii) comparison of Na^+ content measured with a technique other than NMR.

1. Comparison of the evolution of ^{23}Na integrals: This approach is based on the fact that extracellular sodium is usually 100% NMR visible; therefore, it is possible to compare intracellular and extracellular ^{23}Na signals under at least two different conditions where Na^+ ions are taken up or expelled by the cell. If intracellular sodium is 100% visible, then the sum of intracellular and extracellular ^{23}Na integrals remains constant; if not, its variation can be related to the visibility factor as follows:

$$\text{First spectrum: } I^{eff}_{ex} + I^{eff}_{in} = x \quad (1)$$

$$\text{Second spectrum (after influx of Na}^+\text{): } I'^{eff}_{ex} + I'^{eff}_{in} = y \quad (2)$$

$$I^{eff}_{in} = I_{in} \times Vis_{in}$$

$$I^{eff}_{ex} = I_{ex} \times Vis_{ex}; \text{ if } Vis_{ex} = 1,\ I^{eff}_{ex} = I_{ex}$$

So (1) can be written:

$$I^{eff}_{ex} + I_{in} \times Vis_{in} = x$$

And (2) can be written:

$$I'^{eff}_{ex} + I_{in} \times Vis_{in} + (I^{eff}_{ex} - I'^{eff}_{ex}) \times Vis_{in} = y$$

$(I^{eff}_{ex} - I'^{eff}_{ex})$ corresponds to the variation of the integrals of the external sodium due to the influx of Na^+ into the cell.

$$(1) - (2) = (I^{eff}_{ex} - I'^{eff}_{ex}) - (I^{eff}_{ex} - I'^{eff}_{ex})\ Vis_{in} = x - y$$

Finally:

$$Vis_{in} = 1 - [(x - y)/\ (I^{eff}_{ex} - I'^{eff}_{ex})]$$

The limitation of this method is the precision of the determination of $(I^{eff}_{ex} - I'^{eff}_{ex})$ and $(x - y)$; these variations must be large enough to be far from error measurements. It has been applied successfully with erythrocytes ***(29,39)*** and tissues *(23)*. In the case of microorganisms, the intracellular sodium signal is very small compared to that of the extracellular one, as the total cell volume is in the range of 1–10% of the total volume; consequently, this approach cannot be applied under physiological conditions. Nevertheless this strategy was used for measuring the sodium visibility in *S. cerevisiae* and *E. coli*, as the authors used concentrated sodium-loaded cells ***(9,19–20)*** and made the measurements on the cell pellet.

2. Comparison with another analytical method: The basis of this approach is straightforward: the intracellular sodium concentration is measured by ^{23}Na NMR in parallel to another analytical method, typically flame photometry or atomic absorption. The assumption is made that the latter methods quantify all the Na^+ cations; therefore, any lower concentration measured by NMR is attributed to sodium invisibility, and the comparison of the two values allows the invisibility factor to be calculated. This approach has been used in the case of *S. faecalis* ***(13)***, and in eucaryotic cells ***(23,38)***. However, because the intracellular volume was small and because the sodium gradients were usually $[Na^+_{ex}] >> [Na^+_{in}]$, it was difficult to determine $[Na^+_{in}]$, and thus the exact visibility factor, accurately.

3.4.3. Use of Ionophores

As shown in **Subheadings 3.4.1.** and **3.4.2.**, the determination of intracellular volume and sodium NMR visibility factor is time consuming, and accurate measurements are difficult. Consequently, an alternative method has recently been described in the case of *F. succinogenes,* using ionophores ***(18,25)***. This methodology is based on the assumption that the ionophores used are efficient in equalizing intra- and extracellular sodium concentrations. Sodium gradient is only collapsed when K^+, Na^+, and H^+ transmembrane gradients are completely dissipated, so a mixture of three membrane effectors must be used: monensin (Na^+/H^+ exchange), valinomycin (selective for K^+ transport), and CCCP (protonophore). In the case of *F. succinogenes* ***(18,25)***, ^{23}Na NMR spectra were collected every 5 min for 15 min, then ionophores (monensin at 90 μ*M* and valinomycin and CCCP at 30 μ*M* each) were added and three extra spectra were collected (*see* **Fig. 1B**).

In the presence of ionophores, ^{23}Na NMR signal integrals can be used to access intracellular volume, as in that case the intracellular Na^+ concentration is known to be equal to the external one:

$$C'_{in} = C'_{ex} = [(k \times I'_{in} \times Vis_{in})/I'_{cap}]/Vol_{in} \quad (2)$$

C'_{in} = intracellular sodium concentration (m*M*) when ionophores were added to the cells; C'_{ex} = extracellular sodium concentration (m*M*) when ionophores were added to the cells; Vol_{in} = intracellular volume (μL); I^{pr}_{cap} = integral of the reference capillary when ionophores were added to the cells; k = calibration factor of capillary (μmol); I'_{in} = integral of intracellular ^{23}Na NMR signal when ionophores were added to the cells; Vis_{in} = visibility factor of Na^+_{in}

$$C'_{ex} = [(k \times I'_{ex} \times Vis_{ex})/I'_{cap}]/Vol_{ex} \quad (3)$$

I'_{ex} = integral of extracellular ^{23}Na NMR signal when ionophores were added to the cells; Vis_{ex} = visibility factor of Na^+_{ex}.

Under these conditions, extracellular sodium visibility was checked to be 100% by comparing the real Na^+ concentration and that measured from ^{23}Na NMR signals of external sodium, consequently $Vis_{ex} = 1$.

Vol_{ex} = external volume; it can be determined as follows for 1 mL of measurement volume (Vol_{tot}) (*see* **Fig. 2**):

$$Vol_{ex} = Vol_{tot} - Vol_{in} - Vol_{cap}$$

Under these conditions Vol_{cap} (volume of the capillary in the measurement volume) = 116 μL and Vol_{in} can be neglected, as $Vol_{in} << Vol_{ex}$ (approx 1%); so finally, Vol_{ex} = 1000 μL – 116 μL = 884 μL.

From **Eq. 3**: $C'_{ex} = [(k \times I'_{ex} \times 1)/I'_{cap}]/884$

Putting C'_{ex} into **Eq. 2**,

$[(k \times I'_{ex} \times 1)/I'_{cap})]/884 = [(k \times I'_{in} \times Vis_{in})/I'_{cap}]/Vol_{in}$ and $Vol_{in} = [(k \times I'_{in} \times Vis_{in})/I'_{cap}] \times [884/(I'_{ex} \times 1/I'_{cap}) \times k] = [(I'_{in} \times Vis_{in} \times 884)/I'_{ex}]$

Putting Vol_{in} into **Eq. 1**:

$$C_{in} = [(k \times I_{in} \times Vis_{in})/I_{cap}]/[(I_{in} \times Vis_{in} \times 884)/I_{ex}]$$

And finally:

$$C_{in} = (I_{in} \times I'_{ex} \times k)/(I_{cap} \times I'_{in} \times 884) \quad (4)$$

It is worth noting that in **Eq. 4** the visibility factor of intracellular sodium (Vis_{in}) is no longer present, as it is similar in the two spectra.

In conclusion, this methodology using ionophores avoids having to measure both the visibility factor of intracellular sodium and the intracellular volume.

3.5. General Conclusion

^{23}Na NMR spectroscopy is a very convenient method to measure sodium gradients in microorganisms, directly in vivo, under various environmental conditions. Some specific problems linked to this technique can be easily overcome—namely, the toxicity of shift reagents, which can be checked, and the NMR visibility, which can be measured or even avoided by using ionophores. The main question is to choose the best shift reagent. It is difficult to give general conclusions about the relative toxicity of these reagents as well as their efficiency to separate intra- and extracellular ^{23}Na NMR signals, because these reagents have not been tested systematically on the same biological model. However comparative studies have been made for the three reagents on *F.*

succinogenes **(*25*)** and on isolated perfused rat hearts **(*27*)**. The conclusion of these two studies were in agreement and confirmed other independent observations: $Dy(TTHA)^{3-}$ is the less efficient reagent; $Dy(PPPi)_2{}^{7-}$ creates the largest chemical shift, but its use is limited because it brings 10 Na^+ ions and because it is sensitive to pH and interacts with Ca^{2+} cations. $TmDOTP^{5-}$ seems to be the most efficient shift reagent. Concerning toxicity, some problems have been reported for $Dy(TTHA)^{3-}$ and $Dy(PPPi)_2{}^{7-}$, while cellular metabolism was not disturbed in the case of $TmDOTP^{5-}$.

In conclusion, we suggest, as a first approach, the use of $TmDOTP^{5-}$ in combination with ^{23}Na NMR and ionophores to measure sodium gradients, although other conditions can be satisfactory as well.

4. Notes

1. The addition of tripolyphosphate salt drop by drop is a key point for the good formation of the complex, especially when using the choline$^+$ salt. Indeed, when this salt comes out of the resin column it is very viscous, and it is difficult to assay the exact amount of $choline_{10}$ $(PPPi)_2$ salt and thus to calculate the exact ratio 1:2. By monitoring the formation of the complex, it is not necessary to evaluate precisely the concentration of tripolyphospohate salt; the ratio is controlled by the exact concentration of $DyCl_3$.
2. For unexplained reasons, some batches of $Tm(DOTP)^{5-}$ induced a slow Na^+ ion leakage from the cells. Consequently, every batch was carefully checked for not inducing such cell leakage before use.
3. The internal diameter of the NMR tube must be perfectly calibrated to fit with the capillary. Indeed, we found some NMR tubes that do not fit with the capillary.
4. Note from **Fig. 2** that although the relative volumes of the capillary and the NaCl solution are different, they are exactly the same as those encountered when bacteria are used instead of NaCl; therefore, the calibration factor (the equivalent Na^+ concentration in the capillary) will be directly used in the subsequent calculations (*see* **Subheading 3.4.**).
5. A detailed description of in vivo ^{13}C NMR experiments is presented in Chapter 18 of this book, entitled: "^{13}C and ^1H NMR Study of Glycogen Futile Cycle in *Fibrobacter succinogenes*," by Delort et al.
6. On the ^{13}C NMR spectrum recorded in the presence of the shift reagent, it was possible to distinguish the internal and the external lactate signal due to a strong magnetic susceptibility effect. Other investigators observed two distinct peaks of lactate in ^{13}C NMR spectra of *Saphylococcus aureus* and *Lactococcus lactis* **(*40,41*)**. This property can be used to measure lactate transport in these bacteria.
7. ^{23}Na NMR parameters were as described in **Subheading 3.1.3**. ^2H NMR spectra were performed at 46.07 MHz in one pulse experiment (90° pulse). Direct Fourier transform was applied for measuring deuterium integrals.

References

1. Madigan, M. T. and Oren, A. (1999) Thermophilic and halophilic extremophiles. *Curr. Opin. Microbiol.* **2,** 265–269.
2. Ventosa, A., Nieto, J. J., and Oren, A. (1998) Biology of moderately halophilic aerobic bacteria. *Microbiol. Mol. Biol. Rev.* **62,** 504–544.
3. Sakhnini, A. and Gilboa, H. (1993) Double quantum sodium NMR studies of the halotolerant Ba_1 bacterium. *Biophys. Chem.* **46,** 21–25.
4. Colet, J.-M., Bansal, N., Malloy, C. R., and Sherry, A. D. (1999) Multiple quantum filtered ^{23}Na NMR spectroscopy of the isolated, perfused rat liver. *Magn. Reson. Med.* **41,** 1127–1135.
5. Gupta, R. K. and Gupta, P. (1982) Direct observation of resolved resonances from intra- and extracellular sodium-23 ions in NMR studies of intact cells and tissues using dysprosium(III) tripolyphosphate as paramagnetic shift reagent. *J. Magn. Reson.* **47,** 344–350.
6. Lazar, I., Hrncir, D. C., Kim, W.-D., Kiefer, G. E., and Sherry, A. D. (1992) Optimized synthesis structure, and solution dynamics of 1,4,7,10-tetraazacyclododecane-1,4,7,10-tetrakis (methylenephosphonic acid) (H_8DOTP). *Inorg. Chem.* **31,** 4422–4424.
7. Bansal, N., Germann, M. J., Seshan, V., Shires, G. T,. Malloy, C. R, and Sherry, A. D. (1993)Thulium 1,4,7,10-tetra-azacyclododecane-1,4,7,10-tetrakis (methylenephosphonate) as a ^{23}Na shift reagent for the in vivo rat liver. *Biochemistry* **32,** 5638–5643.
8. Chu, S. C., Pike, M. M., Fossel, E. T., Smith, T. W., Balschi, J. A., and Springer, C. S. (1984) Aqueous shift reagent III, Dy(TTHA)$^{3-}$, and Tm(PPP)$_2^{7-}$. *J. Magn. Reson.* **56,** 33–47.
9. Castle, A. M., Macnab, R. M., and Shulman, R. G. (1986) Measurement of intracellular sodium concentration and sodium transport in *Escherichia coli* by ^{23}Na nuclear magnetic resonance. *J. Biol. Chem.* **261,** 3288–3294.
10. Castle, A. M., Macnab, R. M., and Shulman, R.G. (1986) Coupling between the sodium and proton gradients in respiring *Escherichia coli* cells measured by ^{23}Na and ^{31}P nuclear magnetic resonance. *J. Biol. Chem.* **261,** 7797–7806.
11. Nagata, S., Adachi, K., Shirai, K., and Sano, H. (1995) ^{23}Na NMR spectroscopy of free Na^+ in the halotolerant bacterium *Brevibacterium* sp. and *Escherichia coli. Microbiology* **140,** 729–736.
12. Gilboa, H., Kogut, M., Chalamish, S., Regev, R., Avi-Dor, Y., and Russell, N. J. (1991) Use of ^{23}Na nuclear magnetic resonance spectroscopy to determine the true intracellular concentration of free sodium in a halophilic *Eubacterium. J. Bacteriol.* **173,** 7021–7023.
13. Delort, A. M., Dauphin, G., Guyot, J., and Jeminet, G. (1989) Study by NMR of the mode of action of monensin on *Streptococcus faecalis* de-energized and energized cells. *Biochim. Biophys. Acta.* **1013,** 11–20.
14. Rabaste, F., Dauphin, G., Guyot, J., Jeminet, G., and Delort, A. M. (1991) A phosphate dependent sodium transport in *S. faecalis* investigated by ^{23}Na and ^{31}P NMR. *Biochem. Biophys. Res. Com.* **181,** 74–79.

15. Rabaste, F., Jeminet, G., Dauphin, G., and Delort, A. M. (1992) Conditions modulating the ionic selectivity of transport by monensin examined on *Enterococcus hirae (S. fæcalis)* by ^{23}Na NMR and K^+ atomic absorption. *Biochim. Biophys. Acta.* **1108,** 177–182.
16. Rabaste, F., Dauphin, G., Guyot, J., Jeminet, G., and Delort, A. M. (1992) Monensin-induced modifications of ionic gradients on *Streptococcus faecalis.* A ^{23}Na and ^{31}P NMR study. *J. Chim. Phys.* **89,** 217–225.
17. Rabaste, F., Dauphin, G., Guyot, J., Jeminet, G., and Delort, A. M. (1993) Na^+ and K^+ transport by 4-chlorophenylurethane-monensin in *Enterococcus hirae* de-energized and energized cells studied by ^{23}Na-NMR and K^+ atomic absorption. *Biochim. Biophys. Acta* **1179,** 166–169.
18. Schwaab, V., Matheron, C., Delort, A. M., Gaudet, G., and Forano, E. (2001) In vivo ^{23}Na nuclear magnetic resonance study of maintenance of a sodium gradient in the ruminal bacterium *Fibrobacter succinogenes* S85. *Appl. Environ. Microbiol.* **67,** 4390–4392.
19. Ogino, T., den Hollander, J. A., and Shulman, R. G. (1983) ^{39}K, ^{23}Na, ^{31}P NMR studies of ion transport in *Saccharomyces cerevisiae. Proc. Natl. Acad. Sci.* **80,** 5185–5189.
20. Höfeler, H., Jensen, D., Pike, M. M., Delayre, J., Cirillo, V. P., Springer, C. S. Jr., et al. (1987) Sodium transport and phosphorus metabolism in sodium-loaded yeast: simultaneous observation with sodium-23 and phosphorus-31 NMR spectroscopy in vivo. *Biochemistry* **26,** 4953–4962.
21. Balschi, J. A., Cirillo, V. P., and Springer, C. S. (1982) Direct high-resolution nuclear magnetic resonance studies of cation transport in vivo, Na^+ transport in yeast cells. *Biophys. J.* **38,** 323–326.
22. Greenfield, N. J., Cherapak, C. N., Adebodum, F., Jordan, F., and Lenard, J. (1990) Intracellular sodium content of a wall-less strain of *Neurospora crassa* and effects of insulin: a ^{23}Na-NMR study. *Biochim. Biophys. Acta* **1025,** 15–20.
23. Springer, C. S. (1987) Measurement of metal cation compartmentation in tissue by high resolution metal cation NMR. *Annu. Rev. Biophys. Chem.* **16,** 375–399.
24. Chu, S. C. K., Qiu, H. Z. H., Springer, C. S., and Wishnia, A. (1990) Aqueous shift reagents for high-resolution cation NMR. V. Thermodynamics of interaction of $DyTTHA^{3-}$ with Na^+, K^+, Mg^{2+} and Ca^{2+}. *J. Magn. Reson.* **87,** 287–303.
25. Delort, A. M., Gaudet, G., and Forano, E. (2002) ^{23}Na NMR study of *F. succinogenes* S85: comparison of three different shift reagents and calculation of sodium concentration using ionophores. *Anal. Biochem.* **306,** 171–180.
26. Simor, T., Kim, S. K., Chu, W. J., Prohost, G. M., and Elgavish, G. A. (1993) A selective inversion recovery method for improvements of ^{23}Na NMR spectral resolution in isolated perfused rat hearts. *NMR Biomed.* **6,** 201–208.
27. Simor, T., Lorand, T., Szöllösy, A., Gaszner, B., Digerness, S. B., and Elgavish, G. A. (1999) ^{23}Na NMR shift reagents enhance cardiac staircase effect in isolated perfused rat hearts. *NMR Biomed.* **12,** 267–274.
28. Elgavish, G. A. (1989) Shift-reagent-aided ^{23}Na NMR spectroscopy. *Invest. Radiol.* **24,** 1028–1033.

29. Rochdi, M., Delort, A. M., Guyot, J., Sancelme, M., Dauphin, G., and Jeminet, G. (1994) A new Na^+ and K^+ carrier from chemically modified monensin studied in human erythrocytes by ^{23}Na nuclear magnetic resonance, K^+ atomic absorption and H^+ potentiometry. *Bioelectrochemistry and Bioenergetics* **33,** 83–90.
30. Rochdi, M., Delort, A.-M., Guyot, J., Sancelme, M., Gibot, S., Gourcy, J.-G., et al. (1996) Ionophore properties of monensin derivatives studied on human erythrocytes by ^{23}Na NMR and K^+ and H^+ potentiometry: relationship with antimicrobial and antimalarial activities. *J. Med. Chem.* **39,** 588–595.
31. Gibot, S., Jeminet, G., Juillard, J., Gumila, C., Ancelin, M.-L., Vial, H., et al. (1999) Cationomycin and monensin partition between serum proteins and erythrocyte membrane: consequences for Na^+ and K^+ transport and antimalarial activities. *Arch. Biochem. Biophys.* **363,** 361–372.
32. Gupta, R. K. and Gupta, P. (1982) Direct observation of resolved resonances from intra- and extracellular sodium-23 ions in NMR studies of intact cells and tissues using dysprosium(III) tripolyphosphate as paramagnetic shift reagent. *J. Magn. Reson.* **47,** 344–350.
33. Martin, J. B., Klein, G., and Satre, M. (1987) ^{23}Na NMR study of intracellular sodium ions in *Dictyostelium discoideum* amoeba. *Arch. Biochem. Biophys.* **254,** 559–567.
34. Winter, P. M. and Bansal, N. (2001) $TmDOTP^{5-}$ as a shift reagent for subcutaneously implanted 9L gliosarcoma in rats. *Magn. Reson. Med.* **45,** 436–442.
35. Stock, J. B., Rauch, B., and Roseman, S. (1977) Periplasmic space in *Salmonella typhimurium* and *Escherichia coli. J. Biol. Chem.* **252,** 7850–7861.
36. Cayley, D. S., Guttman, H. J., and Record, M. T. (2000) Biophysical characterization of changes in amounts and activity of *Escherichia coli* cell and compartment water and turgor pressure in response to osmotic stress. *Biophys. J.* **78,** 1748–1764.
37. Shinar, H. and Navon, G. (1985) The determination of intracellular water space by NMR. *FEBS Lett.* **193,** 75–78.
38. Rooney, W. D., and Springer Jr., C. S. (1991) A comprehensive approach to the analysis and interpretation of the resonances of spins 3/2 from living systems. *NMR Biomed.* **4,** 209–226.
39. Ogino, T., Shulman, G. I., Avison, M. J., Gullans, S. R., den Hollander, J. A., and Shulman, R. G. (1985) ^{23}Na and ^{39}K NMR studies of ion transport in human erythrocytes. *Proc. Natl. Acad. Sci.* **82,** 1099–1103.
40. Ezra, F. S., Lucas, D. S., Mustacich, R. V., and Russel, A. F. (1983) Phosphorus-31 and carbon-13 nuclear magnetic resonance studies of anaerobic glucose metabolism and lactate transport in *Staphylococcus aureus* cells. *Biochemistry* **22,** 3843–3849.
41. Neves, R. A. (2001) Metabolic strategies to reroute carbon fluxes in *Lactococcus lactis:* kinetics of intracellular metabolites pools by in vivo nuclear magnetic resonance. Ph.D. Thesis, Universidade Nova de Lisboa.

38

Denaturing Gradient Gel Electrophoresis (DGGE) as a Fingerprinting Tool for Analyzing Microbial Communities in Contaminated Environments

Laurent Eyers, Spiros N. Agathos, and Said El Fantroussi

1. Introduction

1.1. Advantages of DGGE/TGGE

The characterization of microbial communities has been very limited due to the lack of appropriate methods. The traditional selection of pure cultures and further study of their physiological and biochemical properties is not well adapted to the study of microbial communities. Indeed, it is assumed that nearly 99% of the microorganisms present in nature cannot be isolated and cultivated because of our ignorance of their physiological needs *(1)*. Therefore, novel molecular techniques have been developed to compensate for the lack of cultivation methods.

One of the available molecular techniques is the construction of 16S ribosomal DNA (rDNA) libraries. Briefly, the DNA is extracted from the ecosystem under study and its 16S rDNA is amplified by PCR and ligated into plasmid vectors. Further sequencing of 16S rDNA fragments of the library and comparison of the sequences with other 16S rDNA sequences can provide information on the identity of the bacteria of the community. However, this technique is labor-intensive, time-consuming, and quite expensive. Furthermore, the evolution of the microbial composition over time cannot be easily monitored using this technique. For these reasons, a simple, rapid, and relatively cheap molecular method was needed. To meet these criteria, Muyzer and his coworkers *(2)* developed a method combining the amplification power of PCR on 16S rDNA of complex microbial populations together with the possibility of denaturing

From: *Methods in Biotechnology: Environmental Microbiology: Methods and Protocols*
Edited by: J. F. T. Spencer and A. L. Ragout de Spencer

gradient gel electrophoresis (DGGE) to separate DNA fragments of the same size but differing according to their base-pair sequences.

1.2. Principles of DGGE/TGGE

In DGGE systems, double-stranded DNA fragments are run through a polyacrylamide gel containing a linear gradient of denaturing agents (a mixture of urea and formamide) at a constant temperature (generally 60°C). During the migration in the gel, the fragment remains double-stranded until it reaches a concentration of denaturing agents equivalent to a melting temperature that causes the lower melting part (also called "domain") of the molecule to melt. At this stage, the fragment changes from a helical structure to a partially melted molecule, which practically halts its migration through the gel.

In TGGE systems, the gradient of chemical denaturing agents is replaced by a gradient of temperature. The same principle governs the migration of the fragments. In this way, two fragments in which an AT is replaced by a GC can be theoretically separated and visualized. However, it is basically estimated that only 50% of all the single-base-pair changes in a fragment of 50 to several hundred bp can be detected by DGGE *(3)*. To overcome this limitation, the fragments to be separated can be previously amplified by PCR using a pair of "modified" primers. The modification consists in the adjunction of a GC-rich sequence (generally 40 bp) to the 5' end of one of the primers *(4)*. This prevents the complete dissociation of the fragments during migration in the gel and therefore the loss of sequence-dependent gel migration upon complete strand separation *(4)*.

Bacterial and fungal communities can be analyzed by DGGE after PCR amplification of 16S rDNA and 18S rDNA genes, respectively. Primers are designed to specifically hybridize to conserved regions of the bacterial 16S rDNA or fungal 18S rDNA genes. By using group-specific primers, it is also possible to target definite communities, like the *Actinomycetes* ***(5)*** or the *Archaea* ***(6)***. Although 16S rDNA or 18S rDNA are the genes mostly targeted, other specific genes can be amplified. Accordingly, Wawer et al. *(7)* analyzed the sequence diversity of the [NiFe] hydrogenase gene of *Desulfovibrio* species, an important group of sulphate-reducing bacteria.

A common set of primers used to amplify the rDNA of bacterial and fungal communities is listed in **Table 1.**

1.3. General Strategy

The general strategy to obtain DGGE/TGGE fingerprints and exploit the results is as follows (**Fig. 1**): First of all, the DNA has to be isolated from the ecosystem, which can range from natural samples (sediments, soils, drinking water, air, and so on) to artificial systems (cultures, bioreactors, etc.). Because

Table 1
Primers Used for DGGE

Pair of primers[a]	Size of the PCR product (bp)[b]	Target	Sequence	Reference
P63f- P518r	511	*Bacteria*	5'-CAGGCCTAACACATGCAAGTC-3' 5'-ATTACCGCGGCTGCTGG-3'	*(31)*
P338f- P518r	236	*Bacteria*	5'-ACTCCTACGGGAGGCAGCAG-3' 5'-ATTACCGCGGCTGCTGG-3'	*(2)*
Arch340f- Arch519r	233	*Archaea*	5'-CCCTACGGGG(C/T)GCA(G/C)CAG-3' 5'-TTACCGCGGC(G/T)GCTG-3'	*(6)*
Act243f- Act513r	342	*Actinomycetes*	5'-GGATGAGCCCGCGGCCTA-3' 5'-CGGCCGCGGCTGCTGGCACGTA-3'	*(5)*
NS2f- Fungr5r	230	Fungi	5'-GGCTGCTGGCACCAGACTTGC-3' 5'-GTAAAAGTCCTGGTTCCC-3'	*(32)*
GC clamp[c]			5'-CGCCCGCCGCGCGCGGCGGGC GGGGCGGGGGCACGGGGGG-3'	*(2)*

[a]f, forward primer; r, reverse primer.
[b]GC clamp included.
[c]GC clamp attached to the 5' end of the reverse primer.

the DNA has to be amplified by PCR, it has to be clean enough to avoid any inhibition of the polymerase. This problem arises frequently with samples taken from soil matrices. Furthermore, the differences in cell adhesion and cell-wall structure, together with the soil's characteristics, can affect the efficiency of the DNA isolation and purification. To tackle these problems, different protocols of extraction can be found in the literature, especially for soil matrices. Therefore it is recommended, once it has been experimentally confirmed that a specific protocol of extraction is efficient enough (in terms of extraction yield and purification suited for the PCR), to use the same one for all subsequent

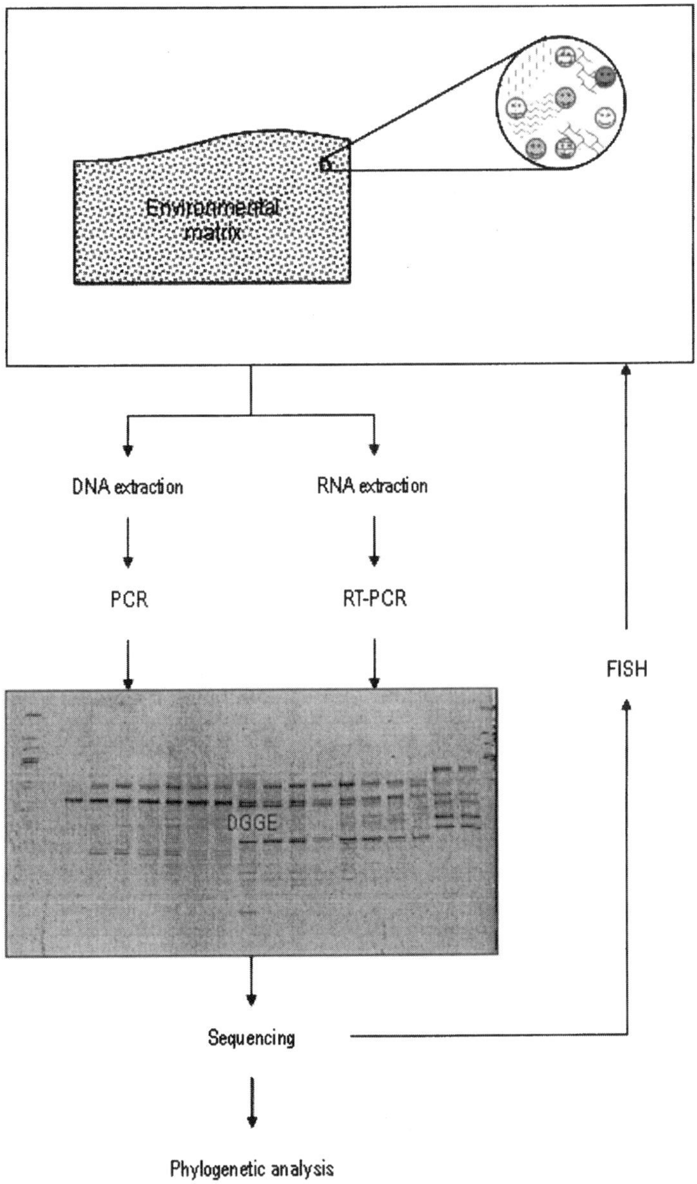

Fig. 1. Flow-chart showing the successive steps required to obtain a DGGE/TGGE fingerprint (sampling, nucleic acid extraction, PCR amplification, and DGGE/TGGE). Each band can be sequenced for further phylogenetic analysis and/or fluorescent *in situ* hybridization.

extractions. Indeed, it has been demonstrated that different extraction protocols can give different DGGE fingerprints and therefore different apparent bacterial community structures ***(8)***.

Alternatively to DNA, RNA can be isolated as an indicator of the level of activity of each microorganism in the environmental matrix under study, since the number of ribosomes per cell is roughly proportional to the growth rate of bacteria ***(9)***.

Once DNA or RNA is extracted and purified, the next step is respectively the PCR or RT-PCR. The primers of interest can be selected based on the list that is given in **Table 1**. During the PCR program, the annealing temperature has to be properly chosen to avoid the amplification of nonspecific sequences. Different increasing annealing temperatures can be evaluated until nonspecific amplifications are avoided. Alternatively to this trial-and-error method, a "touchdown" PCR can be used ***(10)***, which relies on incremental annealing temperature decrease during subsequent PCR cycles.

When sufficient PCR products are available, they can be loaded on a DGGE or TGGE. A detailed method to run DGGE/TGGE can be found in **Subheading 3.**

The microbial identity of each DGGE/TGGE band can be identified after excision of the band of interest and further sequencing. This sequence can be compared with available 16S rDNA in the databases to identify the phylogenetic affiliation of each sequenced band. The Ribosomal Database Project (http://rdp.cme.msu.edu/html) offers the possibility to compare the sequence with other sequences available in the database and contains a useful package of tools for phylogenetic analysis ***(11)*** and probe/primer design ***(12)***. However, it has to be pointed out that these databases are incomplete and still growing. Therefore, if the sequence has a low similarity with known sequences, it is difficult to determine whether the sequence represents a novel microorganism or is part of known taxa for which no 16S rDNA sequences are available or of low quality (i.e., partial sequences or ambiguous sequences). Apart from this aspect of comparison, the sequence can also be used for the design of oligonucleotide probes for *in situ* monitoring of bacterial populations (fluorescent *in situ* hybridization [FISH] ***[13]***).

As mentioned before, DGGE/TGGE is a powerful tool to monitor the response of microbial communities to stimuli and/or stresses by comparison of the fingerprint patterns before and after the external perturbation. Statistical analysis can be performed on these fingerprints (provided a numerical image of the gel is available) to determine whether a significant change occurred in response to the stimulus/stress. Two statistical methods are commonly used—principal coordinate analysis and canonical variate analysis ***(14)***.

1.4. Applications of DGGE/TGGE

Since the first application of DGGE/TGGE in natural ecosystems *(2)*, it has found many applications for the monitoring of microbial populations in complex matrices, from natural to artificial systems. This section will focus on some of these applications.

As mentioned above, molecular techniques avoid the need for cultivation, which is an important selection factor. By using a library of clones containing a fragment of 16S rDNA together with TGGE, Felske and co-workers *(15)* discovered an uncultured member of the class of the *Actinobacteria* in grassland soils.

Analysis of DGGE/TGGE fingerprints is possible on a wide range of natural and artificial matrices (provided that sufficient DNA/RNA can be recovered and purified), from soils *(16)*, wall painting *(17)*, to wastewater systems for the bioremediation of phenol *(18)* or biotrickling filters containing styrene-degrading biofilms *(19)*. Furthermore, as described above, it is possible to study part of the microbial community structure by the use of primers targeting specific groups.

Thanks to the ability of this technique to analyze the evolution of microorganism composition, MacNaughton et al. *(20)* evaluated the microbial population changes that occurred during bioremediation of an experimental oil spill. They discovered that the treatment of the contaminated soil promoted the growth of Gram-negative bacteria in the α-protobacteria and *Flexibacter-Cytophaga-Bacteroides* phylum. As previously mentioned, group-specific populations can be analyzed in very complex communities. In this way, Henckel et al. *(21)* used PCR-DGGE type I and type II to monitor methylotrophs in rice-field soil, along with the functional genes for particulate methane monooxygenase and methanol dehydrogenase, which are important enzymes for the catabolism of all methanotrophs. They found that there was a pronounced shift in the methanotroph community when CH_4 oxidation began.

1.5. Limitations of DGGE/TGGE

Even if DGGE/TGGE is common and one of the most utilized methods to analyze microbial communities, it has to be emphasized that the technique does have some limitations, but also that there are solutions to overcome these limitations.

First of all, the way a sample is stored can affect the microbial community. For example, anaerobic or aerobic storage or direct freezing of the samples can result in the identification of different microbial communities *(22)*. In addition, insufficient or preferential disruption of cells can introduce a bias during subsequent DGGE analysis.

Considerable bias can also occur during the PCR *(23)*. Suzuki and Giovannoni *(24)* detected a preferential amplification of rRNA genes in mixed microbial populations. The authors discovered that this preferential amplification can be limited by an appropriate choice of primers and by decreasing the numbers of cycles of replication. However, they postulated that this phenomenon would be small if the sample contains a high diversity of templates.

Another bias is the formation of chimeric molecules during PCR. Chimeric molecules are composed of parts of two sequences originating from templates with a high sequence similarity. These templates compete with primers during the annealing step of the PCR. The percentage of chimeric molecules can be very important, depending on the number of PCR cycles and sequence similarities between templates. For example, Wang and Wang *(25)* observed up to 30% chimeras after 30 cycles with templates showing a 99.3% sequence similarity. They also suggested increasing the elongation time (2 to 5 min) and diminishing the number of cycles of replication to reduce the formation of chimeras. If chimeric molecules are still formed, the Ribosomal Database Project *(11)* offers the possibility to screen sequenced fragments for possible chimeras.

Chimeric structures are not the only factor that can affect the interpretation of DGGE/TGGE fingerprints and/or lead to an overestimation of the number of microbial species. Indeed, bacteria can have more than one 16S rRNA gene (*rrn* operons) on their chromosome. The number of *rrn* operons can differ widely, from one copy for *Bradyrhizobium japonicum* *(26)* to 10 copies for *Bacillus subtilis* *(27)*. In addition, these copies can show variable sequence heterogeneities, as shown by the detection of up to 10 variants in *Paenibacillus polymyxa* when a 347-bp fragment of its 16S rRNA was amplified by PCR and then loaded on a TGGE *(28)*.

In addition to storage and PCR bias, the interpretation of DGGE/TGGE fingerprints themselves can be ambiguous when very complex communities are analyzed. It is estimated that bacterial populations that compose more than 1% of the total community can be detected by PCR-DGGE *(2)*. This limit can be overcome by bacterial fractionation *(29)*, differential centrifugation of extracted DNA bound to bisbenzimidazole in CsCl gradients according to their G+C content *(30)* or by using group-specific primers.

2. Materials

2.1. DNA Extraction and Purification

There are several protocols available in the literature. However we would highly recommend the kit developed by Mo Bio for environmental samples (Mo Bio UltraClean Soil DNA Kit, Solana Beach, CA). The kit is very rapid and yields relatively high amounts of clean DNA suitable for PCR.

3. Methods

3.1. PCR Conditions

A 2-µL volume of the extracted DNA is amplified by PCR with a 9600 thermal cycler (Perkin-Elmer, Norwalk, CT). The PCR mixture contains 0.5 µ*M* of each primer, 100 µ*M* of each deoxynucleoside triphosphate, 10 µL of 10X PCR buffer, 2 U of Expand high-fidelity DNA polymerase (Boehringer, Mannheim, Germany), 400 ng of bovine serum albumin (Boehringer) per microliter, and sterile water to a final volume of 100 µL. Other polymerases could be used but one should keep all the conditions standard to avoid any biases. Samples are amplified as follows: 94°C for 5 min, followed by 30 cycles of 92°C for 1 min, 55°C for 1 min, and 72°C for 1 min, with a final extension at 72°C for 10 min.

3.2. DGGE

DGGE based on the method cited by Muyzer et al. *(2)* is performed with the D Gene System (Bio-Rad, Hercules, CA). PCR samples are loaded onto 6% (w/v) polyacrylamide gels in 1X TAE (20 m*M* Tris, 10 m*M* acetate, 0.5 m*M* EDTA [pH 7.4]) for the primer set P63f plus P518r, or 8% polyacrylamide gels for the set P338f plus P518r. The polyacrylamide gels are made with a denaturing gradient ranging from 40 to 60% (where 100% denaturant contains 7 *M* urea and 40% formamide). For more details concerning the percentage of denaturants the reader is referred to any basic molecular biology handbook. The electrophoresis could be run overnight at 60°C at 75 V for the primer set P63f plus P518r, or at 35 V for the set P338f plus P518r. Shorter times with higher voltages could be used as well. After the electrophoresis, the gels are soaked for 30 min in SYBR Green I nucleic acid gel stain (1:10,000 dilution; FMC BioProducts, Rockland, ME). The stained gels are immediately photographed on a UV transillumination table with a video camera module (Vilbert Lourmat, Marne la Vallée, France). Alternatively, ethidium bromide can be used.

Statistical comparison of different DGGE patterns, run on the same gel, are done with GelCompar software 4.1 (Applied Maths, Kortrijk, Belgium) or any other available software designed for electrophoresis gel analysis.

3.3. Sequencing of DGGE Fragments

DNA fragments to be sequenced are excised from the gel, placed into sterile Eppendorf tubes containing 25 µL of sterilized water, and incubated overnight at 4°C. A 4-µL volume of the DNA diffused in water serves as a template for PCR amplification. The amplified products are subjected to a new DGGE step to confirm their electrophoretic mobility. The PCR products are purified with a QIAquick PCR purification kit (QIAGEN GmbH, Hilden, Germany) before sequencing. However, it is important to note that this strategy does not always work. Therefore, it is advisable to clone the PCR products into a cloning vector before sequencing. The Topo TA cloning kit from Invitrogen (San Diego, CA) is recommended. The sequences are aligned to 16S rRNA sequences available in the National Center for Biotechnology Information database by using the BLAST 2.0 search program.

References

1. Amman, R. I., Ludwig, W., and Schleiffer, K. H. (1995) Phylogenetic identification and *in situ* detection of individual microbial cells without cultivation. *Microbiol. Rev.* **59,** 143–169.
2. Muyzer, G., De Waal, E. C., and Uitterlinden, A. G. (1993) Profiling of complex microbial populations by denaturing gradient gel electrophoresis analysis of polymerase chain reaction-amplified genes coding for 16S rRNA. *Appl. Environ. Microbiol.* **59,** 695–700.
3. Myers, R. M., Fischer, S. G., Lerman, L. S., and Maniatis, T. (1985) Nearly all single base substitutions in DNA fragments joined to a GC-clamp can be detected by denaturing gradient gel electrophoresis. *Nucleic Acids Res.* **13,** 3131–3145.
4. Sheffield, V. C., Cox, D. R., Lerman, L. S., and Myers, R. M. (1989) Attachment of a 40-base-pair G+C rich sequence (GC-clamp) to genomic DNA fragments by the polymerase chain reaction results in improved detection of single-base changes. *Proc. Natl. Acad. Sci.* **86,** 232–236.
5. Heuer, H., Krsek, M., Baker, P., Smalla, K., and Wellington, E. M. H. (1997) Analysis of actinomycete communities by specific amplification of genes encoding 16S rRNA and gel-electrophoretic separation in denaturing gradients. *Appl. Environ. Microbiol.* **63,** 3233–3241.
6. Ovreas, L., Forney, L., Daae, F. L., and Torsvik, V. (1997) Distribution of bacterioplankton in meromictic lake Saelenvannet, as determined by denaturing gradient gel electrophoresis of PCR-amplified gene fragments coding for 16S rRNA. *Appl. Environ. Microbiol.* **63,** 3367–3373.
7. Wawer, C., Jetten, M. S. M., and Muyzer, G. (1997) Genetic diversity and expression of the [NiFe] hydrogenase large subunit gene of *Desulfovibrio* spp. in environmental samples. *Appl. Environ. Microbiol.* **63,** 4360–4369.
8. Niemi, R. M., Heiskanen, I., Wallenius, K., and Lindstrom, K. (2001) Extraction and purification of DNA in rhizosphere soil samples for PCR-DGGE analysis of bacterial consortia. *J. Microbiol. Methods* **45,** 155–165.
9. Wagner, R. (1994) The regulation of ribosomal RNA synthesis and bacterial cell growth. *Arch. Microbiol.* **161,** 100–106.
10. Don, R. H., Cox, P. T., Wainwright, B. J., Baker, P., and Mattick, J. S. (1991) "Touchdown" PCR to circumvent spurious priming during gene amplification. *Nucleic Acids Res.* **19,** 4008.
11. Maidak, B. L., Cole, J. R., Lilburn, T. G., Parker, C. T., Saxman, P. R., Farris, R. J., et al. (2001) The RDP-II (Ribosome Database Project). *Nucleic Acids Res.* **29,** 173–174.

12. Ashelford, K. E., Weightman, A. J., and Fry, J. C. (2002) PRIMROSE: a computer program for generating and estimating the phylogenetic range of 16S rRNA oligonucleotide probes and primers in conjunction with the RDP-II database. *Nucleic Acids Res.* **30,** 3481–3489.
13. Watanabe, K., Kodama, Y., Syutsubo, K., and Harayama, S. (2000) Molecular characterization of bacterial populations in petroleum-contaminated groundwater discharged from underground crude oil storage cavities. *Appl. Environ. Microbiol.* **66,** 4803–4809.
14. McCaig, A. E., Glover, L. A., and Prosser, J. I. (2001) Numerical analysis of grassland bacterial community structure under different land management regimens by using 16S ribosomal DNA sequence data and denaturing gradient gel electrophoresis banding patterns. *Appl. Environ. Microbiol.* **67,** 4554–4559.
15. Felske, A., Rheims, H., Wolterink, A., Stackebrandt, E., and Akkerman, A. D. L. (1997) Ribosome analysis reveals prominent activity of an uncultured member of the class *Actinobacteria* in grassland soils. *Microbiology* **143,** 2983–2989.
16. Ovreas, L. and Torsvik, V. (1998) Microbial diversity and community structure in two different agricultural soil communities. *Microb. Ecol.* **36,** 303–315.
17. Rolleke, S., Muyzer, G., Wawer, C., Wanner, G., and Lubitz, W. (1996) Identification of bacteria in a biodegraded wall painting by denaturing gradient gel electrophoresis of PCR-amplified gene fragments coding for 16S rRNA. *Appl. Environ. Microbiol.* **62,** 2059–2065.
18. Whiteley, A. S. and Bailey, M. J. (2000) Bacterial community structure and physiology state within an industrial phenol bioremediation system. *Appl. Environ. Microbiol.* **66,** 2400–2407.
19. Tresse, O., Lorrain, M. J., and Rho, D. (2002) Population dynamics of free-floating and attached bacteria in a styrene-degrading biotrickling filter analyzed by denaturing gradient gel electrophoresis. *Appl. Microbiol. Biotechnol.* **59,** 585–590.
20. MacNaughton, S. J., Stephen, J. R., Venosa, A. D., Davis, G. A., Chang, Y.-J., and White, D. C. (1999) Microbial population changes during bioremediation of an experimental oil spill. *Appl. Environ. Microbiol.* **65,** 3566–3574.
21. Henckel, T., Friedrich, M., and Conrad, R. (1999) Molecular analyses of the methane-oxidizing microbial community in rice field soil by targeting the genes of the 16S rRNA, particulate methane monooxygenase, and methanol dehydrogenase. *Appl. Environ. Microbiol.* **65,** 1980–1990.
22. Rochelle, P. A., Cragg, B. A., Fry, J. C., Parkes, R. J., and Weightman, A. J. (1994) Effect of sample handling on estimation of bacterial diversity in marine sediments by 16S rRNA gene sequence analysis. *FEMS Microbiol. Ecology* **15,** 215–226.
23. Wintzingerode, F. V., Gobel, U. B., and Stackebrandt, E. (1997) Determination of microbial diversity in environmental samples: pitfalls of PCR-based rRNA analysis. *FEMS Microbiol. Rev.* **21,** 213–229.
24. Suzuki, M. T. and Giovannoni, S. J. (1996) Bias caused by template annealing in the amplification of mixtures of 16S rRNA genes by PCR. *Appl. Environ. Microbiol.* **62,** 625–630.

25. Wang, G. C. Y. and Wang, Y. (1996) The frequency of chimeric molecules as a consequence of PCR co-amplification of 16S rRNA genes from different bacterial species. *Microbiology* **142,** 1107–1114.
26. Kundig, C., Beck, C., Hennecke, H., and Gottfert, M. (1995) A single rRNA gene region in *Bradyrhizobium japonicum. J. Bacteriol.* **177,** 5151–5154.
27. Stewart, G. C., Wilson, F. E., and Bott, K. F. (1982) Detailed physical mapping of the ribosomal RNA genes of *Bacillus subtilis. Gene* **19,** 153–162.
28. Nubel, U., Engelen, B., Felske, A., Snaidr, J., Wieshuber, A., Amann, R. I., Ludwig, W., and Backhaus, H. (1996) Sequence heterogeneities of genes encoding 16S rRNAs in *Paenibacillus polymyxa* detected by temperature gradient gel electrophoresis. *J. Bacteriol.* **178,** 5636–5643.
29. Jaspers, E. and Overmann, J. (1997) Separation of bacterial cells by isolelectric focusing, a new method for analysis of complex microbial communities. *Appl. Environ. Microbiol.* **63,** 3176–3181.
30. Holben, W. E. and Harris, D. (1995) DNA-based monitoring of total bacterial community structure in environmental samples. *Mol. Ecol.* **4,** 627–631.

Index

H

I

L

M

N

O

P